DISCRETE MATHEMATICS

DISCRETE MATHEMATICS

A Bridge to Computer Science and Advanced Mathematics

Olympia Nicodemi

State University of New York,
College at Geneseo

WEST PUBLISHING COMPANY

St. Paul / New York / Los Angeles / San Francisco

Copyeditor: Constance Day
Designer: Michael Michaud, Unicorn Production Services
Art Studio: ANCO/Boston
Compositor: Weimer Typesetting Co., Inc.
Production Coordinator: Schneider & Company

COPYRIGHT © 1987 by WEST PUBLISHING COMPANY
50 W. Kellogg Boulevard
P.O. Box 64526
St. Paul, MN 55164-1003

Printed in the United States of America

Library of Congress Cataloging-in-Publication Data

Nicodemi, Olympia.
    Discrete mathematics.

    Includes index.
    1. Mathematics   2. Electronic data
processing—Mathematics.   I. Title.
QA39.2.N55   1987   510   86-24616
ISBN 0-314-28503-2

For my sons, Adam and Jacob

CONTENTS

Chapter 1	**LOGIC 2**

1.1 Propositions   *4*
1.2 The Conditional and the Biconditional   *11*
1.3 Disjunctive Normal Form and Simplification   *16*
1.4 Predicates   *25*
1.5 Valid Arguments and Proofs   *30*
1.6 Proofs in Mathematics   *38*
1.7 Bridge to Computer Science: Logic Circuits (optional)   *42*
Key Concepts   *48*
Exercises   *49*

Chapter 2	**SET THEORY 54**

2.1 Sets   *55*
2.2 Set Relations   *57*
2.3 Set Operations   *59*
2.4 Infinite Collections of Sets   *65*
2.5 Power Sets   *68*
2.6 Cartesian Products   *69*
2.7 Inductively Defined Sets   *71*
2.8 Formal Languages   *75*
2.9 Proofs by Induction   *79*

2.10 Bridge to Computer Science: A Compiling Problem (optional)   *84*
         Key Concepts   *94*
         Exercises   *95*

Chapter 3   FUNCTIONS   100

3.1 Functions   *101*
3.2 Injective and Surjective Functions   *107*
3.3 Composition of Functions   *110*
3.4 Inverse Functions   *114*
3.5 Functions and Set Operations   *117*
3.6 Bridge to Computer Science: A Look at Coding Theory (optional)   *121*
         Key Concepts   *127*
         Exercises   *128*

Chapter 4   COUNTING AND COUNTABILITY   132

4.1 Counting Principles   *133*
4.2 Functions and Counting   *142*
4.3 Permutations and Combinations   *146*
4.4 Combinatorial Arguments   *153*
4.5 Infinite Sets and Countability   *157*
4.6 Bridge to Computer Science: An Algorithm Analyzed (optional)   *167*
         Key Concepts   *172*
         Exercises   *173*

Chapter 5   RELATIONS   178

5.1 Relations   *180*
5.2 Composition of Relations   *186*
5.3 Equivalence Relations   *191*
5.4 Equivalence Classes   *196*
5.5 Order Relations   *200*
5.6 Bridge to Computer Science: Two's-Complement Arithmetic (optional)   *212*
         Key Concepts   *215*
         Exercises   *216*

Chapter 6   **GRAPH THEORY   222**

6.1 Basic Concepts   *225*
6.2 Paths and Connectivity   *233*
6.3 Planar Graphs   *244*
6.4 Trees   *249*
6.5 Rooted Trees   *260*
6.6 Bridge to Computer Science: TREESORT (optional)   *273*
     Key Concepts   *281*
     Exercises   *282*

Chapter 7   **DIGRAPHS   292**

7.1 Digraphs   *294*
7.2 Paths and Connectivity   *300*
7.3 Weighted Digraphs   *311*
7.4 Acyclic Digraphs   *320*
7.5 Finite State Machines   *329*
7.6 Bridge to Computer Science: Kleene's Theorem (optional)   *342*
     Key Concepts   *355*
     Exercises   *356*

Chapter 8   **AN INTRODUCTION TO ALGEBRA   366**

8.1 Binary Operations, Semigroups, and Monoids   *367*
8.2 The Integers   *375*
8.3 Groups   *383*
8.4 Subgroups   *394*
8.5 Rings and Fields   *400*
8.6 Bridge to Computer Science: More about Coding Theory (optional)   *408*
     Key Concepts   *413*
     Exercises   *414*

APPENDIX: Matrices and Matrix Operations   421

ANSWERS TO END-OF-SECTION EXERCISES   429

INDEX   485

PREFACE

WE offer discrete mathematics to our computer science and mathematics students to teach them certain aspects of noncontinuous mathematics such as boolean arithmetic, combinatorics, and graph theory, as well as to help them develop the skills of mathematical reasoning: deduction, proof, and recursive thinking. This book is designed to do both in a one-semester course for computer-science majors and mathematics majors in their first or second year. The text assumes no prerequisites beyond the mathematical maturity usually provided by one semester of calculus and a high-level computer language.

For the computer-science major, a course based on this book will

Develop recursive thinking—we emphasize recursively defined sets, functions, proofs by induction, and algorithms

Present boolean algebra within the contexts of set theory, logic, and lattices

Introduce and integrate the theory of formal languages

Present graph theory and its applications (automata are introduced within this context)

Provide an introduction to combinatorics, countability, and number theory

In general, prepare the student for advanced work in the analysis of algorithms, data structures, and compiling

For the student of mathematics, a course based on this book will

Provide a solid foundation in logic, set theory, and functions

Develop the student's skill in both reading and writing technical mathematics

Introduce the student to topics such as combinatorics and graph theory that are not generally covered in the standard calculus sequence

Provide an introduction to number theory and abstract algebraic structures

In general, prepare the student for advanced work in applied and modern algebra, analysis, and probability theory

Exercises and examples abound. Exercises at the end of each section focus on the basics of that section. Answers to all these exercises, including proofs where required, are found at the end of the text. At the end of each chapter there is a large set of exercises. These cover the fundamentals again but also include more challenging exercises. Answers to these are found in the instructor's manual.

The needs and interests of the computer-science major and the mathematics major overlap considerably. A course in discrete mathematics is an excellent vehicle in which to fuse theory and application. To this end we have integrated computer-science applications throughout the text while maintaining a constant focus on the theoretical underpinnings of the topics at hand. To acknowledge just how important the link between discrete mathematics and computer science is, we have ended each chapter with a special section called "Bridge to Computer Science." In each of these sections an extensive application of the mathematics of the chapter to computer science is developed. Several Pascal programs are presented. These illustrate how to implement the concepts under discussion, but they are not necessary to the understanding of these concepts. The bridge sections are optional in the sense that the material is not referred to elsewhere in the text. They vary considerably in their nature and difficulty. For instance, the bridge sections to Chapters 1, 5, and 6 are quite transparent and might be assigned for unassisted independent study or given light coverage within a lecture. The bridges to Chapters 2, 3, 4, 7, and 8 require a bit more effort.

Although mathematics and computer science are integrated within the text, different courses may be developed from this text depending on the interests of the instructor and students, or the demands of the syllabus. Any course should cover almost all of Chapters 1 through 5 and selected topics from the remaining chapters. Here are some suggestions.

For a course serving primarily computer-science students:

(1) Cover Chapters 1 through 5. Emphasize boolean algebra and inductive procedures. Cover Sections 2.4, 3.5, 4.2, 4.5, and 5.4 lightly.

(2) Cover Chapter 6. Emphasize rooted trees and their applications. Cover Section 6.3 lightly (or omit).

(3) Choose one (or more) of the following options:
   (a) Focus on the algorithms in Sections 7.1 through 7.4.
   (b) Cover digraphs briefly in Sections 7.1 and 7.2, and focus on finite state machines in Section 7.5. Cover monoids and semigroups in Section 8.1.
   (c) Cover the elementary number theory presented in Section 8.2. (This material can follow any time after Chapter 5, Section 5.4.)

For a course serving primarily mathematics majors:

(1) Do Chapters 1 through 6 in their entirety. Emphasize proofs, especially proofs by induction.

(2) Choose one of the following options:

    (a) For applied algebra do Chapter 7, focusing on finite state machines. Also do bridge Sections 3.6 and 7.6.

    (b) For an introduction to abstract algebraic structures, do Chapter 8.

In all cases, let us urge our students to learn through thinking, exploring, and doing.

Acknowledgments

I must first thank the members of my family for their boundless support and encouragement.

I wish to thank my colleagues in the departments of mathematics and computer science for their many helpful corrections, comments, and suggestions, and for their cooperation in class testing this text in its numerous preliminary versions. I especially wish to thank Steven West, Donald Trasher, Yi-Neh Gau, Gary Towsley, Jung Tsai, and Sidney Sanders.

I wish to thank my students for their patience in reading and using preliminary copies of the manuscript and for their help with the solutions to the exercises. I would especially like to thank Steven Cole, Mary Ann Huntley, Paul Jackson, Christine MacDonald, and Carin Clement.

The reviewers of this text provided many helpful suggestions, corrections, and much encouragement. I would like to thank them. (Of course, any errors that remain are my own.) They are:

Harold Reiter, *University of Maryland*

E. G. Whitehead, *University of Pittsburgh*

Ken Holladay, *University of New Orleans*

R. C. Entringer, *University of New Mexico*

Thomas Upson, *Rochester Institute of Technology*

Robert Hunter, *Pennsylvania State University*

Davis Cusik, *Marshall University*

G. F. Orr, *University of Southern Colorado*

Peter Jones, *Marquette University*

Frank Cannonito, *University of California at Irvine*

Glenn Hopkins, *University of Mississippi*

Julian Hook, *Florida International University*

Bob Adams, *University of Kansas*

Max Garzon, *Memphis State University*

Thomas Ralley, *Ohio State University*

At West Publishing Company, my sincerest thanks go to Patrick Fitzgerald, my editor; Nancy Hill-Whilton, developmental editor; Patricia Lewis and Barbara Fuller, production editors. Thanks also go to Debbie Schneider of Schneider & Company.

Chapter 1

LOGIC

outline

1.1 Propositions
1.2 The Conditional and the Biconditional
1.3 Disjunctive Normal Form and Simplification
1.4 Predicates
1.5 Valid Arguments
1.6 Proofs in Mathematics
1.7 Bridge to Computer Science: Logic Circuits (optional)
     Key Concepts
     Exercises

Introduction

LOGIC is the study and analysis of the nature of the valid argument, the reasoning tool by which philosophers and mathematicians draw valid inferences from a given set of facts or premises. In this chapter we seek to analyze such arguments as the following:

> All philosophers are immortal.
> Socrates is a philosopher.
> Therefore Socrates is immortal.

> If one is crazy one drills teeth.
> My dentist drills teeth.
> Therefore my dentist is crazy.

The first argument is valid. If we accept as true the immortality of philosophers and the vocation of Socrates, we must logically conclude that Socrates will live forever. (The argument is valid even though its conclusion may not be true.) The second argument, however, is not valid though its conclusion may be true. Even if we accept as true the premise that all crazy people drill teeth, others who are sane might also drill. Thus logic does not force us to indict "my dentist" as crazy. In the sections that follow we shall develop a symbolic representation of arguments such as the preceding, and we shall develop computational devices that, in restricted cases, determine validity.

Logic, a subject of far greater complexity than we shall encounter here, was extensively developed in classical Greece. In the Middle Ages the treatises of Aristotle concerning logic were rediscovered. This event so excited and invigorated the philosophical community that its best thinkers sought to restate all of philosophy in terms of this powerful tool. However, whereas the goal of philosophy is to uncover truth, logic provides only the technology of deduction. The task remained to determine a core of true premises to which logic could be applied.

Mathematics coevolved with logic, its history interwoven with that of logic but not identical to it. For many years, much of mathematics was not so much deduced as inspired from physical models. But in the late nineteenth and early twentieth centuries, mathematicians such as David Hilbert sought to establish a small set of premises or axioms from which all of mathematics could be deduced by means of the tools of logic. (This approach to mathematics is almost universally accepted today.) Yet logic has its limitations. The profound work of Kurt Gödel in the first half of this century demonstrated that there is an intrinsic limit to the deductions possible in any consistent logical system. Computer science has given new impetus to the study of such limitations. Today we ask, "What is the nature of, and what are the limits of, that which can be computed algorithmically?" To answer this complex question thoroughly would take us far beyond the scope of this book. Nonetheless, by the end of Chapter 7 we shall see a connection between language (formal languages) and model computing devices (finite-state machines). Mathematicians and logicians investigate computability in terms of such connections. At the end of this first chapter we shall see how logic is employed in circuit or chip design.

Section 1.1 | Propositions

In our spoken language the declarative sentence is the basic vehicle with which we convey a thought. It is also the basic structure under consideration in logic. In logic it is called a proposition.

> **DEFINITION.** A **proposition** is a declarative sentence to which we can assign a truth value of either *true* or *false*, but not both.

The following are examples of propositions. Each can be established as true or false. Here, propositions (ii) and (iv) are true and propositions (i) and (iii) are false.

(i) George Washington signed the Declaration of Independence.

(ii) $2 + 3 = 5$

(iii) The moon is made of green cheese.

(iv) The set of prime integers is infinite.

"Is the earth round?" and "Go!" are not declarative sentences and hence not propositions. The sentence "The number x is a positive integer" is not a proposition unless the variable x is assigned a specific value. The sentence "This sentence is false" cannot be assigned a truth value of either true or false, because either assignment contradicts the sense of the sentence. Although it is a declarative sentence, it is not a proposition.

We shall use uppercase letters such as A, B, P, Q, and R to represent propositions. These letters are called **logical variables.** We combine simple propositions to form compound propositions by using the connectives "and" and "or." For instance, "John will pass calculus and he will graduate" consists of two simple propositions connected by the word "and." We use the symbol $\wedge$ for the word "and" and the symbol $\vee$ for the word "or." Thus if P and Q are propositions, $P \wedge Q$ is the proposition "P and Q." It is called the **conjunction** of P and Q. Similarly, $P \vee Q$ is the proposition "P or Q" and it is called the **disjunction** of P and Q. An expression involving logical variables and connectives is called a **propositional form.** For example, $P \wedge (Q \vee R)$ is a propositional form.

Example 1. Let A represent the proposition "The roof is red." Let B represent "The wall is white." Then we have

(a) $A \wedge B$: The roof is red and the wall is white.

(b) $A \vee B$: The roof is red or the wall is white.

In proposition (b) the word "or" is used to mean that the roof is red or the wall is white *or both*. In mathematics and in logic, the word "or" is always *inclusive*. We shall usually omit the phrase "or both," but it will always be implied.

Example 2. Let A denote "The earth is round"; let B denote "The sun is cold"; and let C denote "It rains in Spain." Then we have

(a) $A \wedge B$: The earth is round and the sun is cold.

(b) $B \vee C$: The sun is cold or it rains in Spain.

(c) $A \wedge (B \vee C)$: The earth is round and either the sun is cold or it rains in Spain (or both).

A propositional form becomes a true or a false proposition after each of its variables has been assigned a truth value of true or false. In Example 2 let us assume that propositions A and C are true and that proposition B is false. Then the truth value of $A \wedge B$ is false because B is false. The truth value of $B \vee C$ is true because C is true. The truth value of $A \wedge (B \vee C)$ is true because both A and $(B \vee C)$ are true.

An essential concept of symbolic logic is that the truth value of forms such as $A \wedge B$ and $A \wedge (B \vee C)$ depends only on the truth value assigned to each of the variables, not on the particular meaning of the propositions represented by those variables. For example, if P and Q are logical variables with P true and Q false, then $P \wedge Q$ is false and $P \vee Q$ is true. The form $P \wedge Q$ is true only when both P and Q are true. If P or Q (or both) are false, then $(P \wedge Q)$ is false. The form $P \vee Q$ is false only when both P and Q are false.

To summarize, the form $P \wedge Q$ is true when both P and Q are true. It is false when P or Q (or both) are false. The form $P \vee Q$ is true when P or Q (or both) are true. It is false only when both P and Q are false.

Example 3. Let P, Q, and R be logical variables. Suppose that P and Q are true and that R is false. Then we have

(a) $P \vee Q$ is true.

(b) $Q \wedge R$ is false.

(c) $P \wedge (Q \vee R)$ is true.

(d) $P \vee (Q \wedge R)$ is true.

(e) $(P \wedge R) \vee (Q \wedge R)$ is false.

However, if we suppose that P is false and that Q and R are true, then we have

(a) $P \vee Q$ is true.

(b) $Q \wedge R$ is true.

(c) $P \wedge (Q \vee R)$ is false.

(d) $P \vee (Q \wedge R)$ is true.

(e) $(P \wedge R) \vee (Q \wedge R)$ is true.

The negation of the true proposition "All integers are rational numbers" is the false proposition "At least one integer is not a rational number." The negation of the false proposition "The moon is purple" is the true proposition "The moon is not purple." The negation of a proposition always results in a proposition with the opposite truth value. If A is a proposition, then its negation is a proposition denoted by $\sim A$. The propositions A and $\sim A$ always have opposite truth values.

Example 4.

A: It rained on July 4, 1983.	$\sim A$: It did not rain on July 4, 1983.
B: All primes are odd.	$\sim B$: Some primes are not odd.
C: Some even numbers are prime.	$\sim C$: No even number is prime.
D: Jo passed or Mary passed.	$\sim D$: Jo failed and Mary failed.

Note that an assertion about *all* the members of a set is false as soon as that assertion fails for *one* member of the set. Thus the negation of "All Martians are green" is the proposition "Some Martians are not green." Note also that in mathematics the word "some" applied to a set means "at least one" and does not preclude the condition "all." Thus the proposition "Some humans are mammals" is a true proposition, as is "Some primes are even."

A **truth table** is a computational device by which we can determine the truth value of a propositional form once we know the truth values of each of its variables. We let 1 represent truth value "true" and 0 represent truth value "false." Then we tabulate all the possibilities. Here are the truth tables for $\sim P$, $P \wedge Q$, and $P \vee Q$:

Figure 1.1

$\sim P$:

P	$\sim P$
0	1
1	0

$P \wedge Q$:

P	Q	$P \wedge Q$
0	0	0
0	1	0
1	0	0
1	1	1

$P \vee Q$:

P	Q	$P \vee Q$
0	0	0
0	1	1
1	0	1
1	1	1

For more complicated forms, we work in stages from inner parentheses to outer. The values for the form under consideration appear in the last column. We now give the truth tables for the forms $\sim(P \land Q)$ and $\sim P \lor \sim Q$.

Figure 1.2

$\sim(P \land Q)$:

P	Q	$P \land Q$	$\sim(P \land Q)$
0	0	0	1
0	1	0	1
1	0	0	1
1	1	1	0

$\sim P \lor \sim Q$:

P	Q	$\sim P$	$\sim Q$	$\sim P \lor \sim Q$
0	0	1	1	1
0	1	1	0	1
1	0	0	1	1
1	1	0	0	0

Truth tables can be used for forms containing any number of variables. If there are n variables, there will be 2^n rows corresponding to the 2^n possible assignments of truth values to the n variables. We proceed systematically through the possibilities by letting the truth assignments of the variables give successively the binary representation of the integers 0 through $2^n - 1$. Figure 1.3 gives the truth tables for the forms $P \land (Q \lor R)$ and $(P \land Q) \lor (P \land R)$. Note that the first three columns of these truth tables give the binary representation of the integers 0 through 7.

The reader will note that the last columns of the truth tables for the forms $\sim(P \land Q)$ and $\sim P \lor \sim Q$ given in Figure 1.2 are identical. That is, for a given assignment of truth values to P and Q, the forms $\sim(P \land Q)$ and $\sim P \lor \sim Q$ have the same truth value. If two forms have identical truth tables, those forms are said to be **equivalent.** We give an example to show that this equivalence is not merely symbolic but is also an equivalence of meaning. Let P be the proposition "It is snowing." Let Q be the proposition "It is dark outside." Then the proposition $\sim(P \land Q)$ is "It is not the case that it is both snowing and dark outside." This means that either it is not snowing or it is not dark outside; that is, $\sim P \lor \sim Q$.

If A and B are equivalent forms, we write $A = B$; the expression $A = B$ is called a **logical identity.** The preceding paragraph shows that we have the logical identity $\sim(P \land Q) = \sim P \lor \sim Q$. This particular identity is one of DeMorgan's laws, which we shall state in a moment. We can prove such identities by showing that the truth tables for the forms on either side of the equality are identical. We urge the reader to do this. However, here we shall offer an alternative verbal argument. Understanding such verbal arguments is essential if the symbols of symbolic logic are to become tools of reasoning.

Figure 1.3

$P \wedge (Q \vee R)$:

P	Q	R	$Q \vee R$	$P \wedge (Q \vee R)$
0	0	0	0	0
0	0	1	1	0
0	1	0	1	0
0	1	1	1	0
1	0	0	0	0
1	0	1	1	1
1	1	0	1	1
1	1	1	1	1

$(P \wedge Q) \vee (P \wedge R)$:

P	Q	R	$P \wedge Q$	$P \wedge R$	$(P \wedge Q) \vee (P \wedge R)$
0	0	0	0	0	0
0	0	1	0	0	0
0	1	0	0	0	0
0	1	1	0	0	0
1	0	0	0	0	0
1	0	1	0	1	1
1	1	0	1	0	1
1	1	1	1	1	1

Theorem 1.   DeMorgan's Laws.

(a) $\sim(P \wedge Q) = \sim P \vee \sim Q$

(b) $\sim(P \vee Q) = \sim P \wedge \sim Q$

PROOF:

(b) Suppose the proposition $\sim(P \vee Q)$ is true. This means that it is not the case that either P or Q is true. Thus if $\sim(P \vee Q)$ is true, then both P and Q are false. If both P and Q are false, then the proposition $\sim P \wedge \sim Q$ is true. Conversely, if the proposition $\sim P \wedge \sim Q$ is true, then both $\sim P$ and $\sim Q$ are true and so both P and Q are false. If both P and Q are false, then the proposition $P \vee Q$ is also false and so the proposition $\sim(P \vee Q)$ is true.

To prove the preceding identity, we showed that the truth values of the variables that make the right side of the identity true also make the left side true, and vice versa.

Example 5. Let P be the proposition "The grass is wet." Let Q be the proposition "It is raining." Then we have

(a) $\sim(P \vee Q)$: It is not the case that the grass is wet or it is raining. (Thus when $\sim(P \vee Q)$ is true, neither P nor Q holds.)

(b) $\sim P \wedge \sim Q$: The grass is not wet and it is not raining. (Again, when $\sim P \wedge \sim Q$ is true, neither P nor Q holds.)

The truth tables given in Figure 1.3 in this section show that we have the identity $P \wedge (Q \vee R) = (P \wedge Q) \vee (P \wedge R)$. This identity is one of the distributive laws given in the following theorem. We prove part (b) by showing that the same assignment of truth values that makes the left side of the identity true also makes the right side true, and vice versa.

Theorem 2. The Distributive Laws.

(a) $P \wedge (Q \vee R) = (P \wedge Q) \vee (P \wedge R)$

(b) $P \vee (Q \wedge R) = (P \vee Q) \wedge (P \vee R)$

PROOF:

(b) Suppose that $P \vee (Q \wedge R)$ is true. Then either P is true or both Q and R are true. If P is true, then both $P \vee Q$ and $P \vee R$ are true and so $(P \vee Q) \wedge (P \vee R)$ is also true. Now if both Q and R are true, again we have that $(P \vee Q)$ and $(P \vee R)$ are both true. Again it is the case that we have that $(P \vee Q) \wedge (P \vee R)$ is true. Conversely, suppose that $(P \vee Q) \wedge (P \vee R)$ is true. This occurs if P is true, and in this case $P \vee (Q \wedge R)$ is also true. If P is false, then $(P \vee Q) \wedge (P \vee R)$ can be true only if both Q and R are true. Again it is the case that we have that $P \vee (Q \wedge R)$ is true, and we conclude our proof.

Example 6. Let A represent the proposition "It is windy"; let B represent "It is warm"; and let C represent "It is sunny." Then we have

(a) $A \vee (B \wedge C)$: Either it is windy or it is both warm and sunny.

(b) $(A \vee B) \wedge (A \vee C)$: It is windy or warm and it is windy or sunny.

In both propositions (a) and (b) all three conditions may hold, but if it is not windy it must be both sunny and warm.

| Section 1.1 | **Exercises** |

1. Decide which of the following are propositions.

(a) William conquered England in 1066.

(b) 527 is an even integer.

(c) All triangles have three sides.

(d) $(x + 3) < (x + 4)$

(e) If $x = 0.3$, then $x > x$.

(f) I always lie.

(g) Are all circles round?

2. Negate the following propositions.

(a) Mary is tall.

(b) The cat is not black.

(c) All dogs bark.

(d) Some birds have feathers.

(e) Mary is tall and Burt is thin.

(f) John is smart or Fred is not tall.

3. Let P be the proposition "The earth is flat"; let Q be "All birds sing"; and let R be "Manhattan is an island." Write out the following propositions.

(a) $P \lor Q$ (b) $\sim Q \land R$ (c) $\sim(Q \land P)$

(d) $R \lor (P \land Q)$ (e) $P \land \sim(Q \lor R)$

4. Suppose that P is true and that Q and R are both false. Determine the truth value of the following propositional forms.

(a) $P \land (\sim Q \lor R)$ (b) $(P \land Q) \lor R$ (c) $(P \lor \sim Q) \land \sim R$

5. Repeat Exercise 4, supposing that P and Q are false and that R is true.

6. Write out the truth tables for the following propositional forms.

(a) $P \land \sim Q$ (b) $P \lor (Q \land \sim R)$ (c) $(\sim P \lor Q) \land \sim R$

(d) $(\sim P \land Q) \lor (R \land T)$

7. For each of the following pairs of propositional forms, find an assignment of truth value that shows the members *not* to be equivalent.

(a) $\sim P \land Q$ and $\sim(P \land Q)$

(b) $(P \land Q) \lor R$ and $P \land (Q \lor R)$

(c) $\sim P \land \sim Q$ and $\sim(P \land Q)$

8. Determine in which of the following pairs the members are equivalent.

(a) $\sim(P \land Q)$ and $\sim P \land \sim Q$

(b) $\sim P \lor Q$ and $\sim(P \land \sim Q)$

(c) $P \land (Q \lor R)$ and $(P \land Q) \lor R$

(d) $P \land \sim(Q \land R)$ and $(P \land \sim Q) \lor (P \land \sim R)$

9. Use DeMorgan's laws to rewrite the following expressions using only $\sim$ and $\land$.

(a) $P \vee Q$ (b) $\sim P \vee \sim(Q \vee \sim R)$

10. Use DeMorgan's laws to rewrite the following expressions using only $\sim$ and $\vee$.

(a) $P \wedge Q$ (b) $\sim(P \wedge Q) \wedge \sim R$

11. Use an argument like that used to prove Theorem 1 on page 8 to prove that $\sim(P \wedge \sim Q)$ is equivalent to $\sim P \vee Q$.

| Section 1.2 | **The Conditional and the Biconditional** |

Let P be the proposition "John will graduate" and let Q be the proposition "John will get a job." The proposition "If John graduates, then John will get a job" can be restated as "John's graduating implies that John will get a job" or "P implies Q." The connective "implies" is called the **conditional connective,** and we use the symbol $\Rightarrow$ to represent the word "implies." The form $P \Rightarrow Q$ can be rendered into English as either "P implies Q" or "If P then Q."

Example 1. Let P represent the proposition "It is raining"; let Q represent "the game is canceled"; and let R represent "Butch is sad." Then we have

$P \Rightarrow Q$: If it is raining, then the game is canceled.

$Q \Rightarrow R$: The game's being canceled implies that Butch is sad.

$R \Rightarrow P$: That Butch is sad implies that it is raining.

$Q \Rightarrow P$: If the game is canceled, then it is raining.

We use the implication $P \Rightarrow Q$ to mean that whenever P is true, then Q must also be true. Thus the only way that the implication $P \Rightarrow Q$ can fail to be true is if P is true but Q is not. When P is false, the truth of the implication $P \Rightarrow Q$ is not contradicted by either the truth of Q or the falsity of Q. Thus when P is false, the implication $P \Rightarrow Q$ is true no matter what the truth value of Q. We summarize our reasoning in the following truth table for $P \Rightarrow Q$. Note that $P \Rightarrow Q$ is false only when P is true but Q is not.

P	Q	$P \Rightarrow Q$
0	0	1
0	1	1
1	0	0
1	1	1

Example 2. Let P be the proposition "Mary is wearing green socks" and let Q be the proposition "Mary passed algebra." Then $P \Rightarrow Q$ is the proposition "If Mary is wearing green socks, then Mary has passed algebra." Let us suppose that P is true but Q is false. Then it is the case that Mary is indeed wearing green socks but she has failed algebra. So her wearing green socks does not imply that she has passed algebra. The implication $P \Rightarrow Q$ is false, as the truth table indicates. However, let us suppose that P is false and so Mary is not wearing green socks. She may or may not have passed algebra. Neither situation contradicts the implication $P \Rightarrow Q$. This is what is indicated in the first two lines of the truth table for $P \Rightarrow Q$.

We leave it as an exercise to verify that the truth table for the form $\sim P \vee Q$ is the same as that for $P \Rightarrow Q$ and hence that $P \Rightarrow Q$ is equivalent to $\sim P \vee Q$. The form $\sim P \vee Q$ is false only when P is true but Q is not true; that is, P holds but Q fails to follow. To negate the implication $P \Rightarrow Q$, we negate its equivalent, $\sim P \vee Q$, and apply DeMorgan's laws to obtain $\sim(P \Rightarrow Q) = P \wedge \sim Q$.

Example 3. Let A be the proposition "If John loses, then the sky will fall." Its negation, $\sim A$, is the proposition "John loses and the sky will not fall."

Let $P \Rightarrow Q$ be a proposition. Then we call the related proposition $Q \Rightarrow P$ the **converse** of $P \Rightarrow Q$, and we call the form $\sim Q \Rightarrow \sim P$ the **contrapositive** of $P \Rightarrow Q$.

Example 4. Let A be the proposition "If Mr. Smith is running for office, then he is a Democrat." This is of the form $P \Rightarrow Q$, where P is the proposition "Mr. Smith is running for office" and Q is the proposition "Mr. Smith is a Democrat." The converse of A is "If Mr. Smith is a Democrat, then he is running for office." The contrapositive of A is "If Mr. Smith is not a Democrat, then he is not running for office."

The converse of $P \Rightarrow Q$ is *not* equivalent to it. We see this from Example 4. The converse $Q \Rightarrow P$ is true in the case that Mr. Smith is not a Democrat yet is running for office. However, $P \Rightarrow Q$ is not true in this situation. We give the truth tables for $P \Rightarrow Q$ and $Q \Rightarrow P$ to prove that they are not equivalent. (Note that the variables must be written in the same order in both tables.)

$P \Rightarrow Q$:	P	Q	$P \Rightarrow Q$		$Q \Rightarrow P$:	P	Q	$Q \Rightarrow P$
	0	0	1			0	0	1
	0	1	1			0	1	0
	1	0	0			1	0	1
	1	1	1			1	1	1

If we look again at Example 4, we find that $P \Rightarrow Q$ and its contrapositive, $\sim Q \Rightarrow \sim P$, have the same meaning: If Mr. Smith is running for office, we know that he must be a Democrat; if he is not a Democrat, he is not running for office.

The following truth table for $\sim Q \Rightarrow \sim P$ is identical to the truth table for $P \Rightarrow Q$, which shows that these forms are equivalent.

$\sim Q \Rightarrow \sim P$:	P	Q	$\sim Q$	$\sim P$	$\sim Q \Rightarrow \sim P$
	0	0	1	1	1
	0	1	0	1	1
	1	0	1	0	0
	1	1	0	0	1

Example 5. Let A be the proposition "If it is raining, then Mary is studying," and suppose that A is true. The converse of A, "That Mary is studying implies that it is raining," is not necessarily true, because Mary may study even while the sun shines. However, the contrapositive of A, "If Mary is not studying, then it is not raining," is a true statement.

We saw in the preceding paragraphs that $P \Rightarrow Q$ and $Q \Rightarrow P$ are not equivalent. However, if we let A be the true proposition "The number 6 is even," and we let B be the true proposition "The number 6 is divisible by 3," then both $A \Rightarrow B$ and $B \Rightarrow A$ are true. That is, $(A \Rightarrow B) \wedge (B \Rightarrow A)$ is a true proposition. We denote the form $(P \Rightarrow Q) \wedge (Q \Rightarrow P)$ by the symbols $P \Leftrightarrow Q$. It has truth value true exactly when both P and Q have the same truth value, either both true or both false. We call the connective $\Leftrightarrow$ the **biconditional** and render the symbols $P \Leftrightarrow Q$ into English as "P if and only if Q." Thus for A and B as just given, we have $A \Leftrightarrow B$: "The number 6 is even if and only if the number 6 is divisible by 3."

Example 6. Let P represent the proposition "The moon is purple," and let Q represent "Socrates lives." Then $P \Leftrightarrow Q$ is the proposition "The moon is purple if and only if Socrates lives." Because $P \Leftrightarrow Q$ is the same as $(P \Rightarrow Q) \wedge (Q \Rightarrow P)$, we can restate this proposition as "That the moon is purple implies that Socrates lives, and conversely that Socrates lives implies that the moon is purple." Assigning truth value false to both P and Q, we have that both $P \Rightarrow Q$ and $Q \Rightarrow P$ are true implications. Hence, in this case $P \Leftrightarrow Q$ is true.

Example 7. Let A be the proposition "There is snow falling if and only if the temperature is at or below 0 Celsius." This can be represented by $P \Leftrightarrow Q$, where P is "There is snow falling" and Q is "The temperature is at or below 0 Celsius." Let us assume that $P \Rightarrow Q$ is true but $Q \Rightarrow P$ is false. (It might be cold yet clear.) So, assigning a truth value of false to P and true to Q, we have that $P \Leftrightarrow Q$ is false.

The truth table for $P \Leftrightarrow Q$ (also abbreviated P iff Q) is as follows:

$P \Leftrightarrow Q$:	P	Q	$P \Rightarrow Q$	$Q \Rightarrow P$	$P \Leftrightarrow Q$
	0	0	1	1	1
	0	1	1	0	0
	1	0	0	1	0
	1	1	1	1	1

We note again that $P \Leftrightarrow Q$ is true exactly when both P and Q have the same truth value, either both true or both false. In the preceding section we defined two propositional forms A and B as equivalent whenever their truth tables are identical. That is, A is equivalent to B if, for a given assignment of truth value to the variables appearing in A and B, A and B are either both true or both false. We can redefine equivalence in terms of the biconditional as follows: Two forms A and B are equivalent if $A \Leftrightarrow B$ is true for every assignment of truth value to the variables appearing in A and B.

Example 8. The form $(P \wedge Q) \Rightarrow R$ is equivalent to $P \Rightarrow (Q \Rightarrow R)$. We show this by showing that $[(P \wedge Q) \Rightarrow R] \Leftrightarrow [P \Rightarrow (Q \Rightarrow R)]$ is always true. The truth table follows.

P	Q	R	$(P \wedge Q)$	$(P \wedge Q) \Rightarrow R$	$Q \Rightarrow R$	$P \Rightarrow (Q \Rightarrow R)$	$[(P \wedge Q) \Rightarrow R] \Leftrightarrow [P \Rightarrow (Q \Rightarrow R)]$
0	0	0	0	1	1	1	1
0	0	1	0	1	1	1	1
0	1	0	0	1	0	1	1
0	1	1	0	1	1	1	1
1	0	0	0	1	1	1	1
1	0	1	0	1	1	1	1
1	1	0	1	0	0	0	1
1	1	1	1	1	1	1	1

A propositional form that is always true is called a **tautology.** For example, the form $P \vee \sim P$ is a tautology. The preceding truth table shows that $[(P \wedge Q) \Rightarrow R] \Leftrightarrow [P \Rightarrow (Q \Rightarrow R)]$ is a tautology. In fact, two propositional forms A and B are equivalent if and only if $A \Leftrightarrow B$ is a tautology. The propositional form $P \wedge \sim P$ is always false. A propositional form that is always false is called a **contradiction** or an **absurdity.** The form $P \wedge \sim P$ is an absurdity, as is the form $(P \wedge \sim Q) \Leftrightarrow (P \Rightarrow Q)$. (We leave the verification to the reader.) A propositional form that is neither a tautology nor a contradiction is called a **contingency**: Its truth value is contingent on the truth values assigned to each of its variables.

The important logical identities are summarized in Table 1.1. Each form listed is a tautology.

Table 1.1   **IMPORTANT LOGICAL IDENTITIES**

1. (i) $P \Leftrightarrow (P \vee P)$ idempotent laws
      (ii) $P \Leftrightarrow (P \wedge P)$

2. $[\sim(\sim P)] \Leftrightarrow P$ double negation

3. (i) $(P \vee Q) \Leftrightarrow (Q \vee P)$ commutativity
      (ii) $(P \wedge Q) \Leftrightarrow (Q \wedge P)$

Table 1.1 (Continued)

4.　(i) $[P \wedge (Q \vee R)] \Leftrightarrow [(P \wedge Q) \vee (P \wedge R)]$　　　　distributive laws
　　(ii) $[P \vee (Q \wedge R)] \Leftrightarrow [(P \vee Q) \wedge (P \vee R)]$

5.　(i) $[P \vee (Q \vee R)] \Leftrightarrow [(P \vee Q) \vee R]$　　　　associativity
　　(ii) $[P \wedge (Q \wedge R)] \Leftrightarrow [(P \wedge Q) \wedge R]$

6.　(i) $[\sim(P \vee Q)] \Leftrightarrow [\sim P \wedge \sim Q]$　　　　DeMorgan's laws
　　(ii) $[\sim(P \wedge Q)] \Leftrightarrow [\sim P \vee \sim Q]$

7.　$(P \Rightarrow Q) \Leftrightarrow (\sim P \vee Q)$　　　　implication

8.　$(P \Rightarrow Q) \Leftrightarrow (\sim Q \Rightarrow \sim P)$　　　　contrapositive

(In the identities that follow, we shall denote a proposition that is always true by 1 and a proposition that is always false by 0.)

9.　(i) $(P \vee 0) \Leftrightarrow P$
　　(ii) $(P \wedge 1) \Leftrightarrow P$

10.　(i) $(P \vee 1) \Leftrightarrow 1$
　　(ii) $(P \wedge 0) \Leftrightarrow 0$

11.　(i) $(P \vee \sim P) \Leftrightarrow 1$
　　(ii) $(P \wedge \sim P) \Leftrightarrow 0$

Section 1.2　**Exercises**

---

1. Let P be the proposition "John passes calculus"; let Q be the proposition "John is happy"; and let R be the proposition "John has the job." Write out the following propositions.

   (a) $P \Rightarrow Q$

   (b) $Q \Rightarrow (P \wedge R)$

   (c) $R \vee (P \Rightarrow Q)$

   (d) $(R \vee P) \Leftrightarrow Q$

   (e) $\sim(P \Rightarrow Q)$

   (f) $\sim(P \Rightarrow \sim(Q \wedge R))$

   (g) $P \Leftrightarrow (Q \wedge \sim R)$

   (h) $\sim(P \Leftrightarrow Q)$

2. Suppose that P is true and Q and R are false. Determine the truth of each of the forms appearing in Exercise 1.

3. Repeat Exercise 2, assuming that P and Q are false and R is true.

4. Write out the truth tables for each of the following propositions.

    (a) $(P \wedge Q) \Rightarrow R$ (b) $(P \wedge Q) \Leftrightarrow \sim Q$ (c) $(P \vee Q) \Leftrightarrow (R \vee Q)$

5. Negate the following in such a way that the symbol $\Rightarrow$ does not appear.

    (a) $P \Rightarrow (Q \wedge R)$ (b) $(P \wedge Q) \Rightarrow R$ (c) $(P \Rightarrow Q) \Rightarrow R$

6. Write out the converse, contrapositive, and negation of each of the following sentences.

    (a) If Sally wins, then Mary loses.
    (b) If 9 is odd, then the square of 9 is odd.
    (c) If all cats meow, then some dogs bark.
    (d) If John wins, then Mary loses and the school closes.

7. Determine whether each of the following is a tautology, a contradiction, or a contingency.

    (a) $(P \vee \sim Q) \Rightarrow (P \wedge Q)$

    (b) $(\sim P \wedge (P \vee Q)) \Rightarrow Q$

    (c) $(P \Rightarrow Q) \Leftrightarrow (P \wedge \sim Q)$

8. Negate $P \Leftrightarrow Q$ without using either the symbol $\Leftrightarrow$ or the symbol $\Rightarrow$.

9. Prove that $\sim P \Rightarrow Q$ is equivalent to $\sim Q \Rightarrow P$.

Section 1.3   **Disjunctive Normal Form and Simplification**

We saw before that the logical operation $P \Rightarrow Q$ can be written using only the operators $\sim$ and $\vee$. The following examples show that any propositional form that can be defined by a truth table can be expressed using only $\sim$, $\vee$, and $\wedge$. The resulting equivalent form is called its disjunctive normal form.

Example 1. Consider the "exclusive or," $P \oplus Q$, defined in Table A. We look in the last column and find that $P \oplus Q$ is true if and only if one of P or Q is true, but not both. Thus $P \oplus Q$ is equivalent to the form $(\sim P \wedge Q) \vee (P \wedge \sim Q)$, which is its disjunctive normal form.

Table A

P	Q	$P \oplus Q$
0	0	0
0	1	1
1	0	1
1	1	0

In general, we obtain the disjunctive normal form for an n-variable propositional form $f(P_1, P_2, \ldots, P_n)$ from its truth table as follows. For each row in which $f(P_1, P_2, \ldots, P_n)$ assumes the value 1, we form the conjunction $P_1 \wedge P_2 \wedge \cdots \wedge \sim P_k \wedge \cdots \wedge \sim P_n$, where we take P_k if there is a 1 in the kth position in the row and $\sim P_k$ if there is a 0 there. This conjunction is called a **minterm.** Then we form the disjunction of the minterms.

$$(P_1 \wedge \sim P_2 \wedge \cdots \wedge P_n) \vee (\sim P_1 \wedge P_2 \wedge \cdots \wedge \sim P_n) \vee \cdots \vee (P_1 \wedge P_2 \wedge \cdots \wedge P_n)$$

The disjunction of the minterms is what we call the **disjunctive normal form.** It is equivalent to the original propositional form. When $f(P_1, P_2, \ldots, P_n)$ is an absurdity, its disjunctive normal form is 0.

Example 2. We find the disjunctive normal form for the propositional form $f(P, Q, R)$ defined by Table B.

Table B	P	Q	R	$f(P, Q, R)$	Minterms
	0	0	0	0	
	0	0	1	0	
	0	1	0	1	$\sim P \wedge Q \wedge \sim R$
	0	1	1	0	
	1	0	0	1	$P \wedge \sim Q \wedge \sim R$
	1	0	1	1	$P \wedge \sim Q \wedge R$
	1	1	0	0	
	1	1	1	1	$P \wedge Q \wedge R$

The disjunctive normal form is thus

$$(\sim P \wedge Q \wedge \sim R) \vee (P \wedge \sim Q \wedge \sim R) \vee (P \wedge \sim Q \wedge R) \vee (P \wedge Q \wedge R)$$

The disjunctive normal form does not in general yield the simplest expression for a given form in terms of $\sim$, $\vee$, and $\wedge$. For example, $(\sim P \wedge \sim Q) \vee (\sim P \wedge Q) \vee (P \wedge Q)$ is the disjunctive normal form for $P \Rightarrow Q$. But we already know that $P \Rightarrow Q$ is equivalent to $\sim P \vee Q$. To simplify a given expression, we use the distributive laws and the following facts:

    (i) $\sim P \vee P$ is always true.

    (ii) If we denote an expression that is always true by 1, then $P \wedge 1$ is equivalent to P. If we denote an expression that is always false by 0, then $P \vee 0$ is equivalent to P.

We now simplify the disjunctive normal form for $P \Rightarrow Q$ to obtain $\sim P \vee Q$. We use the distributive laws within the square parentheses to see that

(1) $[(\sim P \wedge \sim Q) \vee (\sim P \wedge Q)] \vee (P \wedge Q)$

is equivalent to

(2) $[\sim P \wedge (Q \vee \sim Q)] \vee (P \wedge Q)$

Because $Q \vee \sim Q$ is always true, expression (2) is equivalent to

(3) $\sim P \vee (P \wedge Q)$

Using the distributive law again (in reverse), we find that expression (3) is equivalent to

(4) $(\sim P \vee P) \wedge (\sim P \vee Q)$

which in turn is equivalent to $\sim P \vee Q$.

So far we have seen how to obtain the disjunctive normal form of a given propositional form and how to simplify that expression by using logical identities. In what follows we will provide a criterion by which we can decide when one of two equivalent expressions for a propositional form is simpler than the other. Then we will provide a method for obtaining the simplest possible expression for a propositional form from its disjunctive normal form. The method we present is a pictorial method called the **Karnaugh map method.** It is very useful for simplifying expressions involving relatively few logical variables. (Our discussion will focus on expressions involving only three or four variables.) Other methods are better suited for expressions involving many variables. The Quine-McClusky procedure is one such method that has been implemented on the computer. A discussion of it and other simplification procedures can be found in the book by Taylor Booth, *Digital Networks and Computer Systems,* 2d edition (New York: John Wiley & Sons, 1978).

Before we turn our attention to the Karnaugh map method, we must first present some new terminology. We call the disjunction of two propositional forms A and B the "Boolean sum of A and B," and we call their conjunction the "Boolean product of A and B." A **literal** is either a logical variable or its complement. We shall denote the negation of a literal P by P' (rather than $\sim P$). Thus the expression $P \wedge Q' \wedge S$ is the Boolean product of the literals P, Q', and S. The expression $(P \wedge Q) \vee (R \wedge T)$ is the Boolean sum of the product terms $P \wedge Q$ and $R \wedge T$. In what follows we shall omit the conjunction symbol $\wedge$ between literals so that an expression such as $P \wedge Q \wedge R'$ will appear as PQR'. An expression such as $(P \wedge Q) \vee (R \wedge T)$ will be written as $PQ \vee RT$. We say that an expression for a propositional form is written as "the sum of products" if that expression is the Boolean sum of terms that are the Boolean product of literals. Thus the expression $PQR' \vee PR$ is written as the sum of products whereas the expression $P \wedge (QR' \vee R)$ is not.

Suppose that A and B are equivalent propositional forms and that both are expressed as the sum of products. We say that A is **simpler** than B if one of the following conditions holds:

1. The number of summands occurring in A is less than the number of summands appearing in B.

2. The number of summands occurring in A is the same as the number of summands appearing in B, but the total number of literals appearing in A is less than the total number of literals appearing in B.

For example, the propositional form $PQR \vee P'QR \vee PQ'R'$ is equivalent to $QR \vee PQ'R'$, but the latter is simpler because it contains fewer summands—2 rather than 3. The expressions $P \vee P'Q$ and $P \vee Q$ are equivalent and contain the same number of summands, but the latter is simpler because the total number of literals appearing in it is smaller. A sum of products expression for a propositional form is called **minimal** if no other sum of products expression is simpler.

The Karnaugh map method finds a minimal sum of products expression for a propositional form from its disjunctive normal form. The method is based on the following identity: If A and B are propositional forms, then $(A \wedge B) \vee (A' \wedge B) = B$. In particular, if two minterms differ in exactly one literal, the Boolean sum of those two minterms can be replaced by just one term. That term is the Boolean product of the literals common to both minterms. For example, the terms $PQR'T$ and $PQ'R'T$ are identical except for the literals Q and Q'. Their Boolean sum, $PQR'T \vee PQ'R'T$, is equivalent to $PR'T$. When two minterms differ in exactly one literal, they are said to be **adjacent.**

To make a Karnaugh map for a propositional form involving n variables, we first draw a rectangular grid of 2^n squares, one square for each possible minterm in n variables. We label each square in the grid with a minterm in such a way that if two squares are adjacent—that is, they share a common edge—then the minterms that label them are also adjacent. Also, the minterms that label the first and last squares in a row of the grid must be adjacent, and the minterms that label the top and bottom squares of a column in the grid must be adjacent. Grids for propositional forms involving 2, 3, and 4 variables are shown in Figures 1.4(a), 1.4(b), and 1.4(c), respectively. To find the label of a particular square, form the conjunction of the expressions labeling the row and column of that square. For instance, in Figure 1.4(c), the minterm that labels the entry in the first row, second column of the grid is $PQ'RT$.

Figure 1.4

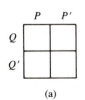

(a)

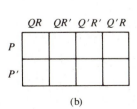

(b)

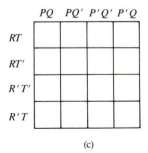

(c)

To find the Karnaugh map for propositional form A involving n variables, we enter a 1 in each square of the grid that is labeled by a minterm appearing in the disjunctive normal form of A. The Karnaugh map of the two-variable expression $P \lor Q$ appears in Figure 1.5(a), and the Karnaugh map for the three-variable expression $P \lor (Q \land R)$ appears in Figure 1.5(b).

Figure 1.5

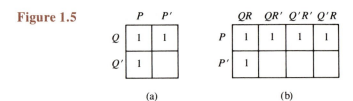

(a)   (b)

The Karnaugh map for the propositional form $A = PQ'R' \lor PQ'R \lor P'Q'R \lor P'QR'$ appears in Figure 1.6(a). There we have circled pairs of adjacent 1's. These pairs correspond to adjacent minterms in the original expression. We know from our previous discussion that each of these pairs is equivalent to a single product term. When we rewrite the expression for A as $(PQ'R' \lor PQ'R) \lor (PQ'R \lor P'Q'R) \lor P'QR'$, grouping adjacent minterms by repeating them where necessary, we see that A can be readily simplified to $A = PQ' \lor Q'R \lor P'QR$. In Figure 1.6(b), we have also circled the isolated minterm so that each minterm appears in some circle. The simplification of A can be carried out schematically on Figure 1.6(b) as follows: In each circle, take the product of each of the literals common to all minterms in that circle; then take the Boolean sum of these product terms.

Figure 1.6

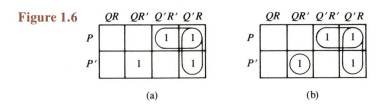

(a)   (b)

When a rectangular block of four or eight 1's occurs in a Karnaugh map, all the minterms in such a block can be replaced by a single term. That term is the product of the literals common to all minterms in the block. For example, the Karnaugh map for the propositional form $B = PQRT \lor PQ'RT \lor PQR'T \lor PQ'R'T \lor P'Q'RT$ appears in Figure 1.7. Note how we have circled the square block of four minterms appearing in the first and last rows of the grid. These correspond to the first four terms occurring in our original expression for B. The Boolean sum of these four terms is equivalent to the single term PT. Schematically we see that the minterms appearing in this block can be replaced by the single term PT—that is,

Figure 1.7

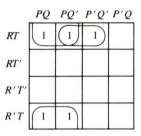

the product of the literals common to all four terms. We have also circled the adjacent pair $PQ'RT$ and $P'Q'RT$. The sum of these terms is equivalent to $Q'RT$. All terms are circled and we see that we can write B more simply as $PT \vee Q'RT$. This expression is a minimal sum of products expression for B. Note that simplifying the sum of all three terms in the top row of the grid would not lead to a single term. In fact, only rectangular blocks of two, four, or eight squares can be replaced by a single term.

The following six steps constitute an algorithm by which we use the Karnaugh map of a propositional form in four or fewer variables to find an equivalent minimal sum of products expression.

Algorithm.

1. Circle all isolated 1's.

2. Find all 1's that are adjacent to exactly one other 1. Encircle each such 1 together with its unique neighboring 1.

3. Find all 1's that are members of exactly one rectangular block of four adjacent 1's. Encircle each such 1 together with the other 1's in its block of four only if at least one member of the block has not already been included in some circle.

4. Find all 1's that belong to exactly one rectangular block of eight 1's. Encircle each such 1 together with the other 1's in its block of eight only if at least one member of that block has not already appeared in some circle.

5. Encircle each remaining 1 together with the largest possible rectangular block of two, four, or eight 1's to which it belongs. Stop as soon as each 1 has appeared in some circle.

6. From each circle appearing on the map after steps 1 through 5 have been completed, form the term that is the product of the literals common to all the minterms appearing in that circle. Form the Boolean sum of these product terms.

We apply the algorithm in the next few examples.

Example 3. The Karnaugh map for the expression $P'QR \lor PQ'R' \lor PQ'R \lor P'Q'R'$ appears in Figure 1.8. From step 1, we must circle the isolated 1 in the second row, first column. From step 2, we must encircle the last 1 in the first row together with its unique adjacent 1. Also from step 2, we must encircle the bottom 1 in the third column together with its unique adjacent 1. After steps 1 and 2 are completed, all terms are circled. From each circle we form the products of the literals common to that circle and take the Boolean sum of these products. The resulting minimal expression is $P'QR \lor PQ' \lor Q'R'$.

Figure 1.8

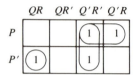

Example 4. The Karnaugh map for the expression $PQ'RT \lor P'Q'RT \lor PQR'T \lor PQ'R'T \lor P'Q'R'T \lor P'QR'T$ appears in Figure 1.9. There are no isolated 1's and there are no 1's that are adjacent to only one other 1. So there is nothing to do from steps 1 and 2. The first 1 in the bottom row is a member of only one rectangular block of four 1's and that block of four is circled in the bottom row from step 3. Similarly, the first 1 in the top row is a member of exactly one square block of four 1's. Note how we have circled that block in the top and bottom rows of the grid. All 1's are circled, and the resulting minimal expression is $R'T \lor Q'T$.

Figure 1.9

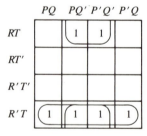

Example 5. The Karnaugh map for the expression $PQR'T \lor PQ'RT \lor P'Q'RT \lor P'Q'RT' \lor P'Q'R'T' \lor P'QR'T'$ appears in Figure 1.10. From step 1, we circle the isolated 1. From step 2, we circle the pair of adjacent 1's appearing in the first row and we circle the pair of 1's appearing in the third row. There are no blocks of four or eight 1's. Now there is a single 1 remaining in the second row. From step 5, we must circle it together with the largest block possible. There are two alternatives. These are designated in Figures 1.10(a) and 1.10(b). Either choice is valid and, depending on our choice, we obtain the minimal expression $PQR'T \lor Q'RT \lor P'R'T' \lor P'Q'T'$ or the minimal expression $PQR'T \lor Q'RT \lor P'R'T' \lor P'Q'R$.

Figure 1.10

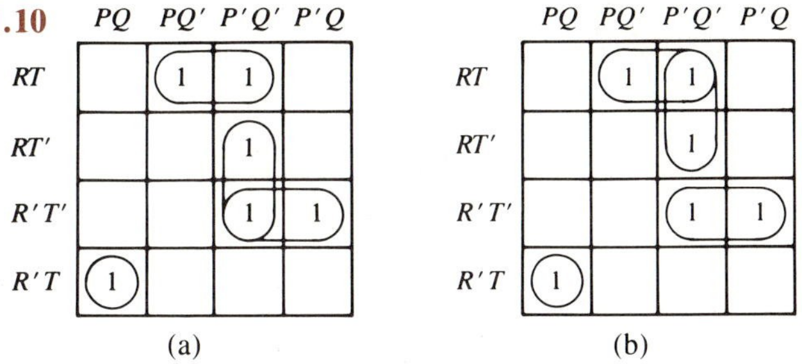

(a)   (b)

Example 6. We simplify the expression represented by the Karnaugh map appearing in Figure 1.11. There are no isolated 1's. From step 2, we must circle the adjacent pair in the third row. From step 3, we must circle the block of four 1's appearing in the top row. From step 4, we circle the block of eight 1's appearing in the first two columns. The minimal sum of products expression is thus $P \vee RT \vee Q'R'T'$.

Figure 1.11

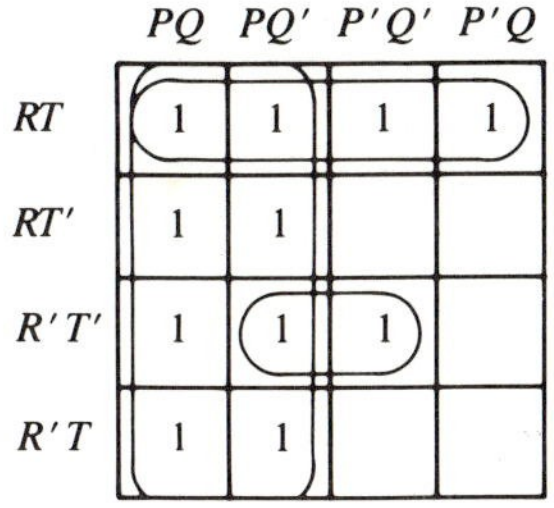

Example 7. We obtain a minimal sum of products representation of the propositional form represented by the Karnaugh map appearing in Figure 1.12. From step 2, we must circle the adjacent pair of 1's appearing in the top and bottom squares of the second column. Similarly, we must circle the adjacent pair in the last column. From step 3, we must circle the two squares of four adjacent 1's. The resulting minimal expression is $PQ'T \vee P'QT \vee Q'R \vee P'R$.

Figure 1.12

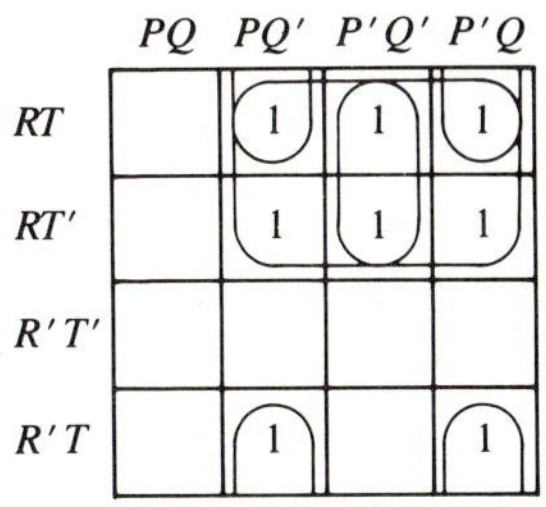

Exercises

1. Find the disjunctive normal form for each of the following.
   (a) The form $f(P, Q)$ defined by the following table:

P	Q	$f(P, Q)$
0	0	1
0	1	0
1	0	1
1	1	1

   (b) The form $f(P, Q, R)$ defined by the following table:

P	Q	R	$f(P, Q, R)$
0	0	0	1
0	0	1	0
0	1	0	1
0	1	1	0
1	0	0	1
1	0	1	0
1	1	0	0
1	1	1	1

2. Find the disjunctive normal form of each of the following propositional forms.
   (a) $P \Leftrightarrow Q$
   (b) $P \Rightarrow (Q \vee R)$
   (c) $(P \wedge Q \wedge \sim R) \vee (\sim P \wedge \sim Q \wedge R)$

3. Simplify the left side of each of the following to obtain the given equivalence.
   (a) $[(P \wedge Q) \vee (P \wedge \sim Q) \vee (\sim P \wedge Q)] \Leftrightarrow P \vee Q$
   (b) $[(P \wedge Q \wedge R) \vee (P \wedge \sim Q \wedge R) \vee (P \wedge Q \wedge \sim R)] \Leftrightarrow P \wedge (\sim Q \Rightarrow R)$

4. The *nand* (not–and) *operator*, or *Sheffer Stroke*, $P \mid Q$ is defined as $\sim(P \wedge Q)$.
   (a) Show that $P \mid P$ is equivalent to $\sim P$.
   (b) Write $P \wedge Q$ in terms of the nand operator alone.
   (c) Write $P \vee Q$ in terms of the nand operator alone.

   Parts (a), (b), and (c) show that any propositional form can be written in terms of the nand operator alone.
   (d) Write $P \Rightarrow Q$ in terms of the nand operator alone.

5. The *nor operator*, or *Pierce Arrow*, $P \downarrow Q$ is defined as $\sim(P \vee Q)$.

   (a) Write $\sim P$, $P \wedge Q$, and $P \vee Q$ in terms of the nor operator alone.

   (b) Write $P \mid Q$ in terms of the nor operator alone.

   (c) Write $P \vee (Q \wedge R)$ in terms of the nor operator alone.

6. Determine the propositional form represented by each of the following Karnaugh maps.

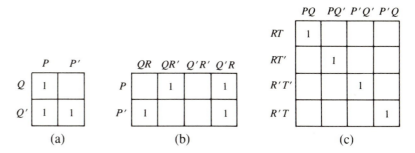

        (a)   (b)   (c)

7. Use a 2×2 Karnaugh map to find a minimal expression for each of the following propositional forms.

   (a) $PQ' \vee P'Q \vee P'Q'$ (b) $PQ \vee P'Q'$

8. Use a 2×4 Karnaugh map to find a minimal expression for each of the following expressions.

   (a) $PQ'R \vee P'QR \vee P'QR' \vee PQR'$

   (b) $PQ'R \vee P'Q'R \vee PQ'R' \vee P'Q'R' \vee P'QR'$

   (c) $PQR \vee P'QR \vee PQR' \vee P'QR'$

9. Use a 4×4 Karnaugh map to find a minimal expression for each of the following propositional forms.

   (a) $P'QRT \vee PQ'RT' \vee P'Q'RT' \vee PQ'R'T'$

   (b) $P'QRT \vee P'Q'RT' \vee PQ'R'T' \vee PQR'T \vee PQ'R'T \vee P'Q'R'T \vee P'QR'T$

   (c) $P'QRT \vee P'Q'RT' \vee P'QRT' \vee PQR'T' \vee PQ'R'T' \vee P'Q'R'T' \vee PQ'R'T \vee P'Q'R'T$

Section 1.4   **Predicates**

Sentences such as "x was the president of the United States in 1803" and "$x + y = 10$" are not propositions when the symbols x and y are used as variables. They become true or false (and hence propositions) only after the values of the variables have been specified. By setting x equal to Jefferson in the first sentence, we obtain a true proposition; by setting x equal to 1 and y equal to 5 in the second sentence, we obtain a false proposition.

An assertion that contains one or more variables is called a **predicate**: its truth value is predicated on the values assigned to its variables. A predicate P containing n variables, $x_1, x_2, \ldots, x_n$, is called an n-place predicate and is denoted by $P(x_1, x_2, \ldots, x_n)$. Before we can assign values to the variables appearing in a predicate, we must specify a **universe of discourse**—that is, a nonempty set of values in which the variables may assume their values.

Example 1. Let the universe of discourse be the set of all integers, and let $P(x, y)$ be the two-place predicate "$x \cdot y = 12$." By setting $x = 3$, we obtain the one-place predicate $P(3, y)$—namely, "$3 \cdot y = 12$." By further setting $y = 4$, we obtain the proposition $P(3, 4)$ or "$3 \cdot 4 = 12$," which is true. However, $P(2, 9)$ is false because "$2 \cdot 9 = 12$" is a false proposition.

When we specify a value for a variable appearing in a predicate, we say that we **bind** that variable. (Variables that are not bound are said to be **free**.) A predicate becomes a proposition only after each of its variables has been bound.

Example 2. Let $Q(x, y, z)$ be the three-place predicate "$x + y = z$," where x, y, and z can assume any real values. (The universe of discourse is the set of real numbers.) Then $Q(2, 5, z)$ is the one-place predicate "$2 + 5 = z$." (It has one free variable and two that are bound.) $Q(z, 3, x)$ is the two-place predicate "$z + 3 = x$." (Note the order in which z and x appear in the latter predicate.) Binding all three variables in $Q(2, 0.3, -0.66)$, we obtain the false proposition "$2 + 0.3 = -0.66$."

Let $P(x)$ be the predicate "$4 \cdot x$ is an integer multiple of 2," where x can be any integer. No matter what value is assigned to x, the resulting proposition is true. Thus the sentence "for all values of x, $P(x)$ is true" is itself a true proposition. In general, for any given predicate $P(x)$, the sentence "For all values of x, $P(x)$ is true" is a proposition. We say that the variable x is bound by **universal quantification,** because we obtain this proposition from the predicate $P(x)$ by an assertion about all the values that x can assume in its universe of discourse. We denote the words "for all" by the symbol $\forall$. Thus the expression $\forall x P(x)$ denotes the proposition "For all values of x, $P(x)$ is true" or, more simply, "For all x, $P(x)$."

Example 3.

(a) Let $Q(x)$ be the predicate "$x \cdot 1 = x$," where x can assume any real value. Then the proposition $\forall x Q(x)$ is true. Let $P(x)$ be the predicate "$x \cdot 0 = x$." Then $\forall x P(x)$ is a false proposition.

(b) Let $Q(x)$ be the predicate "x is an even integer," let $R(x)$ be the predicate "x^2 is a multiple of 4," and assume that the universe of discourse is the set of integers. Then the proposition $\forall x[Q(x) \Leftrightarrow R(x)]$ is true.

Let $Q(x)$ again represent the predicate "$x \cdot 0 = x$." Although $\forall x Q(x)$ is false, there is at least one value of x for which $Q(x)$ is true—namely, $x = 0$. Thus the sentence "There exists a value of x for which $Q(x)$ is true" is a true proposition. In this proposition the variable x is bound by **existential quantification.** We have a statement about the existence of a value of x in the universe of discourse. We denote the words "there exists" by the symbol $\exists$. The expression $\exists x Q(x)$ denotes the sentence "There exists a value of x (in the universe of discourse) for which $Q(x)$ is true" or, more simply, "There exists a value of x such that $Q(x)$."

Example 4. Let the universe of discourse be the set of real numbers. Let $P(x)$ be the predicate "$x^2 \geq 0$" and let $Q(x)$ be the predicate "$3 \cdot x > 10$." Then both the propositions $\forall x P(x)$ and $\exists x P(x)$ are true. The proposition $\exists x Q(x)$ is true (as we can see by letting $x = 5$), but $\forall x Q(x)$ is false (as we can see by letting $x = -5$).

Let $P(x, y)$ be the predicate "$x > y$," where x and y can assume any real value. Then $\exists x \forall y P(x, y)$ is the proposition "There exists a value of x such that, for all values of y, it is true that $x > y$." We can restate this as follows: "We can find a value of x such that, no matter what the value of y, we have $x > y$." This is a false proposition because x is never greater than $x + 1$. However, $\forall y \exists x P(x, y)$ is the proposition "For each value of y, there exists an x such that $x > y$." This is a true proposition because, given any value of y, we may take x to be $y + 1$. The difference is explained as follows: The proposition $\exists x \forall y P(x, y)$ asserts the existence of a specific value of x, let us say x_0, such that $P(x_0, y)$ is true no matter what the value of y; that is, $\forall y P(x_0, y)$ is true. However, the proposition $\forall y \exists x P(x, y)$ is true if for each y we can find some value of x, x perhaps changing as we change y, for which $P(x, y)$ is true. When two or more variables in a predicate are bound by quantification, they are bound in the order in which they appear before the predicate from left to right. Thus $\forall y \exists x P(x, y)$ is $\forall y [\exists x P(x, y)]$. As we have seen, the binding order is crucial to the meaning of the resulting proposition.

Example 5. We assume that the universe of discourse is the set of integers for each of the following predicates, and we transcribe them into logical notation.

(a) For each value of x, we can find a value of y such that $x + y = 0$:
    $\forall x \exists y [x + y = 0]$.

(b) There is some value of x such that, for all values of y, we have $x \cdot y = 0$:
    $\exists x \forall y [x \cdot y = 0]$.

(c) There is some value of x and some value of y for which $x \cdot y = 15$:
    $\exists x \exists y [x \cdot y = 15]$.

(d) For all values of x and y, it is true that $x \cdot y > 0$: $\forall x \forall y [x \cdot y > 0]$.

The propositions given in (a), (b), and (c) are true, whereas that given in (d) is false. We note that the order of x and y in $\forall x \forall y$ and in $\exists x \exists y$ can be changed without affecting the meaning of the resulting proposition.

Example 6. Let $Q(x, y, z)$ be the predicate "$x + y = z$," where x, y, and z can assume any real values. Then we have

(a) $\forall x \forall y \forall z Q(x, y, z)$: For every value of x, y, and z, it is true that $x + y = z$. (Letting $x = 1$, $y = 2$, and $z = 33$, we find that this is false.)

(b) $\forall x \forall y \exists z Q(x, y, z)$: For every value of x and y, we can find a value of z such that $x + y = z$. (This is a true proposition.)

(c) $\exists z \forall y \forall x Q(x, y, z)$: There exists a value of z such that, for all values of x and y, it is true that $x + y = z$. (This is false.)

If $P(x)$ is a predicate, the negation of the proposition $\forall x P(x)$ is the statement "It is not the case that $P(x)$ is true for all values of x." This is the same assertion as "There is some value of x for which $P(x)$ is not true" or, in symbols, $\exists x \sim P(x)$. Thus we have the equivalence of the propositions $\sim[\forall x P(x)]$ and $\exists x \sim P(x)$. Similarly, the negation of the proposition $\exists x P(x)$ is the statement "It is not the case that there is a value of x for which $P(x)$ is true." This is the same assertion as "For every value of x, $P(x)$ is not true" or, in symbols, $\forall x[\sim P(x)]$. Thus we have the equivalence of the propositions $\sim[\exists x P(x)]$ and $\forall x \sim P(x)$.

Example 7. Let $Q(x)$ be the predicate "$x^2 > 1$," where x can be any integer. Then the proposition $\forall x Q(x)$ is false and we have that its negation $\exists x \sim Q(x)$—namely, "There exists a value of x for which $x^2 \leq 1$"—is true. (It is satisfied by letting x equal 1 or 0.) Now let $P(x)$ represent the predicate "$2x$ is odd," where x can assume any integer value. The proposition $\exists x P(x)$ is false, so its negation $\forall x \sim P(x)$—namely, "For all values of x, $2x$ is not odd"—is true.

Example 8.

(a) Let $P(x, y)$ be the predicate "$x \cdot y = 1$," where x and y can assume any real value. The proposition $\forall x \exists y P(x, y)$ is false. (If we let $x = 0$, there is no value of y for which $x \cdot y = 1$.) Thus its negation, $\exists x \forall y[x \cdot y \neq 1]$, is true.

(b) (In this example we use the fact that $P \Rightarrow Q$ is equivalent to $\sim P \vee Q$ and hence that its negation is equivalent to $P \wedge \sim Q$ by DeMorgan's laws.) Let $Q(x)$ be the predicate "x is a prime integer" and let $R(x)$ be the predicate "x^2 is an odd integer." The proposition $\forall x[Q(x) \Rightarrow R(x)]$ is false. Its negation is the true proposition $\exists x[Q(x) \wedge \sim R(x)]$; that is, there exists a value of x that is a prime integer and whose square is not an odd integer. (The latter requirement is satisfied by letting x equal 2.)

Suppose that the universe of discourse is the set of integers and that $E(x)$ is the predicate "x is even" and $O(x)$ is the predicate "x is odd." The proposition $\forall x[E(x) \vee O(x)]$ is true, because it asserts that every integer that is not even is odd.

But the proposition $(\forall x E(x)) \vee (\forall x O(x))$ is false, because it is not the case that either all integers are even or all integers are odd. Thus, given two predicates $P(x)$ and $Q(x)$, we find that the propositions $\forall x P(x) \vee \forall x Q(x)$ and $\forall x[P(x) \vee Q(x)]$ are not equivalent. However, the propositions $\forall x P(x) \wedge \forall x Q(x)$ and $\forall x[P(x) \wedge Q(x)]$ are equivalent. They both assert that, for every value of x in the universe of discourse, both the propositions $P(x)$ and $Q(x)$ are true. The proposition $\exists x[E(x) \wedge O(x)]$ is false (it asserts the existence of an integer that is both even and odd), whereas $\exists x E(x) \wedge \exists x O(x)$ is true. Thus, given any two predicates $P(x)$ and $Q(x)$, the propositions $\exists x P(x) \wedge \exists x Q(x)$ and $\exists x[P(x) \wedge Q(x)]$ are not equivalent. However, we have that the propositions $\exists x P(x) \vee \exists x Q(x)$ and $\exists x[P(x) \vee Q(x)]$ are equivalent. They both assert that there is at least one value of x for which one of the propositions $P(x)$ or $Q(x)$ is true.

Example 9. Suppose that the universe of discourse is the set of integers. Let $Q(x)$ be the predicate "$x^2 \leq 0$" and let $P(x)$ be the predicate "x is an even prime integer." Then the proposition $\exists x Q(x) \wedge \exists x P(x)$ is true, because we may satisfy Q by letting $x = 0$ and P by letting $x = 2$. However, the proposition $\exists x[Q(x) \wedge P(x)]$ is false, because no one integer satisfies both Q and P. Now let $R(x)$ be the predicate "$x^2 > 0$" and let $S(x)$ be the predicate "$x < 2$." Then the proposition $\forall x[R(x) \vee S(x)]$ is true, because the only integer that does not satisfy R is 0, but 0 does satisfy S. However, the proposition $\forall x R(x) \vee \forall x S(x)$ is false, because it asserts that either the square of every integer is strictly greater than 0 or every integer is less than 2.

| Section 1.4 | Exercises |

1. Suppose that the universe of discourse is the set of integers. Let $P(x, y)$ be the predicate "$x - y = 0$." Transcribe each of the following.

   (a) $P(2, 3)$ (b) $P(3, 3)$ (c) $P(y, x)$

   (d) $\forall x \exists y P(x, y)$ (e) $\exists x \forall y P(y, x)$

2. Suppose that the universe of discourse is the set of real numbers. Let $P(x, y)$ be the predicate "$x \cdot y = 4$" and let $Q(x, y)$ be the predicate "$x > y$." Transcribe the following and indicate which of the propositions are true and which are false.

   (a) $P(8, 0.5)$ (b) $\exists y P(2, y)$

   (c) $\forall x \exists y P(x, y)$ (d) $\exists x \forall y P(x, y)$

   (e) $\forall x \forall y[(P(x, y) \Rightarrow Q(x, y)]$ (f) $\exists x \exists y[P(x, y) \wedge Q(x, y)]$

3. Transcribe the following into logical notation. Let the universe of discourse be the real numbers.

   (a) For any value of x, x^2 is nonnegative.

   (b) For every value of x, there is some value of y such that $x \cdot y = 1$.

   (c) There are positive values of x and y such that $x \cdot y > 0$.

   (d) There is a value of x such that, if y is positive, then $x + y$ is negative.

4. Negate the following in such a way that the symbol $\sim$ does not appear outside the square brackets.

   (a) $\forall x[x^2 > 0]$ (b) $\exists x[x \cdot 2 = 1]$

   (c) $\forall x \exists y[x + y = 1]$ (d) $\forall x \forall y[x > y \Rightarrow x^2 > y^2]$

5. Suppose the universe of discourse is the integers. Let $P(x)$ be the predicate "$x > 1$" and let $Q(x)$ be the predicate "$x < 6$." Determine which of the following propositions are true and which are false.

   (a) $\forall x[P(x) \vee Q(x)]$ (b) $\forall x P(x) \vee \forall x Q(x)$

   (c) $\exists x(P(x) \wedge Q(x))$ (d) $\exists x P(x) \wedge \exists x Q(x)$

| Section 1.5 | Valid Arguments |

Consider the following line of reasoning:

> If the butler is guilty of the crime, his shoes will be covered with mud.
> The butler's shoes are indeed covered with mud.
> Therefore the butler is guilty.

It is an example of a logical **argument.** It contains two propositions called premises: "If the butler is guilty of the crime, his shoes will be covered with mud" and "The butler's shoes are indeed covered with mud." It also contains a third proposition called a conclusion: "The butler is guilty." It is, however, an *invalid* argument. If he were guilty the butler would have muddy shoes, but the converse—that muddy shoes imply his guilt—is not necessarily true. The butler might have muddied his shoes in some activity other than the execution of the crime. In a valid argument, the premises must logically force the conclusion. We now define a **valid argument.**

> **DEFINITION.** A **valid argument** is a finite set of propositions P_1, $P_2, \ldots, P_n$ called **premises,** together with a proposition C called the **conclusion,** such that the implication $P_1 \wedge P_2 \wedge \cdots \wedge P_n \Rightarrow C$ is a tautology.

Thus an argument is valid if, whenever each of its premises $P_1, P_2, \ldots, P_n$ is true, its conclusion C is also true. (Note that if one of the premises is false, then $P_1 \wedge P_2 \wedge \cdots \wedge P_n$ is also false, and thus the implication $P_1 \wedge P_2 \wedge \cdots \wedge P_n \Rightarrow C$ is true no matter what the truth value of C.)

According to the preceding definition, the argument as to the butler's guilt is not valid. To see this, let P be the proposition "The butler is guilty of the crime" and let Q be the proposition "The butler's shoes are muddy." Then the premises of the argument are P_1: $P \Rightarrow Q$ and P_2: Q. The conclusion is the proposition P. The following line from the truth table shows that the implication $[(P \Rightarrow Q) \wedge Q] \Rightarrow P$ is not a tautology.

P	Q	$P \Rightarrow Q$	$(P \Rightarrow Q) \wedge Q$	$[(P \Rightarrow Q) \wedge Q] \Rightarrow P$
0	1	1	1	0

Both the premises $(P \Rightarrow Q)$ and Q are true, but the conclusion P is false. (Letting P have truth value 0 while Q has truth value 1 reflects our thinking that the butler's shoes might be muddy for reasons other than his guilt.)

A logical argument with premises $P_1, \ldots, P_n$ and conclusion C can be denoted in either its tautological form, $P_1 \wedge \cdots \wedge P_n \Rightarrow C$, or its inferential form:

$$P_1$$
$$\vdots$$
$$\underline{P_n}$$
$$\therefore C$$

Example 1. Each of the arguments that follow is valid, and each represents a very common form of argument. The inferential form will be given here. We leave it to the reader to verify the appropriate tautology.

(a) If John has a B in calculus, he will graduate.
John does have a B in calculus.
Therefore he will graduate.

Let P be the proposition "John has a B in calculus" and Q the proposition "He will graduate." Then the premises are $(P \Rightarrow Q)$ and P. The conclusion is Q. The inferential form is thus

$$P \Rightarrow Q$$
$$\underline{P}$$
$$\therefore Q$$

This form of valid argument is known as *modus ponens*, or the *law of detachment*.

(b) If Harvey is a dentist, then Harvey drills teeth.
Harvey does not drill teeth.
Therefore Harvey is not a dentist.

Let P be the proposition "Harvey is a dentist" and Q the proposition "Harvey drills teeth." Then the inferential form of the foregoing argument is

$$P \Rightarrow Q$$
$$\frac{\sim Q}{\therefore \sim P}$$

This form of argument is called *modus tollens,* or the *law of contraposition.*

(c) Either elephants are blue or monkeys are green.
Elephants are grey (not blue).
Therefore monkeys are green.

Let P be the proposition "Elephants are blue" and Q the proposition "Monkeys are green." Then the inferential form of this argument is

$$P \vee Q$$
$$\frac{\sim P}{\therefore Q}$$

This form of argument is called a *disjunctive syllogism.*

(d) If Mary is a senior, then Mary wears a pin.
If Mary wears a pin, then Mary will graduate.
Thus if Mary is a senior, then Mary will graduate.

Let P be the proposition "Mary is a senior," Q the proposition "Mary wears a pin," and R the proposition "Mary will graduate." Then the inferential form of the argument is

$$P \Rightarrow Q$$
$$\frac{Q \Rightarrow R}{\therefore P \Rightarrow R}$$

This form of argument is known as a *hypothetical syllogism.*

The forms of the arguments given in Example 1 are very common and are called **rules of inference.** We shall see later that they can be used to augment the set of premises of an argument and thereby simplify the determination of that argument's validity. We now give several additional rules of inference in symbolic form.

Example 1 (continued). **Additional Rules of Inference.**

(e) $\dfrac{P}{\therefore P \vee Q}$ This rule is called the *addition rule*.

(f) $\dfrac{P \wedge Q}{\therefore \qquad P}$ This rule is called *simplification*.

(g) $\begin{array}{c} P \\ \dfrac{Q}{\therefore P \wedge Q} \end{array}$ This rule is called *conjunction*.

(h) $\begin{array}{c} (P \Rightarrow Q) \wedge (R \Rightarrow S) \\ \dfrac{P \vee R}{\therefore Q \vee S} \end{array}$ This rule is called a *constructive dilemma*.

Example 2. The arguments in this example are all invalid. The forms of the first two invalid arguments given here are so common that they have acquired names.

(a) If my congressman is crazy, he likes Washington.
My congressman likes Washington.
Therefore my congressman must be crazy.

Let P represent the proposition "My congressman is crazy" and let Q represent the proposition "He likes Washington." Then we may represent the preceding argument as follows:

$$\begin{array}{c} P \Rightarrow Q \\ \dfrac{Q}{\therefore P} \end{array}$$

It is invalid because, although we are assuming that $P \Rightarrow Q$, we are not assuming that the converse is true—namely, that his liking Washington implies that he is crazy. The following line from the truth table of the tautological form of the argument proves that it is invalid.

P	Q	$P \Rightarrow Q$	$(P \Rightarrow Q) \wedge Q$	$(P \Rightarrow Q) \wedge Q \Rightarrow P$
0	1	1	1	0

Both the premises Q and $(P \Rightarrow Q)$ are true, but the conclusion P is not true. This form of invalid argument is called *affirming the consequent*. (The preceding argument as to the butler's guilt was also of this form.)

(b) If Mary is wearing green socks, she has passed algebra.
But Mary is wearing blue socks.
Unfortunately, she must have failed algebra.

Let P be the proposition "Mary is wearing green socks" and let Q be the proposition "Mary has passed algebra." Then we may represent this argument as follows:

$$P \Rightarrow Q$$
$$\underline{\sim P}$$
$$\therefore \sim Q$$

This form of invalid argument is known as *denying the antecedent*. Its fallacy again lies in assuming that when $P \Rightarrow Q$ holds, so does its converse. (In this case the contrapositive of the first premise—namely, $\sim Q \Rightarrow \sim P$—is true, but the argument incorrectly assumes that its converse—namely, $\sim P \Rightarrow \sim Q$—is true.)

(c) If interest rates are low, then housing starts will be up.
If housing starts are up, then marriage rates will be high.
Whenever interest rates are low, the economy is good.
However, the economy is not good.
Therefore marriage rates must be low.

Although we may validly conclude that interest rates are not low, we cannot conclude that housing starts are necessarily down. (Again, we would be assuming the converse of the first sentence of the argument.) Similarly, even if housing starts are down, we cannot validly conclude that marriage rates are low.

Example 3. Consider the following argument:

$$A \wedge (B \vee C)$$
$$\sim C$$
$$\underline{B \Rightarrow \sim A}$$
$$\therefore \sim A$$

The three premises of the argument cannot be true simultaneously. If $\sim C$ is true, then both A and B must be true in order that the first premise, $A \wedge (B \vee C)$, be true. However, when both A and B are true, the third premise, $B \Rightarrow \sim A$, is false. If we call the three premises P_1, P_2, and P_3, then the conjunction $P_1 \wedge P_2 \wedge P_3$ is always false. Thus the implication $P_1 \wedge P_2 \wedge P_3 \Rightarrow C$ is always true (hence it is a tautology), and the preceding argument is valid.

The following is a valid argument:

If Judy runs for office (P), she will be elected (Q).
If Judy attends the meeting (R), she will run for office (P).
Either Judy will attend the meeting (R) or she will go to Italy (I).
But Judy cannot go to Italy ($\sim I$).
Thus Judy will be elected (Q).

We can represent this argument as follows:

$$P \Rightarrow Q$$
$$R \Rightarrow P$$
$$R \vee I$$
$$\underline{\sim I}$$
$$\therefore Q$$

We shall show that this argument is valid by using the rules of inference given in Example 1. Using these rules, we can replace, simplify, or augment the set of premises of the argument. We can also use logical identities such as DeMorgan's laws to replace any premise by a logically equivalent form. Any conclusion that can be drawn by applying the rules of inference is valid. We proceed:

1. Since $R \Rightarrow P$ and $P \Rightarrow Q$, we also have that $R \Rightarrow Q$ (hypothetical syllogism).
2. Since $R \vee I$ and $\sim I$, we also have R (disjunctive syllogism).
3. Since $[(R \Rightarrow Q) \wedge R] \Rightarrow Q$ (*modus ponens*), we have that Q is a valid conclusion to the premises of the given argument. Thus this argument is valid.

Example 4. We apply the rules of inference given in Example 1 to show that the following argument is valid:

$$(P \vee Q) \Rightarrow S$$
$$S \Rightarrow R$$
$$\underline{\sim (R \vee Q)}$$
$$\therefore \sim P$$

1. $\sim (R \vee Q)$ is equivalent to $\sim R \wedge \sim Q$ by DeMorgan's laws.
2. Since $\sim R \wedge \sim Q$, we have $\sim R$ (simplification).
3. $S \Rightarrow R$ is equivalent to its contrapositive, $\sim R \Rightarrow \sim S$.
4. Since $\sim R$ and $\sim R \Rightarrow \sim S$, we have $\sim S$ (*modus ponens*).
5. Since $(P \vee Q) \Rightarrow S$ and $\sim S$, we have $\sim (P \vee Q)$ (*modus tollens*).
6. $\sim (P \vee Q)$ is equivalent to $\sim P \wedge \sim Q$ by DeMorgan's laws.
7. Since we have $\sim P \wedge \sim Q$, we have $\sim P$ by simplification.

Thus by applying the rules of inference to the premises of the argument, we can obtain $\sim P$ as a valid conclusion to the argument. The argument is valid.

To show that an argument is invalid, we must be able to assign truth values so as to make each of its premises true but its conclusion false.

Example 5. We shall rewrite in inferential form the argument given in Example 2(c).

P: Interest rates are low.
Q: Housing starts are up.
R: Marriage rates are high.
S: The economy is good.

Our argument is thus

$$P \Rightarrow Q$$
$$Q \Rightarrow R$$
$$P \Rightarrow S$$
$$\underline{\sim S}$$
$$\therefore \sim R$$

If we let P and S be false and Q and R be true, then each of the premises of the argument is true, but the conclusion, $\sim R$, is false. The argument is not valid.

The rules of inference are summarized in Table 1.2.

Table 1.2   **RULES OF INFERENCE**

#			
1. $\dfrac{P}{\therefore P \vee Q}$	$P \Rightarrow (P \vee Q)$		addition
2. $\dfrac{P \wedge Q}{\therefore P}$	$(P \wedge Q) \Rightarrow P$		simplification
3. $\dfrac{P \quad Q}{\therefore P \wedge Q}$	$(P \wedge Q) \Rightarrow (P \wedge Q)$		conjunction
4. $\dfrac{P \Rightarrow Q \quad P}{\therefore Q}$	$[(P \Rightarrow Q) \wedge P] \Rightarrow Q$		*modus ponens* (law of detachment)
5. $\dfrac{P \Rightarrow Q \quad \sim Q}{\therefore \sim P}$	$[(P \Rightarrow Q) \wedge \sim Q] \Rightarrow \sim P$		*modus tollens* (law of contraposition)
6. $\dfrac{P \vee Q \quad \sim Q}{\therefore P}$	$[(P \vee Q) \wedge \sim Q] \Rightarrow P$		disjunctive syllogism
7. $\dfrac{P \Rightarrow Q \quad Q \Rightarrow R}{\therefore P \Rightarrow R}$	$[(P \Rightarrow Q) \wedge (Q \Rightarrow R)] \Rightarrow (P \Rightarrow R)$		hypothetical syllogism
8. $\dfrac{(P \Rightarrow Q) \wedge (R \Rightarrow S) \quad P \vee R}{\therefore Q \vee S}$	$\{[(P \Rightarrow Q) \wedge (R \Rightarrow S)] \wedge (P \vee R)\} \Rightarrow (Q \vee S)$		constructive dilemma

Section 1.5	Exercises

1. Give the tautological form of *modus ponens, modus tollens,* and the hypothetical syllogism, and prove that each is a valid argument.

2. Prove that "affirming the consequent" and "denying the antecedent" are invalid by giving the appropriate lines from the truth tables of their tautological forms.

3. Rewrite the following arguments symbolically and determine the validity of each.

  (a) If John graduates he gets the job.
      John does not get the job.
      Therefore John does not graduate.

  (b) If Mary wears the green hat she leads the band.
      Mary does not wear the green hat.
      Therefore Mary does not lead the band.

  (c) Either Fred will sing or the cat will bark.
      The cat won't bark.
      Therefore Fred will sing.

4. Determine the validity of each of the following arguments.

  (a) Either John will run or Mary will speak.
      If Mary speaks, then Fred will fly and the moon is purple.
      The moon is not purple.
      Thus John will run.

  (b) If the cow jumps the cat will dance.
      If the dog barks the cow will jump.
      The dog will not bark.
      Thus the cat will not dance.

5. Determine the validity of each of the following.

(a) $P \Rightarrow \sim Q$
    $R \Rightarrow Q$
    $T \Rightarrow R$
    P
    _____
    $\therefore \sim T$

(b) $P \vee R$
    $Q \Rightarrow R$
    $T \vee \sim R$
    $\sim T$
    _____
    $\therefore P$

Proofs in Mathematics

Before the twentieth century, mathematics was not a fully rigorous subject. Its theorems were accepted as true by argument, intuition, or physical modeling. However, early in this century mathematicians saw the need to establish a rigorous criterion by which the proof of a theorem would be accepted—or not accepted—as valid. Today it is almost universally accepted that the body of facts that constitutes mathematics is those propositions that can be deduced, using the rules of inference of logic, from a set of nine premises (or axioms) known as the *Zermelo-Frankel Axiom System*. (Many of these propositions have yet to be deduced by future mathematicians!) Yet it is seldom that a proof in mathematics refers directly to these axioms. Rather, the proof of a theorem generally uses as its premises a large body of facts accepted as ultimately derivable from this axiom system and assumed to be known to the reader of the proof. How the rules of inference are applied is not always made explicit. We now give an example of the proof of a simple theorem— first as it might appear in an elementary algebra text and then in a slightly expanded form in which the logical argument is more transparent.

Theorem.   Assume that x, y, and z are real numbers and that $x \neq 0$. Assume also that x and z are rational and that $x \cdot y = z$. Then y is also rational.

PROOF 1: Since $y = z/x$ and z/x is rational, y is also rational.

PROOF 2: The premises (or hypotheses) of the theorem are as follows:

P_1: x, y, and z are real numbers.

P_2: x and z are rational numbers.

P_3: $x \neq 0$

P_4: $x \cdot y = z$

The conclusion of the theorem is

C: y is a rational number.

The theorem can be reformulated as $P_1 \wedge P_2 \wedge P_3 \wedge P_4 \Rightarrow C$.

Let Q_1 be the proposition that $y = z/x$. The preceding set of premises is augmented by the following set of premises:

Q_2: $P_1 \wedge P_3 \wedge P_4 \Rightarrow Q_1$

(Q_2 says that if x, y, and z are real numbers with $x \neq 0$, and if $x \cdot y = z$, then $y = z/x$.)

Q_3: $P_2 \wedge Q_1 \Rightarrow C$

(Q_3 says that if x and z are rational and y is the quotient x/z, then y is also rational.)

The proposition C can be validly concluded from the set of augmented premises as follows:

1. Since $P_1 \wedge P_3 \wedge P_4$ is assumed true, and $P_1 \wedge P_3 \wedge P_4 \Rightarrow Q_1$ by premise Q_2, we have that Q_1 is true—namely, that $y = z/x$. (*modus ponens*)
2. Since $P_2 \wedge Q_1$ is true, and $P_2 \wedge Q_1 \Rightarrow C$, we have that C is also true—namely, that y is a rational number. (*modus ponens*)

Thus C can be validly deduced from our augmented set of premises, and our proof is a valid argument.

We cannot provide the reader with an algorithm by which a proof can be obtained for any given theorem in mathematics. It is a fact of our logical system that no such algorithm can be formulated. However, there are several approaches that often make the steps necessary for a proof more obvious. We offer a few examples.

Example 1.   The Direct Proof. Here we simply proceed from the hypotheses or premises of the theorem (perhaps augmented) to its conclusion by using the rules of inference as Theorem 1 illustrates.

Theorem 1.   If x is an even integer, then x^2 is also even.

Note that, as written, the statement of Theorem 1 is a predicate rather than a proposition. The statement should be preceded by the words "for all values of x." The universal quantifier is often omitted in the statements of mathematical theorems and should be provided by the reader.

PROOF: Let x be an even integer. Then $x = 2m$ for some integer m. Thus $x^2 = 4(m^2)$. Since $4(m^2)$ is also a multiple of 2, we conclude that x^2 is also even. (Note that we augmented the hypothesis of the theorem with the fact that an integer is even if and only if it is a multiple of 2.)

Example 2.   The Contrapositive. If a theorem has the form $P \Rightarrow Q$, it is often easier to prove its equivalent contrapositive, $\sim Q \Rightarrow \sim P$, as illustrated by Theorem 2.

Theorem 2.   If x is an integer and x^2 is even, then x is also even.

PROOF: (We assume that x is not even and show that x^2 is not even.) Suppose x is an odd integer. Then x is of the form $(2n + 1)$ for some integer n. Squaring $(2n + 1)$, we have that $x^2 = 4n^2 + 4n + 1$, which is also an odd integer. Thus if x is not even, then x^2 is not even and the theorem is proved.

Example 3.   The Biconditional. If a theorem is of the form $P \Leftrightarrow Q$, then one must prove both the implications $P \Rightarrow Q$ and $Q \Rightarrow P$.

Theorem 3.   Let x be an integer. Then x is even if and only if x^2 is even.

PROOF: First we must prove that if x is even, then x^2 is even. But this was proved in Theorem 1. Then we must prove that if x^2 is even, then x itself is even. But this was proved in Theorem 2. Thus we have proved that x is even if and only if x^2 is even.

Example 4.   The Contradiction. Suppose that the premises or hypotheses of a theorem are $P_1, P_2, \ldots, P_n$, and suppose that the theorem is of the form $P_1 \wedge P_2 \wedge \cdots \wedge P_n \Rightarrow Q$. To prove the theorem by contradiction, we assume that $P_1, P_2, \ldots, P_n$ and $\sim Q$ are true. We then deduce the negation of one of the premises P_k and thus obtain the contradiction $P_k \wedge \sim P_k$. Since $P_k \wedge \sim P_k$ cannot be true, then it must be that $\sim Q$ cannot be true. Hence we conclude that Q is true.

Theorem 4.   There is no rational number x such that $x^2 = 2$.

PROOF: Suppose that x is a rational number that, in lowest terms, is written s/t. (Thus s and t share no common factor.) Suppose also that $(s/t)^2 = 2$. Then, multiplying both sides of this equality by t^2, we have that $s^2 = 2(t^2)$. Since its square is even, s must be even, so $s = 2m$ for some integer m. Substituting $2m$ for s, we obtain that $4(m^2) = 2(t^2)$. Canceling a factor of 2 from both sides of this last equation, we can conclude that t is also even. But if both t and s are even, we have negated the assumption that t and s share no common factor. Thus we have obtained a contradiction. We conclude that there is no rational number x such that $x^2 = 2$.

Example 5.   The Counterexample. Sometimes we wish to show that a proposition of the form $\forall x P(x)$ is false. We know from Section 1.4 that we need find only one example x_0 for which the proposition $P(x_0)$ is false. Such an example is called a **counterexample.** What follows are four false statements and the counterexamples that prove them false.

(1) STATEMENT: All multiples of 3 are odd.
COUNTEREXAMPLE: The number 6 is a multiple of 3, but 6 is not odd.

(2) STATEMENT: All triangles are equilateral.
COUNTEREXAMPLE: The 3-4-5 right triangle is not equilateral.

(3) STATEMENT: All propositional forms are tautologies.
COUNTEREXAMPLE: The propositional form $\sim P \vee Q$ is not a tautology.

(4) STATEMENT: The product of any pair of irrational numbers is itself irrational.
COUNTEREXAMPLE: Let $x = \sqrt{12}$ and let $y = \sqrt{3}$. Both x and y are irrational, but the product $x \cdot y = \sqrt{36} = 6$ is not irrational.

Section 1.6	Exercises

1. Prove directly: The product of two odd integers is odd.

2. Prove using the contrapositive: If the product of two integers x and y is odd, then both x and y are odd.

3. Prove: The product of two integers x and y is odd if and only if both x and y are odd.

4. Prove by contradiction: If x is a rational number and y is an irrational number, then the sum $x + y$ is an irrational number.

5. Prove that each of the following statements is false by giving a counterexample.

    (a) The square root of any integer is irrational.

    (b) All odd numbers are divisible by 3.

    (c) If p is an odd prime number, then $p + 2$ is also prime.

Bridge to Computer Science

1.7 Logic Circuits (optional)

Electronic devices that model logical propositional forms are easy to construct and are the basis of computer technology. Such devices are called **gates.** The leads into a gate (we go from left to right) model the logical variables P, Q, R, and so on. Each lead can assume one of two states, either "on" or "off." When a lead labeled P is on, the variable P has truth value 1, and when the lead labeled P is off, the variable P has truth value 0. The state of the outgoing lead depends on the states of the incoming leads. If the outgoing lead is on, the truth value of the propositional form modeled by the gate is 1. It is 0 if the outgoing lead is off. The output lead of any gate can be used as an input lead to any other gate.

Example 1.

   (a) The AND gate.

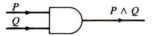

The outgoing lead of the gate is on if and only if both incoming leads are on.

   (b) The OR gate.

The outgoing lead is on if and only if one (or both) of the incoming leads is on.

   (c) The NOT gate (often called an inverter).

The outgoing lead is on if and only if the incoming lead is off.

   Since any propositional form can be expressed using only the operators $\sim$, $\wedge$, and $\vee$, we can build a circuit to model any propositional form, using only the gates described in Example 1. We shall begin with a few simple examples.

Example 2. We build a circuit to model the exclusive or, $P \oplus Q$. In Example 1 of Section 1.3, we showed that $P \oplus Q$ is equivalent to $(\sim P \wedge Q) \vee (P \wedge \sim Q)$. We work from the inner parentheses out.

    1. $P \wedge \sim Q$

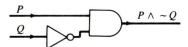

    2. $\sim P \wedge Q$

    3. $(\sim P \wedge Q) \vee (P \wedge \sim Q) = P \oplus Q$

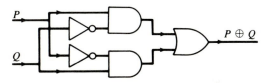

The EXCLUSIVE-OR gate is depicted as follows:

Example 3. The NAND (not–and) gate is simply

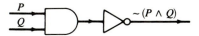

It is depicted as follows:

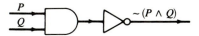

In light of Exercise 4 of Section 1.3, we can build any logical circuit using only NAND gates.

Example 4. The biconditional $P \Leftrightarrow Q$ is equivalent to the form $(P \wedge Q)$ $\vee (\sim P \wedge \sim Q)$. A logic circuit that models the biconditional can be built as follows:

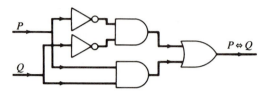

The biconditional circuit or COINCIDENCE gate is depicted as follows:

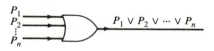

Example 5. The AND and OR gates take only two input leads. To obtain the logic circuit for $P_1 \wedge P_2 \wedge P_3 \wedge P_4$, we frame our expression as $\{[(P_1 \wedge P_2) \wedge P_3] \wedge P_4\}$ and realize the circuit as follows:

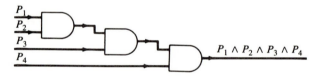

For simplicity, we depict the logic circuit for $P_1 \wedge P_2 \wedge \cdots \wedge P_n$ as follows:

Similarly, we depict the circuit for $P_1 \vee P_2 \vee \cdots \vee P_n$ as follows:

Example 6. **A Half-Adder.** We shall design a circuit to model the addition of two binary digits. For example, $1 + 1 = 10$ or $a + b = cd$. We require two outputs, the sum digit d and the carry digit c. The following truth table summarizes the required output.

a	b	c	d
0	0	0	0
0	1	0	1
1	0	0	1
1	1	1	0

The disjunctive normal form for c is $a \wedge b$. The disjunctive normal form for d is $(a \wedge {\sim}b) \vee ({\sim}a \wedge b)$, which is the same as $a \oplus b$, the exclusive or. Thus we can construct a half-adder with an AND gate and an EXCLUSIVE-OR gate as follows:

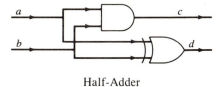

Half-Adder

When we use the half-adder later, we shall depict it as follows:

$$
\begin{array}{c}
a \\
b
\end{array}
\;
\boxed{\text{HA}}
\;
\begin{array}{c}
c \\
d
\end{array}
$$

The carry digit c will always be represented by the top outgoing lead and the sum digit d by the bottom outgoing lead.

Example 7.   A Full-Adder. When adding two binary numbers such as $1011 + 1111 = 11010$, we must have the capacity to add three digits at a time to accommodate a carried digit. Thus $a + b + c = Cd$, where C is the new carry digit. The following table summarizes the results of adding a, b, and the carried digit c.

c	a	b	d	C
0	0	0	0	0
0	0	1	1	0
0	1	0	1	0
0	1	1	0	1
1	0	0	1	0
1	0	1	0	1
1	1	0	0	1
1	1	1	1	1

We can build a full-adder directly by finding the disjunctive normal forms for C

and d directly and using AND, OR, and NOT gates, or we can build it from half-adders as shown in Figure 1.13(a). We shall depict a full-adder as in Figure 1.13(b).

Figure 1.13

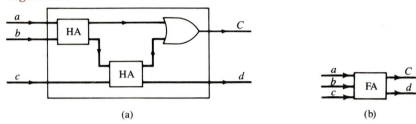

(a)   (b)

We can combine a half-adder and several full-adders to build a circuit whereby we can add two binary numbers of the form $a_n \cdots a_2 a_1$, and $b_n \cdots b_2 b_1$ to obtain $s_{n+1} s_n \cdots s_1$. A circuit that adds two 3-digit numbers can be constructed as shown in Figure 1.14.

Figure 1.14

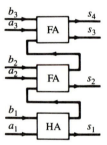

Example 8. Let $a = a_n a_{n-1} \cdots a_1$ and $b = b_n b_{n-1} \cdots b_1$ be two n-bit binary numbers. In this example we shall build a circuit where the outgoing lead g will be 1 if $b > a$ and 0 otherwise. We shall compare the bits of a and b by pairs from right to left. Let g_1 be the result of comparing the i-bit numbers $b^i = b_1 b_{1-i} \cdots b_1$ and $a^i = a_i a_{i-1} \cdots a_1$ so that $g_i = 1$ if $b^i > a^i$ and 0 otherwise. Let $g_0 = 0$. For $i = 0$ to $n - 1$, we can compute g_{i+1} by the following rule:

Rule. $g_{i+1} = 1$ if $b_{i+1} = 1$ and $a_{i+1} = 0$ or if $g_i = 1$ and $a_{i+1} = b_{i+1}$. Otherwise the value of $g_{i+1} = 0$.

The logic circuit for computing g_{i+1} when we know a_{i+1}, b_{i+1}, and g_i can be realized as follows:

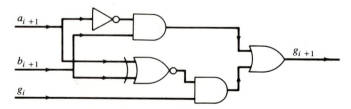

We shall denote this circuit as follows:

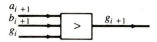

To determine whether $b > a$, we set $g = g_n$. To obtain g_n, we "hard-wire"—that is, permanently set—g_0 to 0 and combine our "$<$" gates as in Figure 1.15.

Figure 1.15

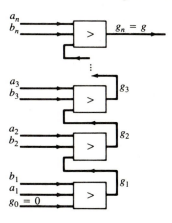

Section 1.7	**Exercises**

1. Use AND, OR, and NOT gates to build the following logic circuits.

   (a) $\sim P \vee (P \wedge R)$ (b) $(P \wedge R) \vee (P \wedge \sim Q)$ (c) $P \wedge Q \wedge R$

   (d) $P \Rightarrow Q$ (e) $P \Rightarrow (Q \Rightarrow R)$

2. Repeat parts (b) and (d) of Exercise 1, first using only OR gates and inverters and then using only NAND gates.

3. Build a full-adder using OR gates, AND gates, and inverters.

4. Suppose that all the data in a certain file are written in n-digit binary numbers $x_1 x_2 x_3 \cdots x_n$ for some fixed value of n. A "parity bit" is an extra digit x_{n+1} added to ensure that the total number of 1's is always even. Thus if $n = 4$, the parity bit added to the number 1011 is 1 and the parity bit added to the number 1010 is 0. Develop the logic and the circuitry for a parity-bit generator for $n = 3$.

5. Build a logic circuit for a three-way branch. Compare two n-bit binary numbers a and b. Set outgoing lead $s = 1$ only if $a < b$. Set outgoing lead $e = 1$ only if $a = b$. Set outgoing lead $g = 1$ only if $a > b$.

key concepts

1.1 Propositions	proposition truth value logical variable logical connectives: ∧ ("and"), ∨ ("or") negation: ~ propositional form truth table logical equivalence logical identities: DeMorgan's laws and the distributive laws
1.2 The Conditional and the Biconditional	conditional connective: ⇒ (implication) converse and contrapositive biconditional: ⇔ (if and only if) tautology, contradiction, and contingency
1.3 Disjunctive Normal Form and Simplification	disjunctive normal form literal Boolean sums and products minimal sum of products expression adjacent terms Karnaugh map
1.4 Predicates	predicate universe of discourse binding quantification: universal ($\forall x$) and existential ($\exists x$)
1.5 Valid Arguments	valid argument rules of inference (See Table 1.2.) invalid arguments (especially "affirming the consequent" and "denying the antecedent")
1.6 Proofs in Mathematics	direct proof contrapositive biconditional contradiction counterexample

Chapter 1	Exercises

1. Let P be the proposition "It is raining." Let Q be the proposition "All cats swim." Let R be the proposition "Socrates lives on." Transcribe each of the following into an English sentence.

   (a) $P \vee Q$ (b) $R \wedge {\sim}Q$ (c) ${\sim}(P \vee Q)$

   (d) ${\sim}R \wedge (Q \vee P)$ (e) $Q \Rightarrow P$

   (f) ${\sim}(R \Rightarrow Q)$ (g) $P \vee ({\sim}Q \Leftrightarrow R)$

2. Let P be the proposition "All integers are divisible by 2." Let Q be the proposition "Some integers are prime." Let R be the proposition "All primes are positive." Transcribe each of the following into an English sentence.

   (a) ${\sim}P$ (b) ${\sim}(P \wedge Q)$ (c) ${\sim}Q \wedge {\sim}R$

   (d) $(P \vee Q) \Rightarrow R$ (e) ${\sim}[P \Rightarrow (Q \wedge R)]$ (f) ${\sim}Q \vee (P \Rightarrow R)$

3. Let P be the proposition "Calculus is hard." Let Q be the proposition "All programs run." Let R be the proposition "Elephants forget." Represent each of the following sentences symbolically.

   (a) Either elephants forget or both calculus is hard and all programs run.

   (b) If elephants do not forget, then all programs run.

   (c) Some programs fail to run, and if calculus is hard, then elephants forget.

4. Let P be the proposition "All primes are odd." Let Q be the proposition "All even integers are divisible by 2." Let R be the proposition "Some integers are positive." Represent each of the following symbolically.

   (a) If all primes are odd, then some even integer is not divisible by 2.

   (b) Either no integers are positive or it is the case that all even numbers are divisible by 2 and all primes are odd.

   (c) Some primes are even if and only if all even numbers are divisible by 2.

5. Determine the truth value of the forms given in parts (a) through (g) of Exercise 1 if we assume that P and R are true and that Q is false.

6. Determine the truth value of the forms given in Exercise 1 if we assume that P and Q are false and that R is true.

7. Write a truth table for each of the forms given in Exercise 1.

8. Write a truth table for each of the following forms, and determine which are tautologies, which contradictions, and which contingencies.

   (a) $(P \vee Q) \wedge {\sim}(P \wedge Q)$

   (b) $(P \wedge {\sim}P) \Rightarrow Q$

   (c) $[P \Rightarrow (Q \wedge R)] \Leftrightarrow [P \wedge ({\sim}Q \vee {\sim}R)]$

(d) $[(P \vee Q) \wedge \sim Q] \Rightarrow P$

(e) $[(P \Rightarrow Q) \wedge \sim P] \Rightarrow \sim Q$

(f) $[(P \Rightarrow Q) \wedge (\sim R \vee Q)] \Rightarrow (P \vee \sim Q)$

9. Determine which of the following pairs of propositional forms are equivalent, and prove your results.

(a) $P \vee (\sim Q \wedge R)$ and $(Q \Rightarrow P) \wedge (\sim R \Rightarrow P)$

(b) $\sim P \wedge (Q \vee R)$ and $(P \wedge Q) \Rightarrow R$

10. Prove DeMorgan's laws by using truth tables.

11. Prove the distributive laws by using truth tables.

12. Prove that $\sim(P \wedge Q) = \sim P \vee \sim Q$, using an argument like that given in Theorem 1 of Section 1.1.

13. Prove that $P \wedge (Q \vee R) = (P \wedge Q) \vee (P \wedge R)$, using an argument like that given in Theorem 2 of Section 1.1.

14. Prove that $P \Rightarrow Q$ is equivalent to $\sim P \vee Q$.

15. Give the converse and the contrapositive of each of the following.

(a) If senators think, then Washington, D.C., is purple.

(b) That Mary is happy implies that she has passed calculus.

(c) If all mice eat cheese, then either the moon is blue or cows fly.

16. Give the converse, contrapositive, and negation of each of the following implications.

(a) If all dogs bark, then some rocks think.

(b) If 2 is an odd number, then all primes are odd.

(c) If all integers are either even or odd, then 0 is not an integer.

17. Find the disjunctive normal form for each of the forms given in Exercise 1.

18. Find the disjunctive normal form for each of the forms given in Exercise 8.

19. In each of the following, simplify the left side to obtain the given identity.

(a) $(P \vee Q) \wedge (\sim P \vee Q) = Q$

(b) $(\sim P \wedge \sim Q) \vee (P \wedge \sim Q) \vee (P \wedge Q) = (P \vee \sim Q)$

(c) $(P \wedge Q \wedge R) \vee (P \wedge \sim Q \wedge R) \vee (P \wedge \sim Q \wedge \sim R) = P \wedge (Q \Rightarrow R)$

20. Use a Karnaugh map to find a minimal expression for each of the following propositional forms.

(a) $P'Q \vee PQ' \vee P'Q$

(b) $PQ'R \vee PQR' \vee P'QR \vee P'Q'R$

(c) $PQR' \vee PQ'R \vee P'QR \vee P'QR'$

(d) $PQRT \vee P'Q'RT \vee PQ'RT' \vee P'Q'RT' \vee PQ'R'T' \vee P'Q'R'T'$

(e) $PQ'RT \lor PQRT' \lor PQ'RT' \lor P'Q'RT' \lor PQ'R'T' \lor PQ'R'T \lor P'Q'R'T$

(f) $PQ'RT \lor P'Q'RT \lor PQR'T' \lor P'Q'R'T' \lor P'QR'T' \lor PQ'R'T \lor P'Q'R'T$

21. Let $P(x, y, z)$ be the predicate "$x \cdot y = z$." Transcribe each of the following into an English sentence.

(a) $P(2, 3, 6)$ (b) $P(6, 3, 2)$

(c) $P(z, 3, x)$ (d) $\forall x \exists y P(x, y, 1)$

(e) $\forall x P(x, 1, x)$ (f) $\exists y \forall x P(x, y, x)$

22. Let the universe of discourse be the set of real numbers. Transcribe each of the following propositions into an English sentence and indicate which propositions are true and which false.

(a) $\exists x[x^2 = 2]$ (b) $\exists x \forall y[x + y = y]$

(c) $\forall y \exists x[x + y = 3]$ (d) $\exists x \forall y[x + y = 3]$

(e) $\exists y \exists z \forall x[x \cdot y = z]$ (f) $\exists z \forall x \forall y[x \cdot y = z]$

23. Assume that the universe of discourse is the set of real numbers, and transcribe the following into logical notation.

(a) There is a real number that is greater than any integer.

(b) If y is a real number greater than 0, then there is a value of x such that $x^2 = y$.

(c) It is not the case that every real number has a square root.

24. Negate each of the following in such a way that the symbol $\sim$ does not appear before a quantifier.

(a) $\exists x \forall y[x > y]$ (b) $\forall y \exists x[x^2 = y]$

(c) $\forall x \forall y[(y > 0) \Rightarrow (x \cdot y > 0)]$

25. Let $P(x)$ be the predicate "$x > 1$" and let $Q(x)$ be the predicate "$x < 3$." Let $R(x)$ be the predicate "$x < 1$." Determine which of the following propositions are true and which false if the universe of discourse is the set of real numbers.

(a) $\forall x[P(x) \lor Q(x)]$ (b) $\forall x P(x) \lor \forall x Q(x)$

(c) $\exists x P(x) \land \exists x R(x)$ (d) $\exists x[P(x) \land R(x)]$

26. Determine which of the following is a valid argument.

(a) If Mary goes to class, she is on time.
But Mary is late.
She will therefore miss class.

(b) If cats are green, then birds go south.
But no cat is green.
Thus no birds go south.

(c) Either it is snowing or it is windy.
If it is windy, John will fly a kite.
If it is snowing, the bird is singing.

But the bird is quiet.
So John will fly a kite.

(d) If I am elected, the streets will be clean.
If the sun shines, the republocrats will vote.
If the republocrats vote, I will be elected.
The sun is not shining.
Thus the streets will not be clean.

27. Determine which of the following are valid arguments.

(a) $P \vee Q$
$P \Rightarrow R$
$\sim(P \wedge Q)$
———
$\therefore Q \vee R$

(b) $P \wedge (Q \vee R)$
$R \Rightarrow \sim Q$
$\sim(P \wedge Q)$
———
$\therefore R$

(c) $P \vee (Q \Rightarrow R)$
$\sim(P \wedge Q)$
———
$\therefore P \vee \sim R$

28. Verify that the rules of inference given in parts (a), (b), and (h) of Example 1 in Section 1.5 are valid arguments.

29. Determine the validity of each of the following arguments and prove your results.

(a) $P \vee Q$
$Q \Rightarrow R$
$P \Rightarrow S$
$\sim S$
———
$\therefore R$

(b) $R \Rightarrow Q$
$R \vee T$
$\sim T \vee S$
$S \Rightarrow P$
———
$\therefore P \vee Q$

(c) $P \Rightarrow Q$
$\sim Q \vee \sim S$
$S \vee \sim P$
$\sim S$
———
$\therefore \sim Q$

(d) $P \vee Q$
$T \Rightarrow \sim Q$
$T \vee R$
———
$\therefore P \vee R$

(e) $\sim P \vee Q$
$\sim(R \vee T)$
$T \vee S$
$\sim(Q \wedge \sim R)$
———
$\therefore P \Rightarrow S$

30. Prove each of the following statements.

(a) The sum of two odd integers is even. (Use a direct proof.)

(b) If x^2 is an odd integer, then x is an odd integer. (Use the contrapositive of this statement.)

(c) It is the case that x is an odd integer if and only if x^2 is an odd integer.

(d) If x is a rational number not equal to 0, y is an irrational number, and $x \cdot y = z$, then z is an irrational number. (Prove this by contradiction.)

31. Find a counterexample to each of the following statements.

(a) The sum of two prime numbers is also prime.

(b) The sum of two irrational numbers is irrational.

(c) The square of a number is greater than the number itself.

SET THEORY

outline

2.1   Sets
2.2   Set Relations
2.3   Set Operations
2.4   Infinite Collections of Sets
2.5   Power Sets
2.6   Cartesian Products
2.7   Inductively Defined Sets
2.8   Formal Languages
2.9   Proofs by Induction
2.10 Bridge to Computer Science: A
      Compiling Problem (optional)
      Key Concepts
      Exercises

Introduction

THE STOCKS used to compile the Dow Jones Industrial Average, all possible seven-card bridge hands, the natural numbers—each of these aggregates of items is a set. Sets are the mental constructs within which we do mathematics. We perform computations on or about the members of sets, or we apply our logical skills to produce theorems about particular sets. So we sum the prices of the stocks on the Dow; we count the number of bridge hands in which all four aces appear; we prove that all natural numbers can be factored into primes.

The need to establish an abstract theory of sets became apparent at the end of the nineteenth century when philosophers and mathematicians sought to clarify the logical underpinnings of mathematics. An axiom system describing sets was laid out, and logical inference was applied. Among the facts and constructs obtained were the natural numbers and their arithmetic. From these we produce almost all of what we familiarly call mathematics.

Two of the many aspects of elementary set theory discussed in this chapter are particularly important. The first aspect is the similarity of the set operations of union, intersection, and complementation to the logical operations of disjunction, conjunction, and negation. The second aspect is induction. We shall study how to define a set inductively. To do so, we establish an **inductive procedure** or program through which we can obtain any member of the set in question in a finite number of steps. Often such an inductive procedure is readily translated into a programming language such as FORTRAN or Pascal. By storing such a program instead of listing the set members per se, we can store a potentially infinite set within the finite capacity of a machine.

Section 2.1	Sets

A **set** is a collection of objects. For example, we may speak of the set of all United States citizens or the set of all integers. All the arrangements of the letters in the word "math" form a set. All real numbers between 0 and 3 form a set. We call the objects in a given set its **elements** or **members.** A set will be denoted by an uppercase letter and its members by lowercase letters. To indicate that an element x is a member of a set A, we write, "$x \in A$." Thus if A is the set of all integers and $x = 3$, then $x \in A$. However, if $x = 3.4$, then $x \notin A$; that is, x is not an element of A. Two sets are equal if they contain the same elements. Stated precisely, the sets A and B are equal if every element x of A is an element of B and every element y of B is an element of A. In this case we write $A = B$.

We may indicate the members of a set in various ways. If the set contains only a small number of elements, we may simply list them in any order within braces. Thus the set $A = \{4, 8, 3\}$ is a set with exactly three elements. The set A may also be denoted $\{8, 4, 3\}$ because its elements may be listed in any order. Furthermore, it may be denoted $\{8, 8, 3, 3, 4\}$ because repetition of the elements 8 and 3 does not change the contents of the set. A set with exactly one element is called a **singleton set.** The set $B = \{q\}$ is a singleton set containing only the element q.

When a set contains a large or perhaps infinite number of elements and the members of the set exhibit an obvious pattern, we may indicate that pattern with ellipses. Thus the set $A = \{2, 4, 6, \ldots\}$ is the set of all even natural numbers. It contains an infinite number of elements. The set $\{7, 8, \ldots, 17, 18\}$ is the finite set of integers between 7 and 18.

Most frequently we shall use **set-builder notation.** To denote the elements of a set A given in this notation, we need a predicate $P(x)$. An element c is then a member of A if and only if $P(c)$ is true. For example, the set of all even integers E is denoted in set-builder notation as follows:

$$E = \{x : x = 2n \text{ for some integer } n\}$$

We read this as "E is the set of all values of x such that $x = 2n$ for some integer n." The predicate follows the colon, and we substitute the words "such that" for the colon. In general, set-builder notation takes the form $S = \{x : P(x)\}$, where $P(x)$ is the predicate satisfied by exactly the members of the set S.

Example 1. $X = \{x : x \text{ is a real number and } 2 < x < 5\}$ is the set of real numbers that lie strictly between 2 and 5.

Example 2. The set $Q = \{x : x \text{ is an integer and } x \text{ is divisible by 3}\}$ can also be denoted by $\{\ldots, -6, -3, 0, 3, 6, 9, \ldots\}$.

Example 3. The set $S = \{x : x \text{ is an integer and } x^2 \text{ is divisible by 4}\}$ contains exactly the same elements as $E = \{x : x \text{ is an integer and } x \text{ is divisible by 2}\}$. Thus we have $S = E$.

We will refer to the following sets frequently:

Z, the set of all integers

N, the set of all natural numbers, $\{1, 2, 3, \ldots\}$

W, the set of all whole numbers, $\{0, 1, 2, 3, \ldots\}$

R, the set of all real numbers

$\mathbf{R^+}$, the set of all real numbers strictly greater than 0

Section 2.1	Exercises

1. List the elements in each of the following sets, using braces and ellipses where necessary.

   (a) $\{x : x$ is a natural number divisible by 5$\}$

   (b) $\{x : x$ is a negative odd integer$\}$

   (c) $\{x : x$ is an even prime number$\}$

   (d) $\{x : x - 1$ is an integer divisible by 4$\}$

   (e) $\{x : x$ is an integer divisible by 2 and by 5$\}$

   (f) The set of all rearrangements of the letters in the word "math"

   (g) The set of all rearrangements of the letters in the word "boot"

2. Redefine each of the following sets, using set-builder notation.

   (a) $\{-2, -4, -6, \ldots\}$

   (b) $\{0, 3, -3, 6, -6, 9, \ldots\}$

   (c) $\{2, 3, 4, 6, 8, 9, 10, 12, 14, 15, \ldots\}$

   (d) $\{1, \frac{1}{2}, \frac{1}{3}, \frac{1}{4}, \ldots\}$

   (e) $\{0, 1, 2, \ldots, 99, 100\}$

3. Let $A = \{x : x$ is rational and $0 \leq x \leq 2\}$, and let $B = \{x : x$ is rational and $1 \leq x \leq 3\}$. Indicate whether each of the following is true or false.

   (a) $0.5 \in A$　　　　　　　　　(b) $2^{1/2} \in B$

   (c) $2.6 \in A$ or $2.6 \in B$　　　(d) $\exists x \{x \in A$ and $x \in B\}$

4. Determine which of the following pairs of sets are equal.

   (a) $S = \{x : x$ is an integer divisible by both 3 and 2$\}$ and $Q = \{6, 12, 18, 24, \ldots\}$

   (b) $X = \{x : x$ is real and $x^2 < x\}$ and $T = \{x : x$ is real and $0 < x < 1\}$

Section 2.2	Set Relations

If A and B are sets, we say that A is **contained** in B if every element of A is also an element of B. We denote the **containment** by $A \subset B$ and call A a **subset** of B. Thus if E is the set of all even integers, then $E \subset \mathbf{Z}$. If A is the set $\{x : x$ is a real number and $0 < x < 5$}, then we have neither $A \subset \mathbf{Z}$ (since $2.5 \in A$ but $2.5 \notin \mathbf{Z}$) nor $\mathbf{Z} \subset A$ (since $6 \in \mathbf{Z}$ but $6 \notin A$).

Example 1. To prove that $A \subset B$, we must prove that if x is an element of A, then x is also an element of B. Let $A = \{x : x$ is an odd integer$\}$, and let $B = \{x : x$ is an integer and x^2 is odd$\}$. We now prove that $A \subset B$.

> PROOF: Let x be an element of A. Then x is an odd integer and we may write x as $x = 2m + 1$ for some integer m. Squaring x, we have that $x^2 = 4m^2 + 4m + 1$, which is also odd. Thus if x is odd, x^2 is also odd, so x must be an element of B. Since any element x of A is also an element of B, we conclude that $A \subset B$.

With A and B as given in Example 1, we can also prove that the relation $A = B$ holds by proving that $B \subset A$. The containment relation does not preclude equality of A and B. When we have sets S and T such that $S \subset T$ but $S \neq T$, we say that S is a **proper subset** of T. In this case every element of S is an element of T, but there is at least one element of T that is not in S.

Example 2. With A and B as given in Example 1, we have that A is a subset of B, but it is not a proper subset of B. The set of natural numbers, however, is a proper subset of the set of integers.

The sets A and B are equal if and only if $A \subset B$ and $B \subset A$. So in general, to show that two sets A and B are equal, we must show that both the containments $A \subset B$ and $B \subset A$ hold.

Example 3. Let $A = \{x : x$ in $\mathbf{R}$ and $0 \leq x < 4\}$ and $B = \{x : x = y^2$ for y in $\mathbf{R}$ and $-1 < y < 2\}$. The predicates defining A and B are quite different, but we shall prove that $A = B$ by showing that $A \subset B$ and then that $B \subset A$.

> PROOF: Let x be any element of A and let $y = x^{1/2}$. Then $x = y^2$ and $0 \leq y < 2$. Thus x is also an element of B and so $A \subset B$. Now let x be any element of B and choose y such that $x = y^2$. (The existence of y is guaranteed by the predicate defining B.) Since $-1 < y < 2$, $0 \leq y^2 < 4$ and thus x is an element of A. We have shown that $B \subset A$. Since both $A \subset B$ and $B \subset A$, we have that $A = B$.

To show that A is not contained in B, we need only show that one element of A fails to be in B. For example, if $A = \{x : x = y^2$ for y in $\mathbf{Z}\}$, then A is not contained in $\mathbf{N}$, the set of natural numbers, since 0 is in A but not in $\mathbf{N}$.

Exercises

1. Let $A = \{x : x$ is an even integer$\}$, let $B = \{x : x$ is an integer divisible by 6$\}$, let $C = \{x : x$ is an integer divisible by 2 or 3$\}$, and let $D = \{x : x$ is an integer divisible by 2 and 3$\}$. Determine which of the following relations holds. If containment holds, determine whether it is proper.

   (a) $A \subset B$ (b) $B \subset C$ (c) $C \subset B$

   (d) $D \subset B$ (e) $A \subset D$ (f) $D \subset C$

   (g) $C \subset D$

2. Repeat the instructions for Exercise 1 taking A, B, C, and D as follows: $A = \{x : x \geq 1\}$, $B = \{x : x \leq 3$ or $x \geq 2\}$, $C = \{x : x \leq 3$ and $x \geq 2\}$, and $D = \{x : x > 0$ and $x^2 \geq 1\}$. Each of the sets A, B, C, and D is a subset of **R.**

3. Determine whether each of the following inclusions is proper.

   (a) $A \subset B$ where $A = \{x : x$ is an odd prime$\}$ and $B = \{x : x$ is an integer not divisible by 2$\}$

   (b) $S \subset T$ where $S = \{x : x$ is a real number with a finite decimal expansion$\}$ and $T = \{x : x$ is a rational number$\}$

   (c) $X \subset Y$ where $X = \{x : x^2$ is an integer divisible by 9$\}$ and $Y = \{x : x$ is an integer divisible by 3$\}$

4. Prove that each of the following relations holds.

   (a) $A \subset B$ where $A = \{x : x$ is an integer multiple of 10$\}$ and $B = \{x : x$ is an integer multiple of 5$\}$

   (b) $A = B$ where $A = \{x : x$ is an even integer$\}$ and $B = \{x : x^2$ is an even integer$\}$

5. Let A and B be subsets of the natural numbers defined as follows:

   $A = \{x :$ if p is a prime and if x is divisible by p, then x is divisible by $p^2\}$ and $B = \{x :$ there is an integer y such that $x = y^2\}$. Prove that $B \subset A$. Show that the containment is proper.

6. Prove that if $A \subset B$ and $B \subset C$, then $A \subset C$.

Section 2.3 **Set Operations**

Suppose $B = \{x : x$ is not a natural number$\}$. It is not clear whether we mean to include in B only negative integers, or all real numbers that are not also natural numbers, or perhaps even such items as the letters of the alphabet. Thus, in order that B be well defined, we must stipulate the **universal set** U of which all sets under discussion are considered to be subsets. If B is defined with set-builder notation, the universal set is the universe of discourse of the predicate defining B. If we take $U = \mathbf{Z}$ and $B = \{x : x \text{ is not a natural number}\}$, then $B = \{x : x \text{ is in } \mathbf{Z} \text{ and } x \leq 0\} = \{\ldots, -3, -2, -1, 0\}$. If we take U to be $\mathbf{R}$, then B contains all real numbers that are less than or equal to 0 and all positive noninteger real numbers.

DEFINITION. Let A be a subset of a universal set U. The set $\{x : x$ is in U and $x \notin A\}$ is called the **complement** of A and is denoted by A'.

Example 1. Let the universal set be the set of real numbers. Let $A = \{x : 1 \leq x \leq 2\}$. Then $A' = \{x : x < 1 \text{ or } x > 2\}$. Let B be the set of all rational numbers. Then B' is the set of all irrational numbers.

Example 2. Let U be the set of integers from 0 to 1000. Let A be the set of integers in which the digit 9 appears. Then A' is the set of numbers in which the digit 9 does not appear. There are 1001 elements in U, 271 elements in A, and 730 elements in A'. We note that every element of U is contained in either A or A' and that A and A' share no common elements. In Chapter 4 we shall develop methods for determining how many elements are in such sets as A and A'.

If we take $\mathbf{Z}$ to be our universal set and define A by $\{x : x \text{ is rational}\}$, then $A = \mathbf{Z}$, and A' has no elements. We denote the set with no elements by the symbol $\emptyset$ and call it the **empty set.** For any set A we have $\emptyset \subset A$ because no element of the empty set can fail to be in A.

Example 3. Let $X = \{x : x > 1 \text{ and } x < 0\}$. Then X is empty; that is, $X = \emptyset$.

Let $A = \{a, b, c, d\}$, $B = \{b, d, e, f\}$, $C = \{a, b, c, d, e, f\}$, and $D = \{b, d\}$. The set C contains only elements that are contained in either A or B, whereas the set D contains only elements that are in both A and B. The set C is called the **union** of A and B and the set D is called the **intersection** of A and B.

DEFINITION. Let A and B be sets. The **union** of A and B, which is written $A \cup B$, is the set $\{x : x \in A \text{ or } x \in B\}$. The **intersection** of A and B, which is written $A \cap B$, is the set $\{x : x \in A \text{ and } x \in B\}$.

For any two sets A and B, we have $A \cap B = B \cap A$ and $A \cup B = B \cup A$. The operations of union and intersection are commutative.

Example 4. Consider the following subsets of the real numbers: $A = \{x : 0 < x < 2\}$, $B = \{x : 1 < x < 3\}$, and $C = \{x : 2 < x < 4\}$. Then $A \cup B = \{x : 0 < x < 3\}$ and $A \cap B = \{x : 1 < x < 2\}$, but $A \cap C$ is empty.

When two sets A and B share no common elements—that is, when $A \cap B = \emptyset$—we say that the sets A and B are **disjoint.** The sets A and C in Example 4 are disjoint. The even integers and the odd integers are disjoint sets.

Proposition 1. Let A be a subset of U. Then

(a) $A \cap A' = \emptyset$ (b) $A \cup A' = U$

We leave the proof as an exercise.

Example 5. Let A and B be subsets of the integers defined by $A = \{x : x$ is divisible by 2$\}$ and $B = \{x : x$ is divisible by 3$\}$. Then $A \cup B = \{x : x$ is divisible by 2 or 3$\} = \{\ldots, -6, -4, -3, -2, 0, 2, 3, 4, 6, \ldots\}$. The intersection $A \cap B = \{x : x$ is divisible by 2 and 3$\} = \{\ldots, -6, 0, 6, 12, 18, \ldots\}$. Also, $A' = \{\ldots, -1, 1, 3, 5, \ldots\}$ and $B' = \{\ldots, -2, -1, 1, 2, 4, 5, \ldots\}$. So we see that $A' \cap B' = (A \cup B)' = \{\ldots, -5, -1, 1, 5, 7, \ldots\}$.

Consider the sets $A = \{a, b, c\}$, $B = \{c, d, e\}$, and $C = \{e, f\}$. Then we have $(A \cup B) \cup C = A \cup (B \cup C) = \{a, b, c, d, e, f\}$. Omitting parentheses, we can form $A \cup B \cup C$ unambiguously because the order in which we take the pairwise unions does not matter. Similarly, $A \cap (B \cap C) = (A \cap B) \cap C$, and we can form $A \cap B \cap C$ unambiguously. (In this case, $A \cap B \cap C$ is empty.) The operations of union and intersection are associative. In general, for any given collection of n sets $A_1, A_2, \ldots, A_n$, we have $A_1 \cup A_2 \cup \cdots \cup A_n = \{x : x \in A_1,$ or $x \in A_2, \ldots,$ or $x \in A_n\}$ and $A_1 \cap A_2 \cap \cdots \cap A_n = \{x : x \in A_1,$ and $x \in A_2, \ldots,$ and $x \in A_n\}$. However, $A \cap (B \cup C) = \{c\}$, whereas $(A \cap B) \cup C = \{c, e, f\}$. Parentheses are indeed necessary in expressions containing both union and intersection. In general, union and intersection are related through the distributive laws given in the following theorem.

Theorem 1.   The Distributive Laws. Let A, B, and C be sets. Then

(a) $A \cap (B \cup C) = (A \cap B) \cup (A \cap C)$
(b) $A \cup (B \cap C) = (A \cup B) \cap (A \cup C)$

PROOF:

(a) First we show that $A \cap (B \cup C) \subset (A \cap B) \cup (A \cap C)$. Let x be any element of $A \cap (B \cup C)$. Then x is an element of A, and x is an element of either B or C. If x is an element of B, then x is an element of $A \cap B$. If x is an element of C, then x is an element of $A \cap C$. Since at least one of these cases occurs, we have that x is an element of $(A \cap B) \cup (A \cap C)$

and we conclude that $A \cap (B \cup C) \subset (A \cap B) \cup (A \cap C)$. Now we show that $(A \cap B) \cup (A \cap C) \subset A \cap (B \cup C)$. Let x be any element of $(A \cap B) \cup (A \cap C)$. Then either x is an element of $A \cap B$ or x is an element of $A \cap C$. Thus x is an element of either B or C, and in either case x is an element of A. Thus x is an element of $A \cap (B \cup C)$ and we have that $(A \cap B) \cup (A \cap C) \subset A \cap (B \cup C)$. We conclude then that $A \cap (B \cup C) = (A \cap B) \cup (A \cap C)$.

We leave the similar proof of part (b) as an exercise.

Example 6. Let A_1, A_2, and A_3 be subsets of **R** defined as follows: $A_1 = \{x : -10 < x < 10\}$, $A_2 = \{x : 0 < x < 20\}$, and $A_3 = \{x : 5 < x < 25\}$. Then we have

(a) $A_1 \cup A_2 \cup A_3 = \{x : -10 < x < 25\}$

(b) $A_1 \cap A_2 \cap A_3 = \{x : 5 < x < 10\}$

(c) $(A_1 \cap A_2) \cup A_3 = (A_1 \cup A_3) \cap (A_2 \cup A_3) = \{x : 0 < x < 25\}$

(d) $A_1 \cap (A_2 \cup A_3) = (A_1 \cap A_2) \cup (A_1 \cap A_3) = \{x : 0 < x < 10\}$

In the following example, we investigate the relation between union, intersection, and complementation.

Example 7. Let the universal set U be the set of all words three letters in length, meaningful or not, formable from the English alphabet. Let A be the subset of U of all words beginning with the letter c, and let B be the subset of all words ending with the letter d. Then $A \cup B$ is the set of words that either begin with c or end with d. Some words in $A \cup B$ are caa, cab, cad, . . ., zxd, zyd, and zzd. Now $(A \cup B)'$ is the set of words neither beginning with c nor ending with d, such as aaa, aab, aac, aae, . . ., bzz, and daa. Thus $(A \cup B)'$ contains only those words not found in A and not found in B. Hence $(A \cup B)' = A' \cap B'$.

The preceding example reflects the general behavior of union, intersection, and complementation as summarized in the following theorem, which is known as DeMorgan's laws for sets.

Theorem 2.   DeMorgan's Laws. Let A and B be sets. Then

(a) $(A \cup B)' = A' \cap B'$ (b) $(A \cap B)' = A' \cup B'$

PROOF:

(b) Let x be any element of $(A \cap B)'$. Since x fails to be in the intersection of A and B, x fails to be in at least one of A or B. Thus x is an element of $A' \cup B'$ and we have that $(A \cap B)' \subset A' \cup B'$. Now let x be any element of $A' \cup B'$. Then either x is not an element of A or it is not an element of

B. In either case, x is not an element of $A \cap B$ and hence is in $(A \cap B)'$. Thus $A' \cup B' \subset (A \cap B)'$ and we conclude that $(A \cap B)' = A' \cup B'$.

We leave the proof of part (a) as an exercise. In the preceding theorem, union, intersection, and complementation are analogous to "or," "and," and "not" in DeMorgan's laws for logical propositions.

If we let $A = \{a, b, c, d, e\}$ and $B = \{a, d, f, r\}$, the elements of A that are not in B form the set $C = \{b, c, e\}$. The set C is called the set difference of A and B.

> **DEFINITION.** Let S and T be sets. The **set difference** of S and T, denoted by $S \backslash T$, is the set $S \cap T' = \{x : x \in S \text{ and } x \notin T\}$.

With A and B as above, we have that $B \backslash A = \{f, r\}$. In general, $S \backslash T \neq T \backslash S$ and, in fact, these two sets are disjoint. (We leave the proof as an exercise.)

Example 8. Let $A = \{x : x > 1\}$, $B = \{x : x > 3\}$, and $C = \{x : x < 2\}$. Then we have that $A \backslash B = \{x : 1 < x \leq 3\}$ and $B \backslash C = B$ whereas $B \backslash A = \emptyset$.

Example 9. We use DeMorgan's laws and the distributive laws to obtain the following identity: $(A \backslash B) \cup (B \backslash A) = (A \cup B) \backslash (A \cap B)$.

(a) $(A \backslash B) \cup (B \backslash A) = (A \cap B') \cup (B \cap A')$ by the definition of set difference.

(b) By the distributive laws, we have first that
$(A \cap B') \cup (B \cap A') = [(A \cap B') \cup B] \cap [(A \cap B') \cup A']$

(c) Distributing again, we have
$[(A \cap B') \cup B] \cap [(A \cap B') \cup A'] = [(A \cup B) \cap (B \cup B')]$
$\cap [(A \cup A') \cap (B' \cup A')]$

(d) Since $A \cup A' = B \cup B' = U$, we may simplify the right-hand side of the expression in step (c) to find that it is equal to
$(A \cup B) \cap (B' \cup A')$

(e) Applying DeMorgan's laws to the expression in step (d), we find that it is equal to
$(A \cup B) \cap [(A \cap B)']$

(f) Since $(A \cup B) \cap [(A \cap B)'] = (A \cup B) \backslash (A \cap B)$, our identity is proved.

(Note that we could also prove the preceding identity using an argument like that used to prove DeMorgan's laws or that used to prove the distributive laws.)

If A and B are sets, we can use the set-difference operation to express $A \cup B$ as the union of three mutually disjoint sets as shown in Proposition 2.

Proposition 2. $A \cup B = (A \backslash B) \cup (B \backslash A) \cup (A \cap B)$.

We leave the proof of Proposition 2 as an exercise. We shall see its importance to counting techniques in Chapter 4.

The important set identities are summarized in Table 2.1.

Table 2.1	IMPORTANT SET IDENTITIES	

1.	$A \cup A = A$	$A \cap A = A$	idempotent laws
2.	$A \cup \emptyset = A$	$A \cap \emptyset = \emptyset$	
3.	$A \cup U = U$	$A \cap U = A$	
4.	$A \cup A' = U$	$A \cap A' = \emptyset$	
5.	$(A \cup B) \cup C = A \cup (B \cup C)$		associativity
	$(A \cap B) \cap C = A \cap (B \cap C)$		
6.	$A \cup B = B \cup A$		commutativity
	$A \cap B = B \cap A$		
7.	$A \cup (B \cap C) = (A \cup B) \cap (A \cup C)$		distributive laws
	$A \cap (B \cup C) = (A \cap B) \cup (A \cap C)$		
8.	$(A \cup B)' = A' \cap B'$		
	$(A \cap B)' = A' \cup B'$		
9.	$(A')' = A$		
10.	$A \cup B = (A \backslash B) \cup (B \backslash A) \cup (A \cap B)$		
11.	$A \backslash (B \cup C) = (A \backslash B) \cap (A \backslash C)$		
12.	$A \backslash (B \cap C) = (A \backslash B) \cup (A \backslash C)$		

Section 2.3 **Exercises**

1. Let U be the set of letters of the alphabet. Let $A = \{a, b, c, \ldots, l\}$, $B = \{h, i, j, \ldots, q\}$, and $C = \{o, p, q, \ldots, z\}$. Find the elements in each of the following sets.

   (a) $A \cap B$ (b) $A \cup C$

   (c) $A \cap (B \cup C)$ (d) $(A \cap B) \cup C$

   (e) $A' \cap B'$ (f) $(A \cap B)'$

   (g) $A \backslash B$ (h) $B \backslash A$

   (i) $A \backslash (B \backslash C)$ (j) $A \backslash (C \backslash B)$

2. Let U be the set of integers. Let $A = \{x : x \text{ is divisible by 3}\}$, let $B = \{x : x \text{ is}$

divisible by 2}, and let $C = \{x : x$ is divisible by 5}. Repeat parts (a) through (j) of Exercise 1.

3. Answer true or false:

   (a) $A' \cup B' = (A \cup B)'$ (b) $A' = U \backslash A$

   (c) $A \cup (B \cup C) = (A \cup B) \cup C$ (d) $A \cup (B \cap C) = (A \cup B) \cap C$

   (e) $A \backslash (B \backslash C) = (A \backslash B) \backslash C$

4. Prove that $A \cup A' = U$ and $A \cap A' = \emptyset$.

5. Prove that $A \backslash B$ and $B \backslash A$ are disjoint sets.

6. Prove Proposition 2 of this section.

7. Let A and B be the following subsets of the real numbers: $A = \{x : 0 < x < 5\}$ and $B = \{x : 2 < x < 8\}$. Express $A \cup B$ as the union of three disjoint sets. (Use Proposition 2 of this section.)

Section 2.4 — Infinite Collections of Sets

For each natural number n, let $A_n = \{x : x$ is real and $-n < x < 1/n\}$. Then for each value of n, we have a different set. For example, when $n = 1$, $A_1 = \{x : -1 < x < 1\}$, and when $n = 3$, $A_3 = \left\{x : -3 < x < \frac{1}{3}\right\}$. Thus we have defined an infinite collection of sets, one set for each natural number.

In general, to speak of an infinite collection of sets we first need an indexing set L with which to name each set. For each w in L, we denote the set indexed or named by w as A_w. We refer to the collection by the notation $\{A_w : w \in L\}$. Usually we can define the members of the set A_w in terms of its index w.

Example 1. We take $\mathbf{R}^+$ to be our indexing set and, for each r in $\mathbf{R}^+$, define a subset Q_r of the real numbers as follows: $Q_r = \{x : x > r\}$. Thus $Q_{3.9} = \{x : x > 3.9\}$. Each member of the collection $\{Q_r : r \in \mathbf{R}^+\}$ is a ray on the real line. The set Q_{-3} is not defined because -3 is not in the index set.

For a collection $\{A_w : w \in L\}$, the union of all sets occurring in the collection is the set $\{x : x \in A_w$ for some w in $L\}$. We denote the union by $\bigcup_{w \in L} A_w$. It contains each of the elements of each of the sets A_w. If we take $\{A_n : n \in \mathbf{N}\}$ as defined in the first paragraph of this section, then $\bigcup_{n \in \mathbf{N}} A_n = \{x : x < 1\}$. When a collection of sets is indexed by the natural numbers, we often denote the union of the collection by $\bigcup_{n=1}^{\infty} A_n$.

Similarly, the intersection of all the sets in a collection of sets $\{A_w : w \in L\}$ is the set $\{x : x \in A_w$ for all w in $L\}$. It is denoted $\bigcap_{w \in L} A_w$. Again taking

$\{A_n : n \in \mathbf{N}\}$ as defined above, we have $\underset{n \in \mathbf{N}}{\cap} A_n = \{x : -1 < x < 0\}$. The intersection of a collection indexed by the natural numbers is often denoted by $\overset{\infty}{\underset{n=1}{\cap}} A_n$.

Example 2. For each n in **N**, let $Q_n = \{x : -1/n < x \leq 1 - (1/n)\}$. Then $\overset{\infty}{\underset{n=1}{\cup}} Q_n$ is equal to $\{x : -1 < x < 1\}$ and $\overset{\infty}{\underset{n=1}{\cap}} Q_n$ is equal to $\{0\}$. Now let $B = \{x : x < 0\}$. Then for each n in **N**, we have $B \cap Q_n = \{x : -1/n < x < 0\}$, and hence $\overset{\infty}{\underset{n=1}{\cup}} (B \cap Q_n)$ is the set $\{x : -1 < x < 0\}$, which also equals $B \cap \left(\overset{\infty}{\underset{n=1}{\cup}} Q_n \right)$. Now for each n in **N**, we have $B \cup Q_n = \{x : x \leq 1 - (1/n)\}$. Thus $\overset{\infty}{\underset{n=1}{\cap}} (B \cup Q_n)$ is equal to $\{x : x \leq 0\}$, which in turn equals $B \cup \left(\overset{\infty}{\underset{n=1}{\cap}} Q_n \right)$.

The preceding is an example of how the distributive laws extend to infinite collections of sets. We summarize this in the following theorem.

Theorem 1. Let $\{A_w : w \in L\}$ be a collection of sets and let B be a set. Then

(a) $B \cap \left(\underset{w \in L}{\cup} A_w \right) = \underset{w \in L}{\cup} (B \cap A_w)$

(b) $B \cup \left(\underset{w \in L}{\cap} A_w \right) = \underset{w \in L}{\cap} (B \cup A_w)$

PROOF:

(a) Let x be any element of $B \cap \left(\underset{w \in L}{\cup} A_w \right)$. Then x is an element of B and x is an element of A_k for some k in L. Thus x is an element of $B \cap A_k$, and x must also be an element of $\underset{w \in L}{\cup} (B \cap A_w)$. We have then that $B \cap \left(\underset{w \in L}{\cup} A_w \right) \subset \underset{w \in L}{\cup} (B \cap A_w)$. Now assume that x is an element of $\underset{w \in L}{\cup} (B \cap A_w)$. Then we can find some k in L such that x is in $B \cap A_k$. Thus x is an element of B, and x is an element of $\underset{w \in L}{\cup} A_w$. So x is an element of $B \cap \left(\underset{w \in L}{\cup} A_w \right)$ and we have that $\underset{w \in L}{\cup} (B \cap A_w) \subset B \cap \left(\underset{w \in L}{\cup} A_w \right)$ to conclude our proof.

We leave the proof of part (b) as an exercise.

Example 3. Suppose that our universal set is the subset of real numbers given by $U = \{x : 0 \leq x \leq 1\}$. For each n in **N**, define A_n to be the set $\{x : 0 \leq x < 1/n\}$. Then for each n, we have $A_n' = \{x : 1/n \leq x \leq 1\}$. Thus $\overset{\infty}{\underset{n=1}{\cap}} A_n' = \{1\} = \left(\overset{\infty}{\underset{n=1}{\cup}} A_n \right)'$ and $\overset{\infty}{\underset{n=1}{\cup}} A_n' = \{x : 0 < x \leq 1\} = \left(\overset{\infty}{\underset{n=1}{\cap}} A_n \right)'$.

The preceding example shows how DeMorgan's laws extend to infinite collections. We summarize this in the following theorem.

Theorem 2.   DeMorgan's Laws. Let $\{A_w : w \in L\}$ be a collection of sets. Then

$$\text{(a)} \quad \bigcap_{w \in L} A_w{}' = \left(\bigcup_{w \in L} A_w \right)' \qquad\qquad \text{(b)} \quad \bigcup_{w \in L} A_w{}' = \left(\bigcap_{w \in L} A_w \right)'$$

PROOF:

(a) Let x be any element of $\bigcap_{w \in L} A_w{}'$. Then, for each w in L, x is not an element of A_w. Thus x is not an element of $\bigcup_{w \in L} A_w$ and x must be an element of $\left(\bigcup_{w \in L} A_w \right)'$. So we have that $\bigcap_{w \in L} A_w{}' \subset \left(\bigcup_{w \in L} A_w \right)'$. Now assume that x is an element of $\left(\bigcup_{w \in L} A_w \right)'$. Thus x is not an element of A_w for any w in L. So, for every element w in L, x is an element of $A_w{}'$. Hence x is an element of $\bigcap_{w \in L} A_w{}'$, and $\left(\bigcup_{w \in L} A_w \right)' \subset \bigcap_{w \in L} A_w{}'$. We conclude that $\bigcap_{w \in L} A'_w = \left(\bigcup_{w \in L} A_w \right)'$.

We leave the proof of part (b) as an exercise.

| Section 2.4 | **Exercises** |

1. For each natural number k, let A_k be the subset of the real numbers defined by $A_k = \{x : -k \leq x \leq k\}$. Find each of the following.

   (a) A_4 (b) $A_5 \cap A_6$ (c) $A_3 \cup A_7$

   (d) $\bigcup_{k=1}^{\infty} A_k$ (e) $\bigcap_{k=1}^{\infty} A_k$

2. Let U be the set of real numbers. For each positive real number $r \in \mathbf{R}^+$, let $A_r = \{x : r \leq x \leq 1 + r\}$. Find $\bigcup_{r \in \mathbf{R}^+} A_r$ and $\bigcap_{r \in \mathbf{R}^+} A_r$.

3. Let U be the set of real numbers. For each natural number n, let $A_n = \{x : -n \leq x \leq 1/n\}$. Find each of the following.

   (a) $\left(\bigcup_{n=1}^{\infty} A_n \right)'$ (b) $\bigcup_{n=1}^{\infty} A_n{}'$

   (c) $\left(\bigcap_{n=1}^{\infty} A_n \right)'$ (d) $\bigcap_{n=1}^{\infty} A_n{}'$

4. For each natural number m, let $A_m = \{x : 1/(m + 2) \leq x \leq 1 - [1/(m + 1)]\}$. Let $B = \left\{ x : x \leq \frac{1}{3} \right\}$. Find each of the following.

   (a) $B \cap \left(\bigcup_{m=1}^{\infty} A_m \right)$ (b) $\bigcup_{m=1}^{\infty} (B \cap A_m)$

   (c) $B \cup \left(\bigcap_{m=1}^{\infty} A_m \right)$ (d) $\bigcap_{m=1}^{\infty} (B \cup A_m)$

Section 2.5	**Power Sets**

Consider the set $A = \{a, b\}$. We list all the subsets of A. They are $\emptyset$, $\{a\}$, $\{b\}$, and $\{a, b\}$. We form a new set $\mathcal{P}(A)$, called the power set of A, the elements of which are each of the subsets of A. Thus $\mathcal{P}(A) = \{\emptyset, \{a\}, \{b\}, \{a, b\}\}$. The set $\mathcal{P}(A)$ has four elements, each of which is a set itself. We note that $\{a, b\}$ is an element of $\mathcal{P}(A)$ but not a subset of $\mathcal{P}(A)$. The elements a and b are not members of $\mathcal{P}(A)$, but the singleton sets $\{a\}$ and $\{b\}$ are. The set $\{\{a\}, \{b\}\}$ is a subset of $\mathcal{P}(A)$.

DEFINITION. Let A be any set. The **power set** of A, denoted $\mathcal{P}(A)$, is the set of all subsets of A. The set $\mathcal{P}(A) = \{x : x \text{ is a subset of } A\}$.

Example 1. Let $A = \{a, b, c\}$. Then $\mathcal{P}(A) = \{\emptyset, \{a\}, \{b\}, \{c\}, \{a, b\}, \{a, c\}, \{b, c\}, \{a, b, c\}\}$. We note that both the empty set and the set A itself are members of the power set.

Example 2. Let $A = \emptyset$. Then $\mathcal{P}(A) = \{\emptyset\}$ and contains one element, the empty set itself.

Example 3. The power set of the set $\mathbf{N}$ is vast. It contains all the finite subsets of $\mathbf{N}$, such as $\{1, 3, 5, 9\}$ and $\{2\}$ and all the infinite subsets of $\mathbf{N}$, such as the set of all odd natural numbers and the set of all primes.

The power set of a finite set A containing n elements contains exactly 2^n elements. We prove this in Chapter 4.

Example 4. Let $A = \{a, b, c, d, e\}$, $B = \{a, b, c\}$, and $Q = \{\{a, b\}, \{d, e\}\}$. Let $\mathcal{P}(A)$ be the power set of A. Then the following relations hold:

$\{c, e\} \in \mathcal{P}(A)$

$B \in \mathcal{P}(A)$

$A \in \mathcal{P}(A)$

$Q \subset \mathcal{P}(A)$

$\{\{a\}, B\} \subset \mathcal{P}(A)$

$Q \cup \{\{a, b\}, \{b\}\} \subset \mathcal{P}(A)$

However,

$a \notin \mathcal{P}(A)$ since a itself is not a member of $\mathcal{P}(A)$.

$B \not\subset \mathcal{P}(A)$ since its members are not elements of $\mathcal{P}(A)$.

$Q \notin \mathcal{P}(A)$ but it is a subset of $\mathcal{P}(A)$.

Exercises

1. List all the members of the power set of each of the following sets.

(a) $A = \{a, b, 2, 3\}$ (b) $B = \{\text{cat, dog, mouse}\}$

(c) $C = \{\{a\}, \{b\}\}$ (d) $D = \{\emptyset, \{\emptyset\}\}$

2. List some of the elements of the power set of the integers. Use set-builder notation for at least two of these elements.

3. Let $A = \{a, b, c, d, e\}$, $B = \{a, b\}$, $C = \{B, \emptyset\}$, and $D = \{a, b, \{a, b\}\}$. Find $A \cap B$, $C \cap D$, $A \cap D$, $C \cap \mathcal{P}(A)$, and $D \cap \mathcal{P}(A)$. Indicate whether each of the following is true or false.

(a) $A \in \mathcal{P}(A)$ (b) $C \subset \mathcal{P}(A)$

(c) $D \subset \mathcal{P}(A)$ (d) $B \subset D$

(e) $B \in D$ (f) $\{a, b\} \in C$

4. Prove that if $A \subset B$, then $\mathcal{P}(A) \subset \mathcal{P}(B)$.

Cartesian Products

> **DEFINITION.** Let A and B be sets. The **cartesian product** of A and B, denoted $A \times B$, is the set of all ordered pairs (x, y) where x is an element of A and y is an element of B. That is, $A \times B = \{(x, y) : x \in A \text{ and } y \in B\}$.

Example 1. Let $A = \{a, b, c\}$ and $B = \{1, 2\}$. Then $A \times B = \{(a, 1), (b, 1), (c, 1), (a, 2), (b, 2), (c, 2)\}$. The set $B \times A = \{(1, a), (2, a), (1, b), (2, b), (1, c), (2, c)\}$.

Example 2. Let $A = \{a, b\}$. Then $A \times A = \{(a, a), (a, b), (b, a), (b, b)\}$. The term "ordered pair" refers to the fact that $(a, b) \neq (b, a)$.

Example 3. Let $A = \{a, b\}$. Then $A \times \mathbf{Z} = \{\ldots, (a, -1), (b, -1), (a, 0), (b, 0), (a, 1), (b, 1), \ldots\}$.

Example 4. The set $\mathbf{N} \times \mathbf{N} = \{(n, m) : n$ and m are natural numbers$\}$. Thus $\mathbf{N} \times \mathbf{N} = \{(1, 1), (1, 2), (1, 3), \ldots, (2, 1), (2, 2), \ldots\}$. The set $D = \{(1, 1), (2, 2), (3, 3), \ldots\}$ cannot be expressed as a cartesian product of two sets A and B. Since 1 appears as the first coordinate of an element of D, it would have to be in A. Since 2 appears as a second coordinate, it would have to be in B. The pair $(1, 2)$ would then have to be in $A \times B$. Thus $D \neq A \times B$.

Example 5. The cartesian product $\mathbf{R} \times \mathbf{R}$ is the cartesian plane. If we let I be the interval $\{x : 0 \leq x \leq 1\}$ and $J = \{x : 0 \leq x \leq 2\}$, then $I \times J$ is the subset of $\mathbf{R} \times \mathbf{R}$ that is the rectangle with corners at $(0, 0)$, $(0, 2)$, $(1, 0)$, and $(1, 2)$ and its interior. We note that whereas the circle $C = \{(x, y) : x^2 + y^2 = 1\}$ is a subset of $\mathbf{R} \times \mathbf{R}$, it is not itself of the form $A \times B$ since $(0, 1)$ and $(1, 0)$ are in C but $(1, 1)$ is not.

We may extend the notion of the cartesian product to collections of more than two sets. Suppose $A_1, A_2, \ldots, A_n$ is a collection of n sets. The cartesian product $A_1 \times A_2 \times \cdots \times A_n$ is the set of all ordered n-tuples of the form $(a_1, a_2, \ldots, a_n)$, where a_k is an element of the set A_k for $1 \leq k \leq n$.

Example 6. Let $A = \{a, b\}$, $B = \{1, 2\}$, and $C = \{x, y\}$. Then $A \times B \times C = \{(a, 1, x), (a, 1, y), (a, 2, x), (a, 2, y), (b, 1, x), (b, 1, y), (b, 2, x), (b, 2, y)\}$.

Example 7. The cartesian product $\mathbf{R} \times \mathbf{R} \times \mathbf{R}$ is the set of all ordered triples of real numbers. It is the familiar 3-space of calculus.

Section 2.6 | **Exercises**

---

1. Let $A = \{1, 2\}$, $B = \{a, b, c\}$, and $C = \{x, y\}$. List all the elements in each of the following sets.

$$A \times C, \quad B \times B, \quad A \times B \times C, \quad (A \times A) \times (C \times C)$$

2. List some of the elements in each of the following sets. Express each as a cartesian product or indicate why this cannot be done.

(a) $\{(x, y) : x$ is a positive integer and y is a negative integer$\}$

(b) $\{(x, y) : x$ is a digit and y is a letter$\}$

(c) $\{(x, y) : x \in \mathbf{N}$ and $y = x + 1\}$

(d) $\{(x, y) : 0 \le x \le 1 \text{ and } y \ge 0, \text{ where } x \text{ and } y \text{ are in } \mathbf{R}\}$

(e) $\{(x, y) : x \text{ is real and } y = x^2\}$

3. List some elements of each of the following sets. Express each as a cartesian product or indicate why this cannot be done.

(a) $\{(x, y, z) : x \text{ and } y \text{ are digits and } z \text{ is any integer}\}$

(b) $\{(x, y, z) : x, y, \text{ and } z \text{ are real and } x^2 + y^2 + z^2 = 1\}$

(c) $\{(x_1, x_2, x_3, x_4) : \text{where } x_k \text{ is real and } 0 < x_k < 1 \text{ for } k = 1, 2, 3, 4\}$

Section 2.7 | **Inductively Defined Sets**

Let S be the set of all words, meaningful or not, that can be formed from the two letters a and b. Consider its subset $L = \{a, b, aa, ab, bb, aaa, aab, abb, bbb, \ldots\}$. The set L is the subset of S of those words in which every a precedes any b. We will now give an inductive definition of the set L. An **inductive definition** of a set first establishes that one or more elements are in that set by listing them explicitly. Then it gives a routine or procedure by which to construct other members of the set from members already known to be in the set. Finally, it ensures that only the elements listed explicitly or obtained through the inductive procedure are in the set. The set L is inductively defined as follows:

(1) The words a and b are in L.

(2) If x is a word in L, then

   (i) ax is in L.

   (ii) xb is in L.

(3) Only words obtained through step (1) and iterations of step (2) are in L.

Step (1) lists a and b explicitly, so we know that they are in L. Step (2) is the inductive step that provides the procedure for building new words in the set from known words. Before applying this inductive procedure, we know only that L contains the words a and b. We apply the inductive procedure to these two words as follows. Letting $x = a$, we obtain the words aa from part (i) of the procedure and ab from part (ii). Letting $x = b$, we obtain the words ab from part (i) and bb from part (ii). (We note that the same word ab can be obtained in two ways.) Now we know that the five words a, b, aa, ab, and bb are in the set L. We reapply the inductive procedure to this set. We let $x = aa$ and obtain aaa and aab. We let $x = ab$ and obtain aab and abb. We let $x = bb$ and obtain abb and bbb. (We did not bother to reapply the procedure to the words a and b, but again there are redundancies.) We now have nine words in *L*. We repeat the procedure *ad infinitum* to generate all the words in *L*. The process may be illustrated with the following diagram:

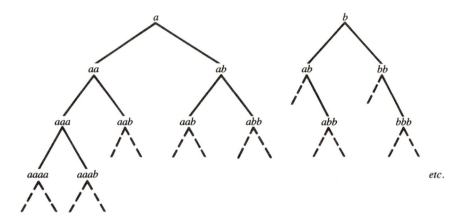

Every inductive definition has the following format:

(1) BASIS CLAUSE: This establishes one or more elements in the set explicitly, ensuring that it is not empty.

(2) INDUCTIVE CLAUSE: This establishes the procedure for obtaining new elements from elements known to be in the set. It is applied interatively (again and again).

(3) TERMINAL CLAUSE: This ensures that no elements except those obtained from the basis clause and iterations of the inductive procedure are included in the set. For brevity, we will usually say, "Only elements so obtained are in the set."

Example 1. Define the set *S* inductively as follows:

(1) $\sqrt{2}$ is in *S*.

(2) If *x* is in S, then $\sqrt{2 + x}$ is also in *S*.

(3) Only numbers so obtained are in *S*.

The basis clause (1) establishes that $\sqrt{2}$ is in *S*. We apply the inductive step (2) to $x = \sqrt{2}$ to obtain that

$$\sqrt{2 + \sqrt{2}}$$

is in *S*. Applying it again, we obtain that

$$\sqrt{2 + \sqrt{2 + \sqrt{2}}}$$

is in S. Using braces and ellipses to denote S, we have

$$S = \{\sqrt{2}, \sqrt{2 + \sqrt{2}}, \sqrt{2 + \sqrt{2 + \sqrt{2}}}, \ldots\}$$

Example 2. Paired Parentheses. In this example we inductively define the set P of all validly paired sequences of left and right parentheses. We want sequences of parentheses like $(( )( ))$ to be in P, but we do not want such sequences as $)( )($. We define P as follows:

    (1) The pair ( ) is in P.

    (2) If x is in P, then so are (x), $(\ )x$, and $x(\ )$.

    (3) Only sequences so formed are in P.

Again we can illustrate the inductive procedure with a diagram:

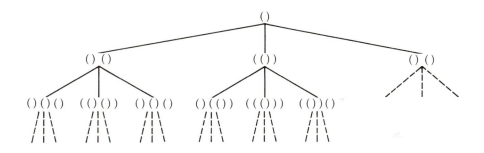

Example 3. Let A be the subset of the integers **Z** defined inductively as follows:

    (1) The numbers 4 and 6 are in A.

    (2) If x and y are any elements of A, then $x + y$ and $x - y$ are also members of A.

    (3) Only numbers so obtained are in A.

From the basis clause we know that the elements 4 and 6 are in A. When we first apply the inductive clause, we may let x equal 4 or 6 and we may let y equal 4 or 6. There are thus four possibilities. We list these and what they generate:

$x = 4$	$y = 4$	$x + y = 8$	$x - y = 0$
$x = 6$	$y = 6$	$x + y = 12$	$x - y = 0$
$x = 4$	$y = 6$	$x + y = 10$	$x - y = -2$
$x = 6$	$y = 4$	$x + y = 10$	$x - y = 2$

Thus after one application of the inductive procedure, we obtain that $-2, 0, 2, 4,$ 6, 8, 10, and 12 are all in A. Upon the second application of the induction procedure, we must let x and y range over each of these 8 elements. (There are 64 possibilities.) Continuing, we find that we may characterize A as follows: $A = \{x : 4m + 6n$ where m and n can assume any integer value$\}$. This is precisely the set of even integers, as we shall prove in Section 2.9.

Section 2.7   Exercises

For Exercises 1 through 4, list at least six elements in each of the inductively defined sets given.

1. (1) $2 \in S$.
   (2) If $x \in S$, then $3x \in S$.
   (3) Only elements so obtained are in S.

2. (1) $5 \in Q$.
   (2) If x and y are in Q, then $x + y$ and $x - y$ are in Q.
   (3) Only elements so obtained are in Q.

3. (1) The symbols ? and ! are in F.
   (2) If x is in F, then x? and !x are in F.
   (3) Only symbol sequences so obtained are in F.

4. (1) $3 \in G$ and $5 \in G$.
   (2) If x and y are in G, then so are $x + y$ and $x - y$.
   (3) Only numbers so obtained are in G.

Give an inductive definition for each of the sets given in Exercises 5 through 9.

5. $\{1, 2, 4, 8, 16, \ldots\}$

6. $\{ab, abab, ababab, \ldots\}$

7. $\{2, 4, 16, 256, \ldots\}$

8. $\{\ldots, -2, 0, 2, 4, 8, \ldots\}$

9. $\{0, 10001, 101000101, \ldots\}$

Section 2.8	**Formal Languages**

In the first paragraph of Section 2.7 we used the symbols a and b to form a set L of words (not necessarily meaningful) spelled with only these symbols. Such a set is called a language over the symbol set $\{a, b\}$. We now generalize from this example.

Let E be a finite set of symbols. (The set E is also called an **alphabet**.) A **word** over E is a finite sequence of symbols of E with possible repetitions. (The term "string" is also used synonymously with the term "word.") For example, if $E = \{a, b, c\}$, the sequence $abbcca$ is a word (or a string) over E. The sequences b, $cabcba$, a, and abc are also words over E. If we let $E = \{0, 1\}$, then $\{000, 101, 110\}$ is a set of three words over E. Given an alphabet E, a set of words over E is called a **language** over E. We use the symbol Λ to denote the empty word, the word with no symbols. When we let $E = \{s, t, u\}$, then $L = \{sts, uvu, \Lambda, s\}$ is a language over E containing four words. The language $Q = \{s, tst, ttstt, tttsttt, \ldots\}$ is a language over E containing an infinite number of words.

Let E be the set $\{a, b\}$, let x be the word aa, and let y be the word bb. We can form the new word xy, called the concatenation of x and y, which is the word $aabb$. We simply juxtapose the symbols of x with those of y in the indicated order xy. More generally, let E be an alphabet; let $a_1, a_2, \ldots, a_n$ and $b_1, b_2, \ldots, b_m$ be symbols in E; let x be the word $a_1 a_2 \ldots a_n$; and let y be the word $b_1 b_2 \cdots b_m$. The **concatenation** of x and y, namely xy, is the word $a_1 a_2 \cdots a_n b_1 b_2 \cdots b_m$. The concatenation of the empty word Λ with any word x is x itself. Thus $\Lambda x = x\Lambda = x$. If z is a word formed by the concatenation of x and y—that is, if $z = xy$— then x is called a **prefix** of z, and y is called a **suffix** of z. If x is a word over E, then we use x^n to denote the word $xxx \ldots x$—that is, x concatenated with itself n times. Thus if $x = aba$, then $x^3 = abaabaaba$. The word $a^5 b^5$ is the word $aaaaabbbbb$, but the word $(ab)^5$ is the word $ababababab$. They are different.

The set of all words formable over an alphabet E is denoted by E^*. (Thus any language over E is a subset of E^*.) If we let $E = \{a\}$ then $E^* = \{\Lambda, a, aa, aaa, \ldots\}$. If $E = \{a, b\}$, then $E^* = \{\Lambda, a, b, aa, ab, bb, aaa, \ldots\}$. If E is empty, then $E^* = \{\Lambda\}$ and contains only one element. Note that for any alphabet E, the set E^* contains the empty word Λ.

We may define E^* inductively as follows:

(1) The empty word Λ is in E^*.

(2) If x is in E^* and y is in E^*, then xy is in E^*.

(3) Only words so formed are in E^*.

Example 1. Let $E = \{a, b\}$. The set $S = \{a, bb, ab\}$ is a language over E containing only three words. The set $L = \{x : x \in E^*$ and x has the same number of a's as b's} is a language over E containing an infinite number of words including the empty word. It contains such words as *abba* and *aaabbb*, but not *abb*.

Example 2. Let $E = \{a, b\}$. We define the set of palindromes P over E (palindromes are words that read the same backwards as forwards) inductively as follows:

(1) The empty word Λ and the words a and b are in P.

(2) If x is in P, then axa and bxb are in P.

(3) Only words so formed are in P.

The set P contains such words as a, b, *aba*, *bab*, *aabbbbaa*, and *abaaaba*.

Example 3. Let $E = \{a, b\}$. We define the langauge Q over E inductively as follows:

(1) The word a is in Q.

(2) (i) If x is in Q and $x = ya$ for some y in E^*, then xbb is in Q.

    (ii) If x is in Q and $x = yb$ for some y in E^*, then xa is in Q.

(3) Only words so formed are in Q.

When we say that $x = ya$ for some y in E^*, we mean that the last symbol of x is a. Similarly, $x = yb$ means that the last symbol of x is b. On the first application of the inductive procedure, we know only that the word a is in the language. Thus only part (i) of the induction procedure is applicable. From this step we obtain the word *abb*. Apply part (ii) to this word to obtain *abba*. Iterating, we find that $Q = \{a, abb, abba, abbabb, \ldots\}$.

Example 4. Let $E = \{s, t, +, -, (,)\}$. We define the set S inductively as follows:

(1) The words s and t are in S.

(2) If x and y are in S, then so are $(+x)$, $(-x)$, $(x + y)$, and $(x - y)$.

(3) Only words so formed are in S.

The set S is a set of well-defined arithmetic expressions using s and t as variables and the operations $+$ and $-$.

Example 5. Let $E = \{1, +, =\}$. Define the set A inductively as follows:

(1) $1 + 1 = 11$ is in A.

(2) If x is a word in A of the form $y + w = z$, then the words $y1 + w = z1$ and $y + w1 = z1$ are also in A.

(3) Only words so formed are in A.

The word $1 + 1 = 11$ is of the form $y + w = z$, and thus $11 + 1 = 111$ and 1

+ 11 = 111 are also in A. If we let a word of the form $x = 11 \ldots 111$ represent the number equal to the number of 1's appearing in itself, then the language A defines the addition of numbers in **N** inductively.

One function of a compiler is to accept or reject the sequences of symbols of a given program as valid "words" in the language the program is written in. The set of compilable programs for Pascal is a language (in the sense of this section) over the symbols allowable in Pascal. In Chapter 7, we will return to the study of formal languages and investigate models for accepting or rejecting a given sequence of symbols as part of a given language.

Let us add that although we usually associate the term "language" with non-numeric symbol processes, Examples 4 and 5 of this section show that numeric processes can also be formulated in terms of words and formal languages.

Since languages over a given alphabet E are simply sets, they are subject to the usual set-theoretic operations of union, complementation, and intersection. There are, however, two additional operations that are applied to languages. We will discuss them now.

Suppose that L is a language over an alphabet E. We may take the words in L to be an alphabet itself. If we do this, then $L*$ may be defined as follows:

(1) The empty word Λ is an element of $L*$.

(2) If x is a word in $L*$ and y is a word in L, then the concatenation xy is contained in $L*$.

(3) Only words so formed are in $L*$.

Obtaining $L*$ from L is called the Kleene-star operation. We note that, given an alphabet E, we obtain the language $E*$ from E itself (regarded as a language) by the Kleene-star operation.

Example 6. Let $E = \{a, b\}$ and let $L = \{aa, aab, bb\}$. Then $L* = \{\Lambda, aa, aab, bb, aabbb, bbbb, aabaabbbbb, \ldots\}$.

Example 7. Let $E = \{0, 1\}$ and let $L = \{1, 10, 100, 1000, \ldots\}$. Then $L* = \{1, 11, \ldots, 10, 110, \ldots, 100, \ldots, 110, 1100, 11000, \ldots, 1010, \ldots\}$.

Let L and S be languages over the alphabet E. The language LS is defined as follows: $LS = \{z : z = xy$ where x is a word in L, and y is a word in $S\}$. The language LS is called the product of L and S.

Example 8. Let $E = \{a, b\}$ and let $L = \{a, aa, aaa, \ldots\}$ and $S = \{bb, bbbb, bbbbbb, \ldots\}$. Then $LS = \{abb, aabb, aaabb, \ldots, abbbb, aabbbb, \ldots\}$. The language LS contains words in which a string of a's of arbitrary length is followed by a string of an even number of b's. Note that $LS \neq SL$.

Exercises

In each of Exercises 1 through 5, a language over $E = \{a, b\}$ is given. List at least 10 words of each, where possible.

1. $L = \{x : x$ is a word that either begins with a or ends with $bb\}$

2. $S = \{x : x$ is a word four letters in length$\}$

3. $Q = \{x : x$ is a word in which every b is preceded immediately by at least two a's$\}$

4. R is defined inductively as follows:

   (1) aa and ba are in R.

   (2) If $x = yaa$ for y in E^*, then xb is in R. If $x = yba$ for y in E^*, then xbb is in R.

   (3) Only words so formed are in R.

5. T is defined inductively as follows:

   (1) $abbb$ is in T.

   (2) If z is a word in T and $z = xy$ where x is a string of a's and y is a string of b's, then the word azy is in T.

   (3) Only words so formed are in T.

In Exercises 6 through 9, give an inductive definition of the indicated language.

6. The language S over $\{a, b\}$ of all words in which a string of an odd number of b's is followed by a string of an odd number of a's

7. The language A over $\{0, 1\}$ of all words with suffix 000

8. Let w be a word in E^* and let L be the language $\{y : y = w^n$ for some n in **N**$\}$

9. The language B of all words over $\{a, b\}$ in which every a is immediately followed by at least one b

10. Let $E = \{0, 1\}$. Find L^* when:

    (a) $L = \{11\}$ (b) $L = \{00, 1\}$ (c) $L = \{10, 100, 1000\}$

11. Let $E = \{0, 1\}$. Let $L = \{x : x$ is in E^* and 10 is a prefix of $x\}$ and let $S = \{\Lambda, 00, 11\}$. Find LS and SL.

12. Let L and S be languages over $E = \{a, b\}$ defined as follows:
    $L = \{x : x$ is a word beginning with $a\}$ and
    $S = \{x : x$ is a word ending with $b\}$.

    (a) Find $L \cap S$ and $L \cup S$.

    (b) Find and characterize LS and SL.

| Section 2.9 | **Proofs by Induction** |

Consider the following propositions:

(1) For each positive integer n, we have $1 + 2 + 3 + \cdots + n = n(n + 1)/2$.

(2) For each real number r where $r \neq 1$, and positive integer n, we have
$$1 + r + r^2 + \cdots + r^n = (1 - r^{n+1})/(1 - r).$$

(3) For each positive integer n and collection of sets $A_1, \ldots, A_n$, we have
$(A_1 \cup A_2 \cup \cdots \cup A_n)' = A_1' \cap A_2' \cap \cdots \cap A_n'$. (DeMorgan's laws)

Each of these propositions has the form of a universally quantified predicate, $\forall n P(n)$, where n assumes its values in **N**. In proposition (1), the predicate $P(n)$ is the sentence "For each n, the sum $1 + 2 + \cdots + n = n(n + 1)/2$." Thus $P(3)$ is the assertion that $1 + 2 + 3 = 3 \cdot 4/2 = 6$, which is indeed true. $P(n + 1)$ is the assertion that $1 + 2 + \cdots + n + (n + 1) = (n + 1)[(n + 1) + 1]/2$. Here we have just replaced each occurrence of the symbol n by the expression $n + 1$. In proposition (3), the predicate is the sentence "For each n we have $(A_1 \cup A_2 \cup \cdots \cup A_n)' = A_1' \cap A_2' \cap \cdots \cap A_n'$." For $n = 1$, the assertion is that $A_1' = A_1'$. For $n = 2$, the assertion is that $(A_1 \cup A_2)' = A_1' \cap A_2'$, which is simply one of DeMorgan's laws for two sets. It was proved in Section 2.3.

We will prove propositions (1), (2), and (3) by **induction.** Such proofs use the following fact about the set of integers—a fact called **the Principle of Induction:**

Suppose that S is a subset of the integers and suppose that

(i) The integer n_0 is in S.

(ii) If any integer x is in S, then the integer $x + 1$ is also in S.

Then S contains all integers that are greater than or equal to n_0.

Clause (i) assures us that the set S is not empty. Clause (ii) enables us to find elements of the set inductively: Since we know that n_0 is in S, we know by clause (ii) that $n_0 + 1$ is also in S. Since we know now that $n_0 + 1$ is in S, we know also that $(n_0 + 1) + 1 = n_0 + 2$ is in the set, and so forth. Upon q applications of clause (ii), we find that $n_0 + q$ is in S. Since every integer n that is greater than or equal to n_0 can be expressed as $n_0 + q$ for some value of q, we know that all integers greater than or equal to n_0 are in S.

To prove that a predicate $\forall n P(n)$ is true by *induction*—that is, by using the Principle of Induction—we let S be the set of integers for which $P(n)$ is true. We establish that S is not empty by proving the assertion true directly for some specific integer n_0, usually $n_0 = 1$ or 0. Then we prove that if we assume $P(n)$ is true, then $P(n + 1)$ must also be true. That is, if n is an element of S, then $n + 1$ is also an element of S, thus satisfying clause (ii). The Principle of Induction then assures us that S contains all values of n greater than or equal to n_0. Thus $P(n)$ is true for all values of n greater than or equal to n_0.

We now prove proposition (1) by induction.

PROOF OF (1): Let $n = 1$. Then $P(1)$ is the proposition that $1 = 1 \cdot 2/2$, which is indeed true. Thus 1 is in the set S of values for which $P(n)$ is true. Now assume that $P(n)$ is true. That is, assume

$$1 + 2 + \cdots + n = n(n + 1)/2$$

(This assumption is called the *induction hypothesis*.) We must show that if $P(n)$ is true, then so is $P(n + 1)$. Now $P(n + 1)$ is the assertion that

$$(1 + 2 + \cdots + n) + (n + 1) = (n + 1)(n + 2)/2$$

We can now use the induction hypothesis and substitute $n(n + 1)/2$ for $(1 + 2 + \cdots + n)$ in the left side of the preceding equation. Thus we obtain $(1 + 2 + \cdots + n) + (n + 1) = [n(n + 1)/2] + (n + 1)$. Simplifying the right side of the latter expression, we obtain that $1 + 2 + \cdots + n + (n + 1) = (n + 1)(n + 2)/2$, which is $P(n + 1)$. Thus we have satisfied clause (ii) of the Principle of Induction by showing that if $P(n)$ is true, then $P(n + 1)$ is also true. We can conclude that $P(n)$ is true for all n that are greater than or equal to 1.

We will now prove propositions (2) and (3). For brevity, we usually omit any direct reference to the set S or to the Principle of Induction.

PROOF OF (2): (We prove the assertion true for all values of $n \geq 0$.) Suppose that r is a real number unequal to 1. For $n = 0$, the proposition is that $1 = (1 - r)/(1 - r)$, which is true. Now assume that the proposition is true for n. That is,

$$1 + r + r^2 + \cdots + r^n = (1 - r^{n+1})/(1 - r)$$

(This is the induction hypothesis.) We show that the proposition is true for $n + 1$. That is,

$$(1 + r + r^2 + \cdots + r^n) + r^{n+1} = (1 - r^{n+2})/(1 - r)$$

Using the induction hypothesis, we can rewrite the left side of the preceding equation as $[(1 - r^{n+1})/(1 - r)] + r^{n+1}$. Upon simplification, we find that the latter expression is equal to $(1 - r^{n+2})/(1 - r)$. We obtain that $1 + r + \cdots + r^{n+1} = (1 - r^{n+2})/(1 - r)$, as was to be shown. Thus the proposition is true for all integers n that are greater than or equal to 0.

PROOF OF (3): For $n = 1$, the assertion is that $A_1' = A_1'$, which is indeed true. Now assume that the proposition is true for the integer n—that is, that $(A_1 \cup A_2 \cup \cdots \cup A_n)' = A_1' \cap A_2' \cap \cdots \cap A_n'$. Now $(A_1 \cup A_2 \cup \cdots \cup A_{n+1})' = [(A_1 \cup A_2 \cup \cdots \cup A_n) \cup A_{n+1}]'$. By DeMorgan's laws for two sets, this is in turn equal to $(A_1 \cup A_2 \cup \cdots \cup A_n)' \cap A_{n+1}'$. Now by the induction hypothesis, the latter is equal to $(A'_1 \cap A'_2 \cap \cdots \cap A'_n)$ $\cap A'_{n+1} = A'_1 \cap A'_2 \cap \cdots \cap A'_{n+1}$, as was to be proved. Thus the proposition is true for all values of n greater than or equal to 1.

In Example 1 that follows, we shall prove that if n is an integer greater than or equal to 2, then n can be factored into a product of prime numbers. The essential part of the proof is the fact that if n is not prime, then n can be factored so that $n = m \cdot q$, where $1 < m < n$ and $1 < q < n$. But neither m nor q will equal $n - 1$, so we cannot apply induction directly. We will use the following equivalent form of induction, which is known as the **Second Principle of Induction:**

Let $P(n)$ be a predicate, where n may assume integer values greater than or equal to n_0.

(1) Suppose that $P(n_0)$ is true.

(2) Suppose that if $P(m)$ is true for all values of m such that $n_0 \leq m < n$, then $P(n)$ is also true.

Then $P(n)$ is true for all values of n such that $n \geq n_0$.

Example 1. We shall use the Second Principle of Induction to prove that if n is an integer greater than or equal to 2, then n can be factored into a product of prime integers.

> PROOF: Let $n = 2$. Since 2 is a prime number, the proposition is true for $n = 2$. Now assume that every integer m such that $2 \leq m < n$ can be factored as the product of primes. Let y be equal to n. If y is a prime number, then the desired factorization has only one factor, y itself. If y is not a prime number, then it can be factored as $y = w \cdot z$, where $2 \leq w < n$ and $2 \leq z < n$. By the induction hypothesis, we can factor both w and z as the product of primes. Multiplying these factorizations, we obtain that y can be factored as the product of primes, thus concluding our proof.

Assertions about an inductively defined set S can also be proved by induction. First we establish that the assertion is true for the elements given in the basis clause directly. Then we prove that if the assertion is true for any subset Q of S, then it is also true for those elements of S obtained from Q through the inductive procedure of the inductive clause.

Example 2. Let E be the alphabet $\{a, b\}$, and define the language S over E inductively as follows:

(1) The empty word Λ is in S.

(2) If x is a word in S, then so are abx, axb, and xab.

(3) Only words so obtained are in S.

We prove inductively the following assertion about S: For any word x in S, the number of a's in x is equal to the number of b's.

> PROOF: The assertion is true about the empty word, because it has no a's or b's. Now assume that x is a word in S and that x has n a's and n b's for some positive integer n. Then all the words that can be obtained from x via the inductive procedure—namely abx, axb, and xab—have $(n + 1)$ a's and b's. Since the assertion is true for all elements established in the basis clause and for all elements obtained from these by iterations of the inductive procedure, it is true for all elements in the set S.

Example 3. Let A be the set defined inductively in Example 3 of Section 2.7. We now prove the assertion that A is the set of even integers.

> PROOF: We prove inductively that every element of A is an even integer. The assertion is true of the elements established in the basis clause—namely 4 and 6. Let Q be a subset of A containing only elements for which the assertion is true; that is, assume that Q contains only even integers. Let x and y be elements of Q. Then, applying the inductive procedure, we have that $w = x + y$ and $z = x - y$ are elements of A. Since x and y are even and since the sum and the difference of even numbers are also even, w and z are even numbers. Thus any elements obtained from the elements of Q through the inductive procedure are even, and we conclude that all the elements of A are even. Thus $A \subset E$.
>
> We must still show that all even integers are in A. We know from the example that the numbers 2, 0, and -2 are all in A. We use induction to prove that $2n$ is an element of A for any positive integer n. The statement is true for $n = 1$ since we know that 2 is in A. Suppose that $2n$ is in A. Then, letting $x = 2n$ and $y = 2$, we obtain that $2n + 2 = 2(n + 1)$ is also in A. We conclude that $2n$ is in A for any positive n. Similarly, since -2 is an element of A, $-2n$ is in A for all positive integers n. Thus $E \subset A$ and we have proved that $E = A$.

Section 2.9   Exercises

1. Let $P(n)$ be the predicate

$$1 + 2 + 3 + \cdots + n = n(n + 1)/2$$

Find $P(5)$, $P(n + 2)$, and $P(2n + 1)$.

2. Let $Q(n)$ be the predicate

$$2 + 4 + 6 + \cdots + 2n = n(n + 1)$$

(a) Find $Q(1)$, $Q(4)$, $Q(n + 1)$, and $Q(2n)$.

(b) Prove by induction that $Q(n)$ is true for all $n > 0$.

3. Let $P(n)$ be the predicate

$$1 + 3 + 5 + \cdots + (2n - 1) = n^2$$

(a) Find $P(1)$, $P(3)$, and $P(n + 1)$.

(b) Prove by induction that $P(n)$ is true for all $n > 0$.

4. Let $Q(n)$ be the predicate

$$1 + 3 + 5 + \cdots + (2n + 1) = (n + 1)^2$$

(a) Find $Q(1)$, $Q(3)$, and $Q(n + 1)$.

(b) Prove by induction that $Q(n)$ is true for all integer values of $n > 0$.

(*Note*: **Exercises 3 and 4 both give formulas for summing consecutive odd numbers. The difference between the formulas is simply which odd number the variable n yields. For instance, if $n = 3$, the formula of Exercise 3 sums odd numbers from 1 to 5, whereas the formula in Exercise 4 sums odd numbers from 1 to 7.)**

5. Prove by induction that

$$1 + 2 + 2^2 + \cdots + 2^n = 2^{n+1} - 1$$

6. Prove by induction that

$$1 + 2^2 + 3^2 + \cdots + n^2 = n(n + 1)(2n + 1)/6$$

7. Prove the following by induction. Let $A_1, A_2, \ldots, A_n$ be a collection of n sets, and let B be a set. Then

$$B \cap (A_1 \cup A_2 \cup \cdots \cup A_n) = (B \cap A_1) \cup (B \cap A_2) \cup \cdots \cup (B \cap A_n)$$

8. Let A be the language over $E = \{a, b\}$ defined inductively as follows:

(1) The empty word Λ is in A.

(2) If x is a word in A, then axa and bbx are also in A.

(3) Only words so formed are in A.

Prove that all words in A have an even number of a's and an even number of b's.

Bridge to Computer Science

2.10 A Compiling Problem (optional)

In Chapter 2 we have studied induction—inductively defined sets and proofs by induction. An inductive definition of a formal language is particularly useful because the inductive procedure provides rules with which to construct or generate words in that language. This is not merely an academic tool. Two very important functions of a compiler are to generate code in a specific formal language (like machine language) and to check that a given sequence of symbols is part of a specified language (such as Pascal). The first task often can be accomplished by writing a program to implement the inductive procedure of an inductive definition. The second task, in turn, often can be accomplished by designing a program that, when reading the string of symbols in a word to be checked, registers which rules of an inductive procedure have been used and thus determines what symbols may follow.

In this, our second bridge to computer science, we look at two simple but useful formal languages—the set of all validly paired strings of parentheses and the set of propositional forms—with the idea of writing an inductive definition for each of them that will enable us to construct an algorithm for each that will determine when a word is part of that language and when it is not. Our small-scale project will reflect some of the key ideas used on a much larger scale in compiler design and in artificial intelligence. On all scales, these ideas are firmly based on the set theory developed in Chapter 2.

We define the set PP of all strings of validly paired parentheses inductively as follows:

(1) The empty word Λ is in PP.

(2) If x and y are words in PP, then

    (i) (x) is in PP.

    (ii) xy is in PP.

(3) Only strings so formed are in PP.

Rule (i) says we can enclose any valid string of parentheses in another pair, and rule (ii) says we can concatenate any two valid strings. If we let the symbol S represent the position into which we can substitute any validly paired string of parentheses, we can paraphrase rules (i) and (ii) by the expressions (S) and SS. With this in mind, we now give an inductive definition of a new language PPG that will include not only the words of PP but also all strings of validly paired parentheses with positions marked into which we can substitute any word from PP. Those positions will be marked by the symbol S. The alphabet for PPG is the set $\{(,), S\}$. It is defined inductively as follows:

(1) S is in PPG.

(2) If *w* is any word in *PPG* in which *S* appears, then a new word *z* in *PPG* may be obtained by taking any one of the following three steps:

   (i) Replace *S* by (*S*).

   (ii) Replace *S* by *SS*.

   (iii) Replace *S* by the empty word.

(3) Only words so obtained are in *PPG*.

A diagram for the words in *PPG* generated by a few iterations of the inductive procedure is given in Figure 2.1. The words of *PP* are those strings in which no *S* appears. They are the terminal leaves in our tree diagram because no further substitutions are possible. We have circled a few.

Figure 2.1

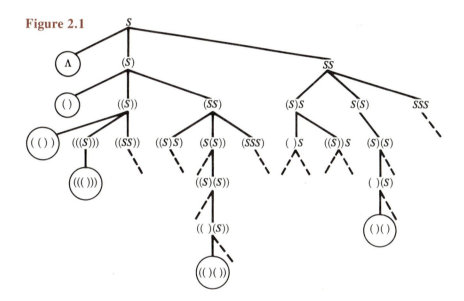

Because *PPG* has symbols for the positions into which we can substitute words, as well as words of *PP* themselves, *PPG* is an example of what is called a **grammar** in the theory of formal languages. Because we can replace *S* no matter where it appears, *PPG* is an example of a **context-free** grammar.

To obtain the sequence ( )(( )), we iterate the inductive procedure of *PPG* as follows:

(1) *S*   Begin

(2) *SS*   Replace *S* by *SS*.

(3) *S*(*S*)   Replace the *S* on the right by (*S*).

(4) (*S*)(*S*)   Replace the *S* on the left by (*S*).

(5) ( )(*S*)   Replace the *S* on the left by the empty word.

(6)   ( )((S))   Replace S by (S).

(7)   ( )(( ))   Replace S by the empty word.

Now we turn our attention to our second compiling task: We wish to develop an algorithm that will read a given string of parentheses once from left to right and determine whether it is validly paired. To do so, we shall construct an auxiliary sequence of symbols called a **stack** in which we shall register what can validly follow that part of the sequence of parentheses that has already been read. We call this sequence a stack because any manipulation of its symbols will be executed only on the last one or two symbols adjoined to it. If q is the last symbol in the stack sequence, we shall say that "q is on the stack." We shall build the stack from right to left so that its last symbol will be its left-most symbol. The symbols), S, and # will appear in the stack. The symbol # will mark the beginning (or bottom) of the stack and will be its right-most symbol. Again, S will give the position where a validly paired sequence of parentheses may occur.

Let $w = x_1 x_2 \ldots x_n$ be a sequence of parentheses. We shall mark the end of w with the symbol # so that $x_{n+1} = \#$. The algorithm to check the validity of w follows.

Algorithm I.   A Parentheses Checker.

(1) Let $i = 1$, and enter $S\#$ onto the stack. (Initially the stack is $S\#$.)

(2)   (i) Suppose $x_i = ($. If S is on the stack, then replace S by S) and go to step 3. Otherwise, stop and reject the string.

(ii) Suppose $x_i =)$. If the last two symbols on the stack are S), then remove S) from the stack, add S to the stack, and go to step 3. Otherwise, stop and reject the string.

(iii) If $x_i = \#$, go to step 4.

(3) Increment i by 1 and go to step 2.

(4) If the entire stack is $S\#$, accept the string and stop. Otherwise, reject the string and stop.

Rule 2, part (i), of the inductive definition of *PPG* says that S can be replaced by (S). This tells us that what can follow a left parenthesis, (, is any validly paired sequence of parentheses, S, and one more right parenthesis,), to close the left parenthesis being read. Step 2, part (i), of the algorithm implements this rule. It tells us that if we have read a left parenthesis, (, then we must register S) on the stack and wait for a valid sequence S and the closing parenthesis,). Rule 2, part (ii), of the inductive definition of *PPG* says that S can be replaced by SS. So any validly paired sequence of parentheses can be followed by another such sequence. Step 2, part (ii), of the algorithm implements this rule as follows. If we are reading a right parenthesis,), then we have completed the reading of a valid sequence S, closed parentheses, and are no longer waiting for S and). So we remove them from the stack. Since another valid sequence may follow, we return S to the stack.

The results of applying the algorithm to the string ( )((( ))) are summarized in Table 2.2. Since the stack is $S\#$ when the symbol $\#$ is read, the sequence ( )((( ))) is accepted.

Table 2.2

Symbol being read	Sequence remaining	Stack
none	( )((( )))$\#$	$S\#$
(	)((( )))$\#$	$S)\#$
)	((( )))$\#$	$S\#$
(	(( )))$\#$	$S)\#$
(	( )))$\#$	$S))\#$
(	)))$\#$	$S)))\#$
)	))$\#$	$S))\#$
)	)$\#$	$S)\#$
)	$\#$	$S\#$
$\#$	none	$S\#$

We now turn to the task of constructing an algorithm that reads a propositional form and checks to determine whether it is a valid form. (Such forms are often called "well-formed formulas," or wff's for short.) We shall restrict our attention to forms in which only the operations $\vee$, $\wedge$, and $\sim$ are used, and we shall take our logical variables from the finite set $V = \{p, q, r, s, t, u\}$. Define the set *WFF* as follows:

(1) If x is an element of V, then x is in *WFF*.

(2) If x and y are in *WFF*, then so are

    (i) $\sim (x)$

    (ii) $(x) \vee (y)$

    (iii) $(x) \wedge (y)$

(3) Only expressions so formed are in *WFF*.

The propositional forms in *WFF* are fully parenthesized so that although we usually write "$p \vee q$," the same form will occur in *WFF* as "$(p) \vee (q)$." Similarly, $\sim p$ will occur as "$\sim(p)$."

Let S represent the position into which we can substitute a valid propositional form (or wff), and let Q represent the position at which we can join one wff to another by $\wedge$ (conjunction) or by $\vee$ (disjunction). Then we may paraphrase rule (i) of the preceding definition as follows:

(i') S may be replaced by $\sim(S)$.

Similarly, rules (ii) and (iii) may be restated as follows:

(ii' and iii') S may be replaced by $(S)Q$, and then Q may be replaced by either $\vee (S)$ or $\wedge (S)$.

Using S and Q, we now define a language (or "grammar") *WFFG* that contains all the strings in *WFF* and also all strings in *WFF* in which places have been marked at which we may insert a wff or join one wff to another by $\wedge$ or $\vee$. The alphabet for *WFFG* is the set $\{(,), \wedge, \vee, \sim, p, q, r, s, t, u, S, Q\}$. The definition of *WFFG* is as follows:

(1) S is in *WFFG*.

(2) (i) If w is any string in *WFFG* in which S appears, then S can be replaced by $\sim(S)$, $(S)Q$, or any symbol in the set $V = \{p, q, r, s, t, u\}$ to form a new word in *WFFG*.

   (ii) If w is a string in *WFFG* in which Q appears, then Q may be replaced by either $\vee (S)$ or $\wedge (S)$ to form a new word in *WFFG*.

(3) Only strings so formed are in *WFFG*.

To derive the wff $[\sim(p)] \vee \{(q) \wedge [\sim(r)]\}$, we iterate the definition of *WFFG* as follows:

(0) S

(1) $(S)Q$

(2) $(S) \vee (S)$

(3) $(S) \vee ((S)Q)$

(4) $(S) \vee ((S) \wedge (S))$

(5) $(S) \vee ((S) \wedge (\sim(S)))$

(6) $(\sim(S)) \vee ((S) \wedge (\sim(S)))$

(7) $(\sim(p)) \vee ((S) \wedge (\sim(S)))$

(8) $(\sim(p)) \vee ((q) \wedge (\sim(S)))$

(9) $(\sim(p)) \vee ((q) \wedge (\sim(r)))$

Let w be a string of symbols over the symbol set $\{(,), \sim, \vee, \wedge, p, q, r, s, t, u\}$. To check that w is a wff, we will again develop an algorithm that uses a stack. The stack will be constructed from right to left from the symbols set $\{(,), S, Q, \#\}$. As we read the symbols of w, we will register in the stack the kinds of expressions that can validly follow that part of w that has already been read. For example, if the symbol $\sim$ is being read, then we know that a valid propositional form enclosed in parentheses must follow and so we enter (S) onto the stack.

Let $w = x_1 x_2 \ldots x_n$ be the propositional form to be read. Mark the end of w by the symbol $\#$ so that $x_{n+1} = \#$. Also mark the bottom of the stack with $\#$. The following algorithm checks that w is a well-formed formula.

Algorithm II. A Wff Checker.

(1) Enter $S\#$ onto the empty stack and let $i = 1$.

(2) Suppose $x_i = \sim$. If S is on the stack, replace S by (S) and go to step 8. Otherwise, stop and reject the string.

(3) Suppose $x_i = ($.

    (i) If S is on the stack, replace S by $S)Q$ and go to step 8.

    (ii) If $($ is on the stack, delete $($ from the top of the stack and go to step 8.

    (iii) Otherwise, stop and reject the string.

(4) Suppose that x_i is an element of V (that is, x_i is a variable). If S is on the stack, remove S from the top of the stack and go to step 8. Otherwise, stop and reject the string.

(5) Suppose that $x_i =)$. If $)$ is on the stack, remove $)$ from the top of the stack and go to step 8. Otherwise, stop and reject the string.

(6) Suppose that $x_i = \vee$ or $x_i = \wedge$. If Q is on the stack, replace Q with (S) and go to step 8. Otherwise, stop and reject the string.

(7) Suppose $x_i = \#$. If $\#$ is on the stack, accept the string and stop. Otherwise, reject the string and stop.

(8) Increment i by 1 and go to step 3.

The results of applying the algorithm to the string $(\sim(p)) \vee ((q) \wedge (\sim(r)))$ are summarized in Table 2.3.

Table 2.3	*Symbol being read*	*Sequence remaining*	*Stack*
	none	$(\sim(p)) \vee ((q) \wedge (\sim (r)))\#$	$S\#$
	$($	$\sim(p)) \vee ((q) \wedge (\sim(r)))\#$	$S)Q\#$
	$\sim$	$(p)) \vee ((q) \wedge (\sim(r)))\#$	$(S))Q\#$
	$($	$p)) \vee ((q) \wedge (\sim(r)))\#$	$S)Q\#$
	p	$)) \vee ((q) \wedge (\sim(r)))\#$	$))Q\#$
	$)$	$) \vee ((q) \wedge (\sim(r)))\#$	$)Q\#$
	$)$	$\vee ((q) \wedge (\sim(r)))\#$	$Q\#$
	$\vee$	$((q) \wedge (\sim(r)))\#$	$(S)\#$
	$($	$(q) \wedge (\sim(r)))\#$	$S)\#$
	$($	$q) \wedge (\sim(r)))\#$	$S)Q)\#$
	q	$) \wedge (\sim(r)))\#$	$)Q)\#$
	$)$	$\wedge (\sim(r)))\#$	$Q)\#$
	$\wedge$	$(\sim(r)))\#$	$(S))\#$
	$($	$\sim(r)))\#$	$S))\#$
	$\sim$	$(r)))\#$	$(S)))\#$
	$($	$r)))\#$	$S)))\#$
	r	$)))\#$	$)))\#$
	$)$	$))\#$	$))\#$
	$)$	$)\#$	$)\#$
	$)$	$\#$	$\#$
	$\#$	none	$\#$

When the last symbol # is read, the stack contains only the symbol #, so the string is accepted.

We end this section with two Pascal programs that implement Algorithms I and II. In the second program, the ASCII symbols ∧ (caret) and ∨ (lowercase vee) are interpreted within an input string as the "and" and "or" operations, respectively.

```pascal
program parentheses(input,output);
{Checks strings of parentheses entered at keyboard for valid pairing.}
const
  maxparen=50;
var
  reject,top,index,nrparen:integer;
  seq,stack:array[1..maxparen] of char;
  rejected,invalidinput:boolean;
begin
  write('Number of parentheses? (maximum ',maxparen:1,'): ');
  readln(nrparen);
  write('Ready for a string composed solely of (''s and )''s : ');
  for index:=1 to nrparen do
    read(seq[index]);
  writeln;
  write('The string received is  ');
  writeln;
  rejected:=false; {Initialize rejection flag}
  invalidinput:=false; {Initialize invalid input sequence flag}
  for index:=1 to nrparen do begin {Echo and check the sequence}
    write(seq[index]);
    if not (seq[index] in ['(',')']) then begin {Invalid char in input}
      invalidinput:=true;
      rejected:=true
      end {Invalid char in input}
    end; {Echo and check the sequence}
  writeln;
  stack[1]:='#'; {Mark "bottom" of stack}
  stack[2]:='S'; {Mark substitution-position on stack}
  top:=2; {Stack pointer indicates latest addition to ("top" of) stack}
  index:=1; {Initialize sequence-scan index}
  while not rejected and (index < nrparen+1) do begin {CHK THE SEQUENCE}
    while (seq[index] = '(') and not rejected do begin {CHECK FOR '('s}
      if (stack[top]='S') then begin {REPLACE 'S' BY 'S)' ON STACK}
        stack[top]:=')'; {Substitute ')' for the 'S' at top of stack}
        top:=top+1;       {Push a new 'S'}
        stack[top]:='S'; {onto the stack}
        end {REPLACE 'S' BY 'S)' ON STACK}
      else
          {ASSERT: Left parenthesis is current character in sequence, }
          {and there is no 'S' at its top. }
          rejected:=true;
      index:=index+1 {Prepare to see if next character in seq is a '('}
      end; {CHECK FOR '('s}
    while(seq[index] = ')') and not rejected do begin {CHECK FOR ')'s}
      if(stack[top] = 'S') and (stack[top-1] = ')') then begin {RID )}
        top:=top-1; {Remove 'S' from stack top}
```

```
        {ASSERT: The ')' is now at stack top}
        stack[top]:='S' {Replace the ')' with an 'S'}
        end {RID )}
      else
          {ASSERT: Right parenthesis is current character in sequence, }
          {and there is no 'S)' on stack.                             }
          rejected:=true;
      index:=index+1 {Prepare to see if next character in seq is a ')'}
      end {CHECK FOR ')'s}
    end; {CHK THE SEQUENCE}
  if invalidinput then
    writeln('Invalid input sequence (contains non-parenthesis)')
  else
    if not rejected and ((stack[top] = 'S') and (stack[top-1] = '#')) then
      write('Accept the sequence.')
    else
      write('Reject the sequence.')
end.
```

program wffg(input,output);
```
{Accepts input string if and only if it is a "Well-Formed Formula"}
var
  inputstring,stack:array[1..50] of char;
  index,nrsymbols,top:integer;
  rejected,invalidinput:boolean;
{---------------------------------------------------------------------}
procedure processnotsymbol;
{Processes '~' in input string}
begin {processnotsymbol}
  if (stack[top] = 'S') then begin {REPLACE 'S' WITH '(S)'}
    stack[top]:=')'; {Replace 'S' at stack top with a ')'}
    stack[top+1]:='S'; {Push a new 'S' onto stack}
    stack[top+2]:='('; {Push a '(' onto stack}
    top:=top+2 {Must always point to last item pushed onto stack}
    end {REPLACE 'S' WITH '(S)'}
  else
    {ASSERT: Symbol '~' encountered in input string,}
    {and the stack does not have 'S' at its top.     }
    rejected:=true {Curr symbol invalid; reject the string}
end; {processnotsymbol}
{---------------------------------------------------------------------}
procedure processopenparen;
{Processes '(' in input string}
begin {processopenparen}
  if (stack[top] = 'S') then begin {REPLACE 'S' WITH 'S)Q'}
    stack[top]:='Q';
    stack[top+1]:=')';
    stack[top+2]:='S';
    top:=top+2
    end {REPLACE 'S' WITH 'S)Q'}
  else
    {ASSERT: Symbol '(' encountered in input string,}
    {and the stack does not have 'S' at its top.     }
    if (stack[top] = '(') then
      top:=top-1 {Delete '(' from stack}
```

```
      else
        {ASSERT: Symbol '(' encountered in input string,}
        {and stack had neither 'S' nor '(' at its top.  }
        rejected:=true {Curr symbol invalid; reject the string}
end; {processopenparen}
{-------------------------------------------------------------------}
procedure processvariable;
{Processes a variable in input string}
begin {processvariable}
  if (stack[top] = 'S') then
    top:=top-1 {Delete 'S' from the stack}
  else
    {ASSERT: Symbol for a variable encountered in input string,}
    {and stack does not have 'S' at its top.                   }
    rejected:=true {Curr symbol invalid; reject the string}
end; {processvariable}
{-------------------------------------------------------------------}
procedure processcloseparen;
{Processes ')' in input string}
begin {processcloseparen}
  if (stack[top] = ')') then
    top:=top-1 {Delete ')' from stack}
  else
    rejected:=true {Curr symbol invalid; reject the string}
end; {processcloseparen}
{-------------------------------------------------------------------}
procedure processlogicaloperator;
{Processes a logical operator ('v' or '^') in input string}
begin {processlogicaloperator}
  if (stack[top] = 'Q') then begin {REPLACE 'Q' WITH '(S)'}
    stack[top]:=')';
    stack[top+1]:='S';
    stack[top+2]:='(';
    top:=top+2
    end {REPLACE 'Q' WITH '(S)'}
  else
    rejected:=true {Curr symbol invalid; reject the string}
end; {processlogicaloperator}
{===================================================================}
begin {main program}
  stack[1]:='#'; {Mark bottom of stack}
  stack[2]:='S'; {Put "substitution marker" on top of stack}
  top:=2; {Stack pointer indicates latest addition to ("top of") stack}
  rejected:=false; {Initialize rejection flag}
  invalidinput:=false; {Initialize invalid-input flag}
  write('Number of symbols in the input string? (Maximum 50): ');
  readln(nrsymbols);
  write('Input string?: ');
  for index:=1 to nrsymbols do begin {Get and check input string}
    read(inputstring[index]);
    if not(inputstring[index] in ['~','(','p'..'u',')','v','^'])
    then begin {Invalid character in input}
      invalidinput:=true;
      rejected:=true
      end {Invalid character in input}
```

```
      end; {Get and check input string}
   writeln;
   if not invalidinput then
      for index:=1 to nrsymbols do
        case inputstring[index] of
           '~':processnotsymbol;
           '(':processopenparen;
           'p'..'u':processvariable;
           ')':processcloseparen;
           'v','^':processlogicaloperator
        end; {Case inputstring[index]}
   if (stack[top] <> '#') then
     {ASSERT: All symbols of input string have been scanned,}
     {and there are still symbols left on the stack.  This   }
     {indicates a premature end-of-string condition.         }
     rejected:=true; {Reject the string}
   if invalidinput then
     writeln('Invalid. Valid symbols are ~, ( , p..u, ), v, ^')
   else
     if rejected then
       writeln('Reject the input string.')
     else
       writeln('Accept the input string.')
end. {program wffg}
```

Section 2.10	Exercises

1. Show how to derive the following sequences of parentheses using the grammar *PPG*.

    (a) (( ))()()   (b) ( )( )( )   (c) ((( )( )))

2. Use Algorithm I and construct the stack used to check whether each of the following sequences is validly paired.

    (a) ( )((( )( )))( )   (b) (( )( )(( )   (c) ( ))(

3. (For Pascal users) Modify the Pascal Parentheses Checker so that it checks the parentheses within a string of characters like $(a + b) * ((x * y) - 5)$.

4. Let L be the language $\{x : x = a^n b^n \text{ where } n \in \mathbf{N}\}$. Write a grammar for L, and write an algorithm that determines whether a string of a's and b's is in L.

5. Show how Algorithm II accepts or rejects the following strings by constructing the stack used in the algorithm.

    (a) $((\sim(r)) \wedge (q)) \wedge (p)$ (b) $(p \wedge (\sim(q))) \wedge (r)$

6. Write a grammar for the set of fully parenthesized arithmetic expressions that use the variables x, y, z, and w and the operations $+$, $-$, $*$, and $/$.

key concepts

2.1 Sets	sets elements or members, $x \in A$ set equality, $A = B$ singleton set set notation: lists, braces and ellipses, set builder
2.2 Set Relations	containment relation, $A \subset B$ subset and proper subset

2.3 Set Operations	universal set
	empty set, $\emptyset$
	set operations:
	    union, $A \cup B$
	    intersection, $A \cap B$
	    complementation, A'
	    set difference, $A \backslash B$
	disjoint sets
	set identities: DeMorgan's laws, the
	    distributive laws, etc. (See Table 2.1.)

2.4 Infinite Collections of Sets	infinite collections of sets
	    index set
	    union, $\bigcup_{w \in L} A_w$
	    intersection, $\bigcap_{w \in L} A_w$

2.5 Power Sets	power set, $\mathscr{P}(A)$
2.6 Cartesian Products	cartesian product, $A \times B$

2.7 Inductively Defined Sets	inductive definition
	    basis clause
	    inductive clause
	    terminal clause

2.8 Formal Languages	Formal languages
	    alphabet
	    word or string
	    concatenation
	    empty word, E^*
	    Kleene-star operation
	    product language LS

2.9 Proofs by Induction	Principle of Induction proof by induction Second Principle of Induction

Chapter 2	Exercises

1. Let $U = \{0, 1, 2, 3, 4, 5, 6, 7, 8, 9, 10\}$ be the universal set, and let $A = \{0, 2, 4, 6, 8, 10\}$, $B = \{1, 3, 5, 7, 9\}$, and $C = \{0, 3, 6, 9\}$. List all the elements in each of the following sets: $B \cup C$, $A \cap C$, C', $A \cap C'$, $(B \cup C)'$, and $B' \cap C'$.

2. With U, A, B, and C as in Exercise 1, find $A \cap (B \cup C)$, $(A \cap B) \cup C$, $A\backslash B$, $A\backslash(B \cup C)$, and $A\backslash(B\backslash C)$.

3. Let U be the set of real numbers. Let $A = \{x : x > 0\}$, $B = \{x : x < 1\}$, and $C = \{x : -2 < x < 2\}$. Find A', C', $A \cap B$, $A \cup B$, $A\backslash B$, and $C\backslash(A \cap B)$.

4. List the elements of each of the following sets. Use ellipses where necessary. Each is a subset of the integers.
   (a) $A = \{x : 0 \le x \le 13\}$
   (b) $B = \{x : x = n^2 \text{ for some element } n \text{ of } \mathbf{N}\}$
   (c) $C = \{x : x \text{ is divisible by } 4\}$
   (d) $D = \{x : x \text{ is divisible by } 3 \text{ or } 5\}$
   (e) $E = \{x : x^2 \text{ is divisible by } 6 \text{ but not by } 4\}$

5. Use set-builder notation to describe each of the following sets.
   (a) $\{-2, -1, 0, 1, 2\}$
   (b) $\{\ldots, -5, -2, 1, 4, 7, \ldots\}$
   (c) {The set of real numbers between 2.4 and 3.9}
   (d) $\{1, 4, 6, 8, 9, 10, 12, \ldots\}$
   (e) $\{-1, 0, 3, 8, 15, 24, \ldots\}$
   (f) $\{\frac{1}{2}, \frac{1}{3}, \frac{1}{4}, \ldots\}$

6. For each of the following pairs of sets A and B, determine whether $A \subset B$ and, if so, determine whether the containment is proper.
   (a) $A = \{x : x \text{ is real and } 0 \le x^2 \le 1\}$ and $B = \{x : -1 \le x \le 1\}$
   (b) $A = \{x : x \text{ is an integer multiple of } 2\}$ and $B = \{x : x \text{ is an integer multiple of } 4\}$
   (c) $A = \{x : 2x \text{ is an integer multiple of } 12\}$ and $B = \{x : x \text{ is divisible by both } 2 \text{ and } 3\}$

7. Prove: $B \subset A$ where $A = \{x : x^2 \text{ is an integer multiple of } 9\}$ and $B = \{x : x \text{ is an integer divisible by } 6\}$.

8. Prove: $A = B$ where $A = \{x : x^2 \text{ is divisible by } 9\}$ and $B = \{x : x \text{ is divisible by } 3\}$.

9. Prove part (b) of Theorem 1 in Section 2.3.

10. Prove part (a) of Theorem 2 in Section 2.3.

11. Prove each of the following set identities with an argument like that given to prove Theorem 1 in Section 2.3.

   (a) $A \cap B = A \backslash B'$

   (b) $A \cup (B \backslash A) = A \cup B$

   (c) $A \backslash (B \cup C) = (A \backslash B) \cap (A \backslash C)$

   (d) $A \backslash (B \cap C) = (A \backslash B) \cup (A \backslash C)$

12. Prove the set identities given in Exercise 11, using DeMorgan's laws, the distributive laws, and other identities.

13. Let A and B be sets. The **symmetric difference** of A and B, denoted by $A \triangle B$, is defined as follows:

$$A \triangle B = (A \backslash B) \cup (B \backslash A)$$

   (a) Let $A = \{1, 2, 3, \ldots, 10\}$, $B = \{2, 4, 6, \ldots, 10, 12\}$, and $C = \{3, 6, 9, 12\}$. Find $A \triangle B$, $B \triangle C$, $(A \triangle B) \triangle C$, and $A \triangle (B \triangle C)$.

   (b) Prove that for arbitrary sets S, T, and R, we have

$$(S \triangle T) \triangle R = S \triangle (T \triangle R)$$

14. For each number $n \in \mathbf{N}$, let $A_n = \{x : x \text{ is real and } 0 < x < 1/n\}$.

   (a) Find A_1 and A_6.

   (b) Find $A_5 \cup A_3$ and $A_2 \cap A_7$ and A_4'.

   (c) Find $A_n \cap A_{n+1}$, $A_n \cup A_{n+1}$, and A_n'.

   (d) Find $A_n \cap A_{n+1}'$ and $A_n' \cap A_{n+1}$.

   (e) Find $\overset{\infty}{\underset{n=1}{\cup}} A_n$ and $\overset{\infty}{\underset{n=1}{\cap}} A_n$.

15. Let L be the subset of real numbers defined as $\{x : 0 < x < 1\}$. For each r in L, let $A_r = \{x : -r < x < 1 - r\}$.

   (a) Find $\underset{r \in L}{\cup} A_r$ and $\underset{r \in L}{\cap} A_r$.

   (b) Let $B = \{x : x > 0\}$. Find $\underset{r \in L}{\cup} (B \cap A_r)$ and $B \cup \left(\underset{r \in L}{\cap} A_r \right)$.

16. Let the universal set U be the set of all real numbers and let $\mathbf{R}^+$ be the index set. For each $r \in \mathbf{R}^+$, let $A_r = \{x : -1/r < x < r\}$. Find each of the following.

   (a) $\underset{r \in \mathbf{R}^+}{\cup} A_r$ and $\underset{r \in \mathbf{R}^+}{\cap} A_r$

   (b) $\underset{r \in \mathbf{R}^+}{\cap} A_r'$ and $\left(\underset{r \in \mathbf{R}^+}{\cup} A_r \right)'$

   (c) $\underset{r \in \mathbf{R}^+}{\cup} A_r'$ and $\left(\underset{r \in \mathbf{R}^+}{\cap} A_r \right)'$

17. Prove part (b) of Theorem 1 in Section 2.4.

18. Prove part (b) of Theorem 2 in Section 2.4.

19. Let $A = \{a, b, c\}$, $B = \{a, b\}$, and $C = \{a, b, \{a, b\}\}$. List all the elements in $\mathcal{P}(A)$, $\mathcal{P}(C)$, and $\mathcal{P}(A) \cap \mathcal{P}(C)$. Indicate whether each of the following statements is true or false.

(a) $B \subset \mathcal{P}(A)$ (b) $B \in \mathcal{P}(A)$ (c) $B \subset C$

(d) $B \in C$ (e) $\mathcal{P}(C) \subset \mathcal{P}(A)$ (f) $\mathcal{P}(B) \subset \mathcal{P}(C)$

(g) $B \in \mathcal{P}(C)$ (h) $A \cap \mathcal{P}(A) = \emptyset$ (i) $C \cap \mathcal{P}(C) = \emptyset$

20. Let $A = \{a, b\}$, $B = \{s, t, v\}$, and $C = \{0, 1, 2, 4\}$. List all the elements of each of the following cartesian products.

(a) $A \times C$ (b) $B \times B$

(c) $\mathcal{P}(A) \times C$ (d) $A \times B \times C$

21. Express each of the following sets as a cartesian product, or show that this is impossible.

(a) The rectangle and its interior in $\mathbf{R} \times \mathbf{R}$ with corners at $(0, 1)$, $(0, 4)$, $(3, 1)$, and $(3, 4)$.

(b) The set of all elements (x, y) where x is an even integer and y is an odd integer.

(c) The set of all elements (n, m) of $\mathbf{N} \times \mathbf{N}$ such that $n = m + 1$.

(d) The first quadrant in $\mathbf{R} \times \mathbf{R}$.

(e) The union of the first and third quadrants in $\mathbf{R} \times \mathbf{R}$.

22. Each of the following sets is defined inductively. Describe each, using set-builder notation.

(a) (1) $\frac{1}{2}$ is in S.
   (2) If x is in S, then $\left(\frac{1}{2}\right)x$ is in S.
   (3) Only numbers so formed are in S.

(b) (1) The numbers 3 and 5 are in Q.
   (2) If x and y are in Q, then $x + y$ and $x - y$ are in Q.
   (3) Only numbers so formed are in Q.

(c) Define the language L over the alphabet $\{a, b\}$ as follows:
   (1) The word b is in L.
   (2) If x is a word in L, then xbb is in L and aax is in L.
   (3) Only words so formed are in L.

(d) Define the language S over the alphabet $E = \{a, b\}$ as follows:
   (1) The empty word Λ is in S.

(2) If x is a word in S, and x is of the form yy for some word y in E^*, then *yaya* and *ybyb* are in S.

(3) Only words so formed are in S.

23. Give an inductive definition for each of the following sets.

(a) The set of all integer multiples of 3.

(b) The set of all elements of $\mathbf{N} \times \mathbf{N}$ of the form $(n, n + 1)$.

(c) The set of all words over the alphabet $\{a, b\}$ for which every a is followed immediately by at least one b.

(d) $\{2, 7, 12, 17, 22, \ldots\}$

(e) The set of all words over $\{0, 1\}$ with an even number of symbols.

24. Prove by induction:

$$1 + 2^3 + \cdots + n^3 = [n(n + 1)/2]^2 \text{ for } n \geq 1$$

25. Prove by induction:

$$1/(1 \cdot 2) + 1/(2 \cdot 3) + \cdots + 1/[n(n + 1)] = n/(n + 1) \text{ for } n \geq 1$$

26. Prove by induction:

$$n^2 > n + 1 \text{ for } n \geq 2$$

27. Prove by induction for $n \geq 1$:

$$1 + 3^2 + 5^2 + \cdots + (2n + 1)^2 = (n + 1)(2n + 1)(2n + 3)/3$$

28. (a) Prove by induction that $2n + 1 \leq 2^n$ for $n \geq 3$.

(b) Use the results of part (a) to prove by induction that $n^2 \leq 2^n$ for $n \geq 4$.

29. Let x be a real number such that $x \geq -1$. Prove by induction that, for $n \geq 0$, $(1 + x)^n \geq 1 + nx$.

30. For $n \geq 1$, prove that $(x - y)$ is a factor of $(x^n - y^n)$. *Hint*: Add $0 = x^n y - x^n y$ to the term $x^{n+1} - y^{n+1}$ and factor.

31. Prove by induction that, for any collection of n sets, $A_1, A_2, \ldots, A_n$, we have $(A_1 \cap A_2 \cap \cdots \cap A_n)' = A_1' \cup A_2' \cup \cdots \cup A_n'$.

32. Let S be the set defined in Example 1 in Section 2.7. Prove by induction that for each x in S, we have $x < 2$.

33. Let Q be the set defined in Example 3 in Section 2.8. Prove that each word x in Q has an even number of b's.

34. Let L be the language over the alphabet $\{a, b\}$ defined as follows:

(1) The word a is in L.

(2) If x is a word in L, then ax and xab are in L.

(3) Only words so formed are in L.

Prove by induction that each word in L has a greater number of a's than b's.

35. Let E be the alphabet $\{1, \cdot, =\}$ and define the language M over E as follows:

(1) $1 \cdot 1 = 1$ is in M.

(2) If w is a word in M of the form $x \cdot y = z$, then $x \cdot y1 = zx$ and $x1 \cdot y = zy$ are also in M.

(3) Only words so formed are in M.

List five different words in M. Prove that if the word $x \cdot y = z$ is in M, and the number of 1's in x is m, and the number of 1's in y is n, then the number of 1's in z is $m \cdot n$.

Chapter 3

FUNCTIONS

outline

3.1 Functions
3.2 Injective and Surjective Functions
3.3 Composition of Functions
3.4 Inverse Functions
3.5 Functions and Set Operations
3.6 Bridge to Computer Science: A
     Look at Coding Theory (optional)
     Key Concepts
     Exercises

Introduction

A CHILD counting raisins assigns a number to each raisin under consideration. An engineer tracking a missile assigns a position to each moment of time the missile is tracked. In each case a correspondence between two sets is established: between raisins and whole numbers, between moments of time and positions in space. Such correspondences are called functions. While the idea of counting occurred to mankind in time immemorial (its basis in functions was not always made explicit), the idea that the physics of bodies in motion is best understood by studying position as a function of time was at the heart of the scientific revolution of the seventeenth century.

Functions are central to the study of physics and enumeration, but they occur in many other situations as well. For instance, the correspondence between the data stored in computer memory and the set of addresses of the memory locations occupied is a function. The encoding of the standard symbols of our written language—$a, b, c, \ldots, z, 0, 1, 2, \ldots, 9, ?, !, +, \ldots$—into strings of 0's and 1's for digital processing and the subsequent decoding of the strings obtained: these are functions. (In the bridge section of this chapter, we take a look at the problem of encoding and decoding in terms of the possibility of error detection and correction.) Thus, to understand the general use of functions, we must study their properties in the general terms of set theory, which is what we do in this chapter. We will turn to the study of counting techniques (and their underlying functions) in Chapter 4, where we shall also extend our notions of enumeration and cardinality to infinite sets (again through functions).

| Section 3.1 | Functions |

The reader is no doubt familiar with functions of the form $y = f(x)$. For instance, if $f(x) = x^2$, $x = 2$, and $y = f(2)$, then the value of y is 4. If $g(x) = x^{1/2}$, we know that $g(0.49) = 0.7$, but whether we can evaluate $y = g(-4)$ depends on whether we allow y to assume complex values. In order to determine a function f, we must stipulate the values of x for which $f(x)$ can be evaluated, the values that $f(x)$ can assume, and the rule or method to be used to determine the value of $f(x)$.

The concept of a function is not restricted to numeric computations. Suppose that A is the set of all last names of U.S. citizens, and suppose that B is the set of all words of length three or less over the English alphabet. We can define a function $y = g(x)$, where x assumes its values in A and y assumes its values in

B: We let $g(x)$ be the first three letters of the name x. Thus $g(\text{Smith}) = \text{Smi}$ and $g(\text{Nu}) = \text{Nu}$.

We now formalize our notion of a function.

> **DEFINITION.** Let A and B be sets. A **function** f from A to B is a rule that assigns to each element x in A exactly one element y in B. We call A the **domain** of f, and B we call the **codomain** of f. We write $f : A \rightarrow B$. If x is an element of A and y is the element of B assigned to x, we write $y = f(x)$ and call y the function value of f at x.

Example 1. Suppose $A = \{a, b, c, d\}$ and $B = \{s, t, u\}$. We define $f : A \rightarrow B$ by the assignment $f(a) = t$, $f(b) = s$, $f(c) = s$, and $f(d) = u$. We can represent f schematically as shown in Figure 3.1. We note that the function f assigns s to both b and c. Another function $g : A \rightarrow B$ may be defined by the assignment $g(a) = g(b) = g(c) = g(d) = u$. When we define a function from A to B, we can assign each element x in A any of the three values in B. Because there are 4 values in A, there are $3^4 = 81$ different functions possible from A to B.

The rule h given by Figure 3.2 does not define a function. There is more than one value assigned to a, and we cannot decide which element of B is equal to $h(a)$.

Figure 3.1   **Figure 3.2**

Example 2. Suppose $A = \{a, b\}$. Figure 3.3 gives all four functions from A to itself.

Figure 3.3

Example 3. We define $f : \mathbf{R}^+ \rightarrow \mathbf{R}$ by the formula $f(x) = (x - 1)/(x + 1)$ for each x in $\mathbf{R}^+$. Then $f(3) = 2/4$ and $f(0.5) = -0.5/1.5$. We also have $f(a - 3) = (a - 4)/(a - 2)$, where a is any real number greater than 3. The value of $f(-2)$ is not defined because -2 is not in the domain of f (even though the formula for f yields a value when $x = -2$).

Example 4. Let $f : \mathbf{R} \rightarrow \mathbf{R}$ be defined as follows: $f(x) = x^2 + 2$ if $x > 1$, and $f(x) = 3x - 5$ if $x \leq 1$. Then $f(2) = 6$ and $f(-2) = -11$. For any real number a,

$f(a + 1) = (a + 1)^2 + 2 = a^2 + 2a + 3$ if $a > 0$, and $f(a + 1) = 3(a + 1)$ $- 5 = 3a - 2$ if $a \leq 0$. The rule for computing $f(a + 1)$ depends on whether the quantity $(a + 1)$ is greater than 1.

Example 5. Let $E = \{a, b\}$ and define the function $f : \mathbf{N} \times \mathbf{N} \to E^*$ as follows: $f(n, m) = a^{n-1}b^{m-1}$. Then $f(3, 2) = aab$ and $f(3, 4) = aabbb$. Also, $f(1, 1) = \Lambda$. In general, $f(n, m)$ is a string of $(n - 1)$ a's followed by $(m - 1)$ b's. Words such as *ababa* are not function values for this function.

Example 6. Let $f : \mathbf{R} \to \mathbf{Z}$ be defined by the rule $y = f(x)$, where y is the greatest integer less than or equal to x. We denote $f(x)$ by $\lfloor x \rfloor$ and call it the **greatest-integer function.** We have $\lfloor 3.46 \rfloor = 3, \lfloor -1.25 \rfloor = -2$, and $\lfloor 45 \rfloor = 45$. If we define $g : \mathbf{R} \to \mathbf{Z}$ by $g(x) = \lfloor x/2 \rfloor$, then $g(5) = 2$ and $g(-3) = -2$. (There are many names and notations for the greatest integer function. It is, for instance, called the *floor function* and denoted by $[x]$ or $[\![x]\!]$. We shall always use $\lfloor x \rfloor$.)

Example 7. If A is any set, we can define a function $f : A \to A$ by letting $f(x) = x$ for each value of x in A. This function is called the **identity function** on A. Thus if f is the identity function on the set of natural numbers $\mathbf{N}$, then $f(4) = 4$ and $f(301) = 301$, whereas $f(-2)$ is not defined (because -2 is not in the domain of f). We use id_A to denote the identity function on A. Thus $\text{id}_\mathbf{N}(4) = 4$.

Example 8. Suppose that X is the set (a, b, c); that $A = \mathcal{P}(X)$, the power set of X; and that $\mathbf{W}$ is the set of whole numbers. Define $h : A \to \mathbf{W}$ by letting $h(x)$ be the number of elements in the set x. (Note that each element of the domain A of h is itself a set.) Thus if x is the set $\{b, c\}$, then $h(x) = 2$, and if x is the empty set, then $h(x) = 0$.

In the examples that follow, we define functions **inductively.** To define a function f inductively, we shall evaluate f explicitly at a few elements in its domain and then give a **recurrence relation** through which we can compute $f(x)$ in terms of previously obtained values.

Example 9. Consider the simple function $f : \mathbf{W} \to \mathbf{W}$ defined by $f(x) = 2x$. We can also define it inductively as follows:

(a) $f(0) = 0$
(b) $f(n + 1) = f(n) + 2$ for $n \geq 0$

The recurrence relation is given by step (b). We know explicitly that $f(0) = 0$ and so $f(1) = f(0) + 2 = 2$. Similarly, $f(2) = f(1) + 2 = 2 + 2 = 4$, and so forth.

Example 10. In this example we define the factorial function $\text{fac}(n) = n!$ inductively.

(a) $\text{fac}(0) = 1$

(b) $\text{fac}(n + 1) = (n + 1)\text{fac}(n)$ for $n \geq 0$

We define $0!$ to be 1 explicitly in step (a). Then from the recurrence relation given in step (b), we have

$$1! = 1 \cdot 0! = 1$$
$$2! = 2 \cdot 1! = 2$$
$$3! = 3 \cdot 2! = 6$$
$$4! = 4 \cdot 3! = 24 \text{ etc.}$$

The factorial function is a very important counting tool, as we shall see in Chapter 4. Its values grow very large very quickly. For example, $5! = 120$ whereas $10! = 3,628,800$.

Example 11. Define $f: \mathbf{W} \to \mathbf{W}$ as follows:

(a) Let $f(0) = 1$ and $f(1) = 1$.

(b) If $x > 1$, let $f(x) = f(x - 1) + f(x - 2)$.

Thus $f(2) = 1 + 1 = 2$, $f(3) = 2 + 1 = 3$, $f(4) = 3 + 2 = 5$, and so forth. The sequence of numbers generated by $f(x)$—namely $\{1, 1, 2, 3, 5, 8, 13, \ldots\}$ —is known as the Fibonacci sequence. It occurs remarkably frequently in mathematical modeling and has important applications in number theory. For Fibonacci (Leonardo of Pisa) himself, the sequence was derived in answer to the following hypothetical problem in Pisan ecology, which was posed to his students in the year 1202. It may be found in Fibonacci's *Liber Abbaci* (Book of the Abacus).

> "How many pairs of rabbits can be bred from one pair in one year?
>
> A man has one pair of rabbits at a certain place entirely surrounded by a wall. We wish to know how many pairs can be bred from it in one year if the nature of these rabbits is such that they breed one other pair every month and begin to breed in the second month after their birth."

If we assume that each pair bears another pair (male and female) of rabbits each month, that there are no deaths, and that the first pair breeds during the first month, then the answer is $f(13) = 377$ pairs of rabbits.

Example 12. Let $f: \mathbf{W} \to \mathbf{W}$ be defined inductively as follows:

(a) $f(0) = 2$

(b) $f(n + 1) = 3f(n) + 5$ for $n \geq 0$

In this example, we prove by induction that for each n in $\mathbf{W}$, $f(n) = (3^{n+2} - 5)/2$. The statement is true for $n = 0$, because $(3^2 - 5)/2 = 2$, and we know that $f(0) = 2$ from step (a). Now let us assume that $f(n) = (3^{n+2} - 5)/2$. We must show that $f(n + 1) = (3^{n+3} - 5)/2$. We know that $f(n + 1) = 3f(n) + 5$. We can use the induction hypothesis to substitute $(3^{n+2} - 5)/2$ for $f(n)$ and obtain that $f(n + 1) = [3(3^{n+2} - 5)/2] + 5$. Simplifying, we have $f(n + 1) = (3^{n+3} - 5)/2$.

If A is an inductively defined set, we may also define functions on it inductively. We must be careful to ensure that the values assigned to the elements in A established by the basis clause are assigned explicitly and that every element in A is assigned exactly one function value.

Example 13. Suppose that E is the set of symbols $\{0, 1\}$ and that the set A is the language defined over E as follows:

(1) Both 0 and 1 are elements of A.

(2) If x is in A and x is not equal to 0, then both $x0$ and $x1$ are in A.

(3) Only words so formed are in A.

We now define $f : A \rightarrow \mathbf{W}$ inductively.

(a) Let $f(0) = 0$ and let $f(1) = 1$.

(b) If x is in A and $q = x0$, then let $f(q) = 2f(x)$. If $q = x1$, let $f(q) = 2f(x) + 1$.

Thus $f(10) = 2$, $f(11) = 3$, and $f(111) = 7$. If we interpret each x in A as the binary expression of a nonnegative integer, then $f(x)$ is that integer's decimal expression.

Example 14. Let $E = \{a_1, a_2, \ldots, a_n\}$ be an alphabet, and let E^* be the set of all words over E. In Section 2.8 we defined E^* inductively. We now offer an inductive definition of the length function $L : E^* \rightarrow \mathbf{W}$, which gives us the length of each word in E^*.

(a) $L(\Lambda) = 0$

(b) If $y = xa_k$ for some a_k in E, then $L(y) = L(x) + 1$

Inductively defined functions are a part of a broader class of functions known as **recursive** functions. A function f is called recursive if the rule used to define f refers to the function f itself.

Exercises

1. Determine which of the following rules define functions.

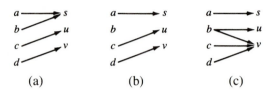

(a)   (b)   (c)

(d) $f : \mathbf{R} \to \mathbf{Z}$, where $f(x)$ is the integer nearest x

(e) $f : \mathbf{R} \to \mathbf{R}$, where $f(x)$ is any real number y such that $y^3 = x$

2. Let $A = \{a, b, c\}$ and let $B = \{s, t\}$. List all the functions from A to B and all the functions from B to A.

3. Let $A = \{x : x \geq 1\}$ and let $f : A \to \mathbf{R}^+$ be defined by $f(x) = (x^2 + 1)/x$. Find $f(2)$, $f(1.5)$, and $f(-1)$ if possible. For what values of z is $f(z + 2)$ defined?

4. Let $g : \mathbf{R} \times \mathbf{R} \to \mathbf{R}$ be defined by $g(x, y) = 2x + y$. Evaluate $g(3, 2)$, $g(2, 3)$, and $g(x, x)$. Let $x = s + t$ and $y = s - t$. Evaluate $g(x, y)$ in terms of s and t. Find the sets in the plane for which $g(x, y) = 0$ and for which $g(x, y) > 1$.

5. Let $f(x) = \lfloor x \rfloor$ be the greatest-integer function. For what values of x is $f(x) = 2$? For what values of x is $f(x) = -3$? In general, given an integer n, for what values of x is $f(x) = n$?

6. Let $g : \mathbf{R} \to \mathbf{Z}$ be defined as follows: $g(x) = n$ where n is the smallest integer that is greater than or equal to x. (This function is called the **least-integer function** and is denoted by $g(x) = \lceil x \rceil$. It is also called the "ceiling" function.) Evaluate $g(4.001)$, $g(-3.2)$, and $g(0)$. Show that $\lceil x \rceil \neq \lfloor x \rfloor + 1$ in general.

7. Each of the following functions on $\mathbf{N}$ is defined inductively. For each, compute the value of the function at x for x between 1 and 7.

(a) $f(1) = 1$ and for $n \geq 1$, $f(n + 1) = 2f(n) + 1$

(b) $g(1) = 1$ and for $n \geq 1$, $g(n + 1) = ng(n)$

(c) $f(1) = 1$, $f(2) = 3$, and for $n \geq 3$, $f(n) = f(n - 1) + f(n - 2)$

8. Let f be the function defined in part (a) of Exercise 7. Prove by induction that $f(n) = 2^n - 1$ for all $n \geq 1$.

9. Give an inductive definition of each of the following functions from $\mathbf{N}$ to $\mathbf{N}$.

(a) $f(n) = 3^n$ (b) $f(n) = 3n + 2$

10. Explain why the answer to Fibonacci's question is $f(13)$. (See Example 11.)

| Section 3.2 | Injective and Surjective Functions |

Suppose $f : \mathbf{R} \to \mathbf{R}$ is defined by $f(x) = 3x - 1$, and let $y = 5$. There is one and only one answer to the question "For what value of x is $f(x) = y$?" The answer is $x = 2$. In fact, given any y in the codomain of f, we can provide a unique answer to the same question by taking x to be $(y + 1)/3$. For example, if we let $y = 101$ and $x = (y + 1)/3 = 34$, then $y = f(34)$. However, if we proceed similarly with the function $g : \mathbf{R} \to \mathbf{R}$ defined by $g(x) = x^2$, we can fail in two ways. If $y < 0$, there is no value of x for which $g(x) = y$; and if $y > 0$, there are two such values.

DEFINITION.

    (a) A function $f : A \to B$ is said to be **injective** or **one-to-one** if, whenever a and b are elements of A such that $a \neq b$, then $f(a) \neq f(b)$.

    (b) We say that f is **surjective** or **onto** if, for any value y in B, there is at least one element x in A for which $f(x) = y$.

Example 1. Define the functions f_1, f_2, f_3, and f_4 by the diagrams shown in Figure 3.4.

Figure 3.4

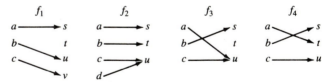

The function f_1 is injective, but it is not surjective because there is no value of x such that $f_1(x) = t$. The function f_2 is surjective, but it is not injective because $f_2(c) = f_2(d)$. The function f_3 is neither injective nor surjective, whereas the function f_4 is both.

DEFINITION. A function that is both injective and surjective is called a **bijection** or a **one-to-one correspondence**.

The function f_4 of Example 1 is a bijection.

Example 2. The function $f : \mathbf{R} \to \mathbf{R}$ defined by $f(x) = 3x - 1$ is a bijection. We first show that f is injective. To do so, we shall use the contrapositive (and hence equivalent) form of the conditional clause with which we defined injectivity:

The function f is injective if, whenever $f(a) = f(b)$, then $a = b$.

So we assume that $f(a) = f(b)$. We must now show that $a = b$. Our assumption is that $(3a - 1) = (3b - 1)$. Adding 1 to both sides, we have $3a = 3b$. Dividing both sides by 3, we have $a = b$. Thus f is injective. To show that a function f is surjective, we let y be any element of the codomain of f and show that we can find a value of x in its domain for which $f(x) = y$. So we let y be a real number. We must find a value of x such that $f(x) = y$. Thus we must find x such that $y = 3x - 1$. Solving for x, we find that if we let $x = (y + 1)/3$, then $f(x) = y$. Thus f is surjective. Because f is both injective and surjective, we have that f is a bijection.

Example 3. Define $f : \mathbf{N} \to \mathbf{N}$ by $f(n) = 2^n$. This function is one-to-one because if n is not equal to m, then 2^n is not equal to 2^m. It is not onto, however, because if we let $y = 3$, there is no integer value of n for which $2^n = 3$.

Example 4. Let $E = \{a, b\}$ and let $A = E^*$, the set of all words over E. Let $L : A \to \mathbf{W}$ be the length function so that $L(x)$ is the length of the word x. The function L is surjective. For any integer $n > 0$, let $x = aaa \cdots a$ (n times). Then $L(x) = n$. If $n = 0$, let x be the empty word and so $L(x) = 0$. The function L is not injective because many words in A have the same length. In fact, for each $n \geq 0$ there are 2^n words over $E = \{a, b\}$ of length n.

Example 5. Define $q : \mathbf{R}^+ \to \mathbf{R}^+$ by $q(x) = x^2$. This is a bijection because for any value of y in $\mathbf{R}^+$, there is exactly one value of x in $\mathbf{R}^+$—namely $x = y^{1/2}$—for which $q(x) = y$. (Note that both the domain and the codomain of q are different from those of the function g defined in the first paragraph of this section. So, although the formulas defining the functions are identical, q is a bijection but g is neither injective nor surjective.)

Example 6. Let $f : \mathbf{R} \to \mathbf{R}$ be defined as follows: $f(x) = x^2 - 2x + 5$ if $x > 1$, and $f(x) = x - 1$ if $x \leq 1$. This function is injective. To see this, we must consider three cases: a and b both greater than 1, a greater than 1 but b less than or equal to 1, and both a and b less than or equal to 1. If a and b are both greater than 1 and $a \neq b$, then $f(a) = a^2 - 2a + 5 = (a - 1)^2 + 4$ and $f(b) = b^2 - 2b + 5 = (b - 1)^2 + 4$. Because $(a - 1)$ and $(b - 1)$ are not equal and are both positive, $f(a) \neq f(b)$. If $a > 1$ and $b \leq 1$, then $f(a) > 4$ but $f(b) \leq 0$. If $a \neq b$ and both $a \leq 1$ and $b \leq 1$, then $a - 1 \neq b - 1$. So in all cases, $f(a) \neq f(b)$. The function f is not surjective, however, because if $x > 1$, then $f(x) > 4$ but if $x \leq 1$, then $f(x) \leq 0$. There are no values of x such that $0 < f(x) \leq 4$. (See Figure 3.5.)

Figure 3.5

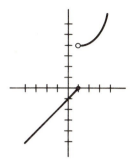

Example 7. Let E be the set of all even integers, and let $r : \mathbf{Z} \to E$ be defined by $r(n) = 2n$. Then r is a bijection or one-to-one correspondence even though E is a proper subset of $\mathbf{Z}$.

Example 8. Let $f : \mathbf{R} \times \mathbf{R} \to \mathbf{R} \times \mathbf{R}$ be defined by $f(x, y) = (x + y, x - y)$. We first show that f is injective. So suppose that $f(x, y) = f(u, v)$; that is, $(x + y, x - y) = (u + v, u - v)$. We have

$$x + y = u + v$$
$$x - y = u - v$$

Adding these two equations, we find that $2x = 2u$ and thus $x = u$. Subtracting, we find that $2y = 2v$ and thus $y = v$. Thus $(x, y) = (u, v)$ and we conclude that f is injective. Now we show that f is surjective. Let (s, t) be any value in the codomain $\mathbf{R} \times \mathbf{R}$. We must find (x, y) such that $f(x, y) = (s, t)$. That is, we must satisfy both of the following equations:

$$x + y = s$$
$$x - y = t$$

Solving these equations simultaneously for x and y in terms of s and t, we find that if $x = (s + t)/2$ and $y = (x - t)/2$, then $f(x, y) = (s, t)$. Thus f is also surjective and hence a bijection. We verify surjectivity with an example. Let $(s, t) = (3, 5)$. Let $x = (3 + 5)/2 = 4$ and $y = (3 - 5)/2 = -1$. Then $f(4, -1) = (3, 5)$.

Section 3.2	Exercises

1. Determine the injectivity and the surjectivity of the following functions. Indicate which is a bijection.

(a) $f : \mathbf{R} \to \mathbf{R}$ defined by $f(x) = (2x + 1)/3$

(b) $f : \mathbf{Z} \to \mathbf{Z}$ defined by $f(n) = 3n + 1$

(c) $f : \mathbf{R} \to \mathbf{R}^+$ defined by $f(x) = x^2 + 2$

(d) $f : \mathbf{N} \to \mathbf{W}$ defined by $f(n) = \lfloor n/2 \rfloor$

(e) $f : \mathbf{R} \to \mathbf{R}$ defined by $f(x) = x(2 - x)$ for $x > 0$ and $f(x) = x$ for $x \leq 0$

2. Define a function from $\mathbf{N}$ to $\mathbf{N}$ that is:

(a) Injective but not surjective

(b) Surjective but not injective

(c) Bijective but not the identity function

3. Prove that each of the following functions is a bijection.

(a) $f : \mathbf{R} \times \mathbf{R} \to \mathbf{R} \times \mathbf{R}$ defined by $f(x, y) = (x + y, 2x - y)$

(b) $f : \mathbf{Z} \times \mathbf{Z} \to \mathbf{Z} \times \mathbf{Z}$ defined by $f(m, n) = (2n + m, n + m)$

4. Is the function $f : \mathbf{Z} \times \mathbf{Z} \to \mathbf{Z} \times \mathbf{Z}$ defined by $f(n, m) = (n + m, n - m)$ a bijection?

5. Determine the injectivity and surjectivity of each of the following functions from $\mathbf{R}$ to $\mathbf{R}$.

(a) $f(x) = x^2 - 4x + 6$ for $x > 0$ and $f(x) = 2x$ for $x \leq 0$

(b) $g(x) = x^2 + 2x + 1$ for $x > 0$ and $g(x) = 2x + 1$ for $x \leq 0$

6. Let A be the language over $\{a, b\}$ defined inductively as follows:

(1) The empty word Λ is a word in A.

(2) If x is a word in A, then ax and xb are words in A.

(3) Only words so formed are in A.

Find a bijection between A and the cartesian product $\mathbf{N} \times \mathbf{N}$.

Section 3.3   **Composition of Functions**

Suppose that $A = \{a, b, c\}$, $B = \{w, y, z\}$, and $C = \{s, t\}$. Suppose further that $f : A \to B$ and $g : B \to C$ are given by the diagrams shown in Figure 3.6. For each element x in A, $f(x)$ is an element of the domain of g, and we can evaluate $g(f(x))$. For example, if $x = b$ then $g(f(b)) = g(z) = t$. We can define a new function $h : A \to C$ by the rule $h(x) = g(f(x))$. The function h is called the composition of f with g and is given by the diagram shown in Figure 3.7.

Figure 3.6   **Figure 3.7**

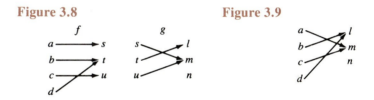

> **DEFINITION.** Suppose that $f : A \to B$ and $g : B \to C$ are functions. The **composition** of f with g is a function from A to C denoted by $g \circ f$, where $g \circ f(x)$ is defined to be $g(f(x))$ for each x in A.

We emphasize that in order for $g \circ f$ to be defined, the codomain of f must be the domain of g.

Example 1. Suppose that the functions f and g are given, respectively, by the diagrams shown in Figure 3.8. Then the function $g \circ f$ is given by Figure 3.9. The function $f \circ g$ is not defined.

Figure 3.8   **Figure 3.9**

Example 2. Suppose that f and g are functions from **R** to **R** defined by $f(x) = x^2 + 3$ and $g(x) = 4x - 5$. Then $f \circ g(x) = f(4x - 5) = (4x - 5)^2 + 3$. But $g \circ f(x) = g(x^2 + 3) = 4(x^2 + 3) - 5$. So although both $f \circ g$ and $g \circ f$ are defined, they are not equal.

Example 3. Let $f : \mathbf{R} \to \mathbf{R}$ be defined by $f(x) = x^2 + 1$ if $x > 0$ and $2x + 1$ if $x \le 0$. Let $g : \mathbf{R} \to \mathbf{R}$ be defined by x^3 if $x > 0$ and $3x - 7$ if $x \le 0$. Then the composition $g \circ f$ is defined as follows:

$$g \circ f(x) = (x^2 + 1)^3 \qquad \text{if } x > 0$$

$$g \circ f(x) = (2x + 1)^3 \qquad \text{if } -\tfrac{1}{2} < x \le 0$$

$$g \circ f(x) = 3(2x + 1) - 7 \qquad \text{if } x \le -\tfrac{1}{2}$$

Example 4. Let A be the set $\{a, b, c, d\}$ and suppose that $g : A \to A$ is defined by Figure 3.10. Then the diagrams for $g \circ g$ and $g \circ (g \circ g)$ are given by Figure 3.11.

Figure 3.10

g

Figure 3.11

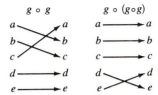

$g \circ g$ $g \circ (g \circ g)$

Given a set S and a function $f : S \to S$, we denote the function f composed with itself n times by f^n. We define f^n inductively as follows:

(i) $f^0 = \mathrm{id}_S$

(ii) $f^{n+1} = f^n \circ f$ for $n > 0$

Example 5. Let $f : \mathbf{R} \to \mathbf{R}$ be defined by $f(x) = 2x + 1$. Then $f^0 = \mathrm{id}_{\mathbf{R}}$ and $f^1 = \mathrm{id}_{\mathbf{R}} \circ f = f$. Also, $f^2(x) = f \circ f(x) = 4x + 3$ and $f^3(x) = 8x + 7$. We now prove (by induction) that $f^n(x) = 2^n x + (2^n - 1)$ for all $n \geq 0$. The statement is true for $n = 0$ because $2^0 x + (2^0 - 1) = x$. Now assume the statement true for n. We shall show that $f^{n+1}(x) = 2^{n+1} x + (2^{n+1} - 1)$. We know that $f^{n+1} = f^n \circ f$. Using the induction hypothesis, we find that $f^{n+1}(x) = f^n(f(x)) = 2^n(f(x)) + (2^n - 1) = 2^n(2x + 1) + (2^n - 1)$. Simplifying, we obtain that $f^{n+1}(x) = 2^{n+1} x + (2^{n+1} - 1)$.

We now relate the concepts of injectivity and surjectivity to composition.

Theorem 1.   Suppose that $f : A \to B$ and $g : B \to C$ are functions.

(a) If both f and g are injective, then $g \circ f$ is also injective.

(b) If both f and g are surjective, then $g \circ f$ is also surjective.

PROOF:

(a) Suppose that s and t are in A and are not equal. Since f is injective, $f(s)$ is not equal to $f(t)$. Since g is injective, $g(f(s))$ is not equal to $g(f(t))$. Thus $g \circ f(t)$ is not equal to $g \circ f(s)$, and $g \circ f$ is injective.

(b) Let z be any element of C. We must find x in A such that $g \circ f(x) = z$. Since g is surjective, there is an element y in B such that $g(y) = z$. Since f is surjective, there is an element x in A such that $f(x) = y$. Thus $g \circ f(x) = g(f(x)) = g(y) = z$, and we have established the surjectivity of $g \circ f$.

The converse of part (a) of Theorem 1 is not true. It is *not* the case that if $g \circ f$ is injective, then both f and g are injective. We give a counterexample. Let f, g, and $g \circ f$ be given by the diagrams shown in Figure 3.12. We see that although

Figure 3.12

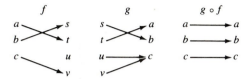

the composition $g \circ f$ is injective, g is not. We leave it to the reader to provide a counterexample to the converse of part (b).

Corollary. If $f : A \rightarrow B$ and $g : B \rightarrow C$ are both bijections, then $g \circ f$ is also a bijection.

| *Section 3.3* | **Exercises** |

1. Let $A = \{a, b, c, d\}$, $B = \{s, t, u\}$, and $C = \{1, 2\}$. Define the functions $f : A \rightarrow B$, $g : B \rightarrow C$, and $h : C \rightarrow A$ as shown in the following diagrams:

Figure for Exercise 1

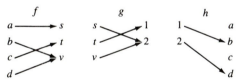

Find $g \circ f$, $h \circ g$, $f \circ h$, and $g \circ h$ if possible. In each case specify the domain and codomain of the resulting function.

2. Let $f : \mathbf{R} \rightarrow \mathbf{R}$ be defined by $f(x) = (3x - 1)/5$, and let $g : \mathbf{R} \rightarrow \mathbf{R}$ be defined by $g(x) = (x^2 + 1)/2$. Find $f \circ g$ and $g \circ f$.

3. Let f and g be functions from $\mathbf{R}$ to $\mathbf{R}$ defined by $f(x) = \lfloor x \rfloor$ and $g(x) = |x|$. Determine whether $f \circ g = g \circ f$.

4. Let $f : \mathbf{R} \times \mathbf{R} \rightarrow \mathbf{R} \times \mathbf{R}$ be defined by $f(x, y) = (x + 2y, y - x)$. Let $g : \mathbf{R} \rightarrow \mathbf{R} \times \mathbf{R}$ be defined by $g(t) = (3t, t^2)$. Let $h : \mathbf{R} \times \mathbf{R} \rightarrow \mathbf{R}$ be defined by $h(x, y) = x + 2y$. Find $f \circ g$, $h \circ f$, and $h \circ (f \circ g)$. In each case specify the domain and codomain of the resulting function.

5. Let $A = \{a, b, c, d\}$ and let $f : A \rightarrow A$ be defined by the diagram at the top of page 114. Find f^k (that is, f composed with itself k times) for $0 \leq k \leq 4$.

Figure for Exercise 5

6. Let $f : \mathbf{R} \to \mathbf{R}$ be defined by $f(x) = 3x + 2$. Find f^k for $0 \le k \le 3$.

7. Give a counterexample to the following statement: If $g \circ f$ is surjective, then both g and f are surjective.

8. Find functions $f : \mathbf{N} \to \mathbf{Z}$ and $g : \mathbf{Z} \to \mathbf{Z}$ such that $g \circ f$ is injective but g is not injective.

9. Find functions $f : \mathbf{N} \to \mathbf{N}$ and $g : \mathbf{N} \to \mathbf{N}$ such that $g \circ f$ is surjective but f is not surjective.

Section 3.4 | Inverse Functions

Suppose that $A = \{a, b, c, d\}$, that $B = \{s, t, u, v\}$, and that f is a bijection defined by Figure 3.13. If we take any value z in B, we can find exactly one value w in A for which $f(w) = z$. For example, if $z = t$ then $w = c$. We can thus define a new function $g : B \to A$ by $g(z) = w$ if $f(w) = z$. The function g is given by Figure 3.14. Both the compositions $f \circ g$ and $g \circ f$ are defined, and the reader will verify that for each x in A, $g \circ f(x) = x$ and that for each x in B, $f \circ g(x) = x$.

We can proceed similarly with any bijection.

Figure 3.13 f

Figure 3.14 g

DEFINITION. Suppose that $f : A \to B$ is a bijection. We define the function $f^{-1} : B \to A$, called the **inverse** of f, by the following rule: If z is an element of B, then $f^{-1}(z) = w$, where $f(w) = z$.

So for f defined in Figure 3.13, we have that g defined in Figure 3.14 is the inverse of f; that is, $g = f^{-1}$.

Proposition 1. Suppose that $f : A \to B$ is a bijection and f^{-1} is its inverse. For each x in B we have $f \circ f^{-1}(x) = x$, and for each x in A we have $f^{-1} \circ f(x) = x$. That is to say, $f \circ f^{-1} = \text{id}_B$ and $f^{-1} \circ f = \text{id}_A$.

PROOF: Let x be an element of B and let $f^{-1}(x) = z$. By the definition of an inverse function, we have $f(z) = x$. Thus $f \circ f^{-1}(x) = f(f^{-1}(x)) = f(z) = x$. Now if x is in A and $f(x) = z$, then $f^{-1}(z) = x$ and so $f^{-1} \circ f(x) = f^{-1}(f(x)) = f^{-1}(z) = x$.

Example 1. Suppose that $f : \mathbf{R} \to \mathbf{R}$ is the bijection given by $f(x) = 3x - 1$. Let y be an element of the codomain of f. To find $f^{-1}(y)$, we must solve for x in the equation $y = 3x - 1$. We find that $x = (y + 1)/3$ and thus $f^{-1}(y) = (y + 1)/3$. Since we usually write the formula for a function using the variable x, we rewrite the formula for f^{-1} as $f^{-1}(x) = (x + 1)/3$. For each x in $\mathbf{R}$, we have that $f \circ f^{-1}(x) = f((x + 1)/3) = 3[(x + 1)/3] - 1 = x$.

Example 2. The function $g : \mathbf{R} \to \mathbf{Z}$ given by $g(x) = \lfloor x \rfloor$ is not injective. We cannot define the inverse of g, because for each value of n in $\mathbf{Z}$ there are infinitely many values of x (rather than just one) for which $g(x) = n$. The function $h : \mathbf{N} \to \mathbf{N}$ given by $h(n) = 2^n$ is not surjective. We cannot define its inverse because, for instance, we cannot define $h^{-1}(3)$ since there is no integer value of n for which $2^n = 3$.

Example 3. The function $r : \mathbf{Z} \to E$ defined in Example 7 in Section 3.2 by $r(x) = 2x$ is a bijection. Its inverse is given by the formula $r^{-1}(x) = x/2$.

Example 4. Let I be the subset of $\mathbf{R}$ given by $\{x : x > 2\}$, and let $g : \mathbf{R}^+ \to I$ be given by $g(x) = x^2 + 2$. Then g is a bijection and we have that its inverse is given by the formula $g^{-1}(x) = (x - 2)^{1/2}$.

Example 5. Let $B = \{x : x \geq 1 \text{ or } x < 0\}$. Define $f : \mathbf{R} \to B$ as follows:

$$f(x) = x + 1 \qquad \text{if } x \geq 0$$
$$f(x) = 2x \qquad \text{if } x < 0$$

Then f is a bijection. Its inverse is defined as follows:

$$f^{-1}(x) = x - 1 \qquad \text{if } x \geq 1$$
$$f^{-1}(x) = x/2 \qquad \text{if } x < 0$$

Example 6. Let $f : \mathbf{N} \to \mathbf{Z}$ be defined as follows:

$$f(n) = n/2 \qquad \text{if } n \text{ is even}$$
$$f(n) = (1 - n)/2 \qquad \text{if } n \text{ is odd}$$

The function f is a bijection between the set of natural numbers and the set of integers. It puts the even natural numbers into one-to-one correspondence with the positive integers and the odd natural numbers into one-to-one correspondence with the integers that are less than or equal to 0. Its inverse $f^{-1} : \mathbf{Z} \to \mathbf{N}$ is defined as follows:

$$f^{-1}(n) = 2n \qquad \text{if } n > 0$$
$$f^{-1}(n) = 1 - 2n \qquad \text{if } n \leq 0$$

Example 7. Let $f : \mathbf{R} \times \mathbf{R} \to \mathbf{R} \times \mathbf{R}$ be defined by $f(x, y) = (x + y, x - y)$. We considered this function in Example 8 in Section 3.2. There we saw that it was a bijection. We also saw that if we let (s, t) be an element of the codomain of f and let $x = (s + t)/2$ and $y = (s - t)/2$, then $f(x, y) = (s, t)$. Thus $f^{-1}(s, t) = ((s + t)/2, (s - t)/2)$. Rewriting in terms of the variables x and y, we have that $f^{-1}(x, y) = ((x + y)/2, (x - y)/2)$.

| *Section 3.4* | **Exercises** |

1. Find the inverse of each of the following bijections.

   (a) $f : \mathbf{R} \to \mathbf{R}$ defined by $f(x) = (1 - 2x)/3$

   (b) $g : \mathbf{R} \to \mathbf{R}$ defined by $g(x) = (x^3 + 1)/3$

   (c) $f : \mathbf{R}^+ \to \{x : x > 1/2\}$ defined by $f(x) = (x^2 + 1)/2$

   (d) $g : \mathbf{R} \times \mathbf{R} \to \mathbf{R} \times \mathbf{R}$ defined by $g(x, y) = (2x + y, x + y)$

   (e) $f : \mathbf{R} \to \mathbf{R}$ defined by $f(x) = x^2$ if $x \geq 2$ and $f(x) = x + 2$ if $x < 2$

2. Define a bijection between each of the following pairs of sets and give its inverse.

   (a) The even integers and the odd integers

   (b) The natural numbers $\mathbf{N}$ and the set of all perfect squares

   (c) $A = \{x : 0 \leq x \leq 1\}$ and $B = \{x : 1 \leq x \leq 3\}$

   (d) The set of even integers and the set of negative integers

3. Let A be the language over $\{a, b\}$ defined inductively as follows:

   (1) The word b is an element of A.

   (2) If x is an element of A, then ax and xbb are in A.

   (3) Only words so formed are in A.

(a) Find a bijection from A into $\mathbf{W} \times \mathbf{N}$ and give its inverse.

(b) Find a bijection from A to $\mathbf{N} \times \mathbf{N}$ and give its inverse.

4. Let $A = \{a, b\}$ and let $B = \{a, b, c\}$. Suppose that S is the subset of $\mathcal{P}(B)$ containing all subsets of B that do not contain the element a. Find a bijection from $\mathcal{P}(A)$ to S and give its inverse.

5. Find sets A and B and functions $f : A \rightarrow B$ and $g : B \rightarrow A$ such that $g \circ f = \mathrm{id}_A$ but $g \neq f^{-1}$.

6. Prove that if f is a bijection, then $(f^{-1})^{-1} = f$.

| Section 3.5 | Functions and Set Operations |

Consider the greatest-integer function $g : \mathbf{R} \rightarrow \mathbf{Z}$ defined by $g(x) = \lfloor x \rfloor$. Let I be the interval $\{x : -1.6 < x < 3.6\}$. We ask, "What values of y in $\mathbf{Z}$ are assumed by $g(x)$ for x in I?" For any value of x in I, $g(x)$ is one of the values in the set $S = \{-2, -1, 0, 1, 2, 3\}$. The set S is called the direct image of I under g and is written $g(I)$. Now consider the set $T = \{-2, 3\}$. We ask, "For what values of x in $\mathbf{R}$ is $g(x)$ in T?" Here we start with a subset of the codomain of g and ask for a subset of its domain. (Before, we proceeded in the opposite manner.) The answer is the set $J = \{x : -2 \leq x < -1 \text{ or } 3 \leq x < 4\}$; these are precisely the values of x in the domain for which $g(x)$ is either -2 or 3. The set J is called the inverse image of T under g and is written $g^{-1}(T)$.

> **DEFINITION.** Let $f : A \rightarrow B$ be a function, and let S be a subset of A and T be a subset of B. The set $f(S) = \{y : y \in B \text{ and } y = f(x) \text{ for some } x \text{ in } S\}$ is called the **direct image** of S under f. The set $f^{-1}(T) = \{x : f(x) \in T\}$ is called the **inverse image** of T under f.

Example 1. Let $A = \{a, b, c, d, e\}$ and $B = \{s, t, u, v\}$. Suppose that $f : A \rightarrow B$ is given by Figure 3.15. Let $S = \{a, c, d\}$ and $T = \{s, t, v\}$. Then $f(S) = \{t, u\}$ and $f^{-1}(T) = \{a, b, c, e\}$.

Figure 3.15

Example 2. Let $f : \mathbf{R} \to \mathbf{R}$ be defined by $f(x) = 2x + 1$. Let $S = \{x : -2 \leq x \leq 2\}$. If x is an element of S, then $-3 \leq f(x) \leq 5$. Thus $f(S) = \{x : -3 \leq x \leq 5\}$. Now let $T = \{x : 0 \leq x \leq 2\}$. Then $f(x)$ is an element of T whenever $0 \leq 2x + 1 \leq 2$. Solving this inequality for x, we find that $f(x)$ is an element of T whenever $-\frac{1}{2} \leq x \leq \frac{1}{2}$. Thus $f^{-1}(T) = \{x : -\frac{1}{2} \leq x \leq \frac{1}{2}\}$.

Example 3. Let $g : \mathbf{R} \to \mathbf{R}$ be defined by $g(x) = x^2$. Let $S = \{x : -1 \leq x \leq 3\}$. Then $f(S) = \{x : 0 \leq x \leq 9\}$. Now $f^{-1}(f(S)) = \{x : -3 \leq x \leq 3\}$ and we see that $f^{-1}(f(S)) \neq S$.

For a function $f : A \to B$ the direct image of the domain A, namely $f(A)$, is called the **image of** f. It is not necessarily B. For example, if we take A to be the interval $\{x : 0 < x < 1\}$ and define $f : A \to \mathbf{R}$ by $f(x) = 1/x$, then $f(A) = \{x : x > 1\}$. This is a proper subset of the codomain $\mathbf{R}$. The function is not surjective. We leave it as an exercise to prove that $f(A) = B$ if and only if f is surjective.

Example 4. Let $g : A \to B$ be a bijection and suppose that y is any element of B. Consider the set $\{y\}$ containing only y. Then $g^{-1}(\{y\}) = \{g^{-1}(y)\}$. That is, the inverse image of the set containing only y is the set containing only $g^{-1}(y)$.

Example 5. Let E be a finite nonempty set of symbols, and let $A = E^*$ be the set of all words over E. Define $L : A \to \mathbf{W}$ by letting $L(x)$ be the length of the word x. Let n be a positive integer and $T = \{n\}$. Then $L^{-1}(T)$ is the set of all words of length n. Thus if $E = \{a, b\}$ and $n = 3$, we have that $L^{-1}(\{3\}) = \{aaa, aab, abb, aba, bbb, bba, baa, bab\}$. Let $S = \{a, aa, aaa\}$. Then $L(S) = \{1, 2, 3\}$.

Example 6. Suppose that $f : \mathbf{R} \to \mathbf{R}$ is defined by $f(x) = x^2$. Let $S = \{x : -2 < x < 3\}$ and $T = \{x : 4 < x < 9\}$. Then $f(S) = \{x : 0 \leq x < 9\}$ and $f^{-1}(T) = \{x : 2 < x < 3 \text{ or } -3 < x < -2\}$.

Example 7. Let $p : \mathbf{R} \times \mathbf{R} \to \mathbf{R}$ be defined by $p(x, y) = x$. Let S be the rectangle and its interior in $\mathbf{R} \times \mathbf{R}$ defined by $\{(x, y) : -1 \leq x \leq 1 \text{ and } 0 \leq y \leq 3\}$. Then $p(S)$ is the interval $\{x : -1 \leq x \leq 1\}$. Let C be the circle of radius 1 that has its center at $(-1, 2)$. It is defined by $\{(x, y) : (x + 1)^2 + (y - 2)^2 = 1\}$. Then $p(C)$ is the interval $\{x : -2 \leq x \leq 0\}$. If we take $x = 1$ to be a point on the real line, then $p^{-1}(1)$ is the line in $\mathbf{R} \times \mathbf{R}$ given by $\{(x, y) : x = 1\}$.

We now consider the relation between the direct and inverse images and the set operations union, intersection, and complementation.

Theorem 1. Suppose that $f : A \rightarrow B$ is a function from A to B and that H and K are subsets of A. Then $f(H \cup K) = f(H) \cup f(K)$.

PROOF: Let y be an element of $f(H \cup K)$. We can find an element x in $H \cup K$ such that $f(x) = y$. Because x is either in H or in K, y is either in $f(H)$ or in $f(K)$. Thus y is in $f(H) \cup f(K)$, and we have shown that $f(H \cup K) \subset f(H) \cup f(K)$.

   Conversely, let y be an element of $f(H) \cup f(K)$. If y is in $f(H)$, then there is some value z in H such that $f(z) = y$. If y is in $f(K)$, then there is some value w in K for which $f(w) = y$. In either case there is a value in $H \cup K$—call it x—for which $f(x) = y$. Thus y is in $f(H \cup K)$, and we have that $f(H) \cup f(K) \subset f(H \cup K)$. We conclude that $f(H \cup K) = f(H) \cup f(K)$.

   Now consider the function $f : A \rightarrow B$ defined by Figure 3.16. Let $H = \{a\}$ and $K = \{c\}$. We have that $f(H) = \{t\}$ and $f(K) = \{t\}$ and so $f(H) \cap f(K) = \{t\}$. However, $H \cap K$ is empty and so $f(H \cap K) = \emptyset$. Thus it is not the case that $f(H \cap K) = f(H) \cap f(K)$. Also $f(H') = f(\{b, c, d, e\}) = \{t, u, v\}$, whereas $(f(H))' = \{s, u, v\}$. Thus neither is it true that $f(H') = (f(H))'$. The direct image does not behave well with respect to intersection and complementation. However, if we let $S = \{s, t\}$ and $T = \{t, u\}$, we can verify that $f^{-1}(S \cup T) = f^{-1}(S) \cup f^{-1}(T)$, $f^{-1}(S \cap T) = f^{-1}(S) \cap f^{-1}(T)$, and $f^{-1}(S') = (f^{-1}(S))'$. This reflects the general behavior of the inverse image.

Figure 3.16

Example 8. Let $f : \mathbf{R} \rightarrow \mathbf{W}$ be defined by $f(x) = \lfloor x^2 \rfloor$, and let $S = \{0, 1\}$ and $T = \{1, 2\}$. Then $f^{-1}(S) = \{x : -\sqrt{2} < x < \sqrt{2}\}$ and $f^{-1}(T) = \{x : -\sqrt{3} < x \leq -1 \text{ or } 1 \leq x < \sqrt{3}\}$. Now $f^{-1}(S \cap T) = f^{-1}(\{1\}) = \{x : -\sqrt{2} < x \leq -1 \text{ or } 1 \leq x < \sqrt{2}\}$ and $f^{-1}(S) \cap f^{-1}(T) = f^{-1}(S \cap T)$.

Theorem 2. Let f be a function from A to B, and let S and T be subsets of B. Then

   (a) $f^{-1}(S \cup T) = f^{-1}(S) \cup f^{-1}(T)$
   (b) $f^{-1}(S \cap T) = f^{-1}(S) \cap f^{-1}(T)$
   (c) $f^{-1}(S') = (f^{-1}(S))'$

PROOF:

(b) Suppose that x is an element of $f^{-1}(S \cap T)$. Then $f(x)$ is an element of both S and T. Thus x is an element of both $f^{-1}(S)$ and $f^{-1}(T)$; that is, x is an element of $f^{-1}(S) \cap f^{-1}(T)$. We have then that $f^{-1}(S \cap T) \subset f^{-1}(S) \cap f^{-1}(T)$. Now let x be an element of $f^{-1}(S) \cap f^{-1}(T)$. Because x is in $f^{-1}(S)$, we have that $y = f(x)$ is in S; and because x is in $f^{-1}(T)$, we have that y is in T. Thus $y = f(x)$ is in $S \cap T$, and so x is in $f^{-1}(S \cap T)$. Thus we have that $f^{-1}(S) \cap f^{-1}(T) \subset f^{-1}(S \cap T)$, and we conclude that $f^{-1}(S \cap T) = f^{-1}(S) \cap f^{-1}(T)$.

We leave the proof of parts (a) and (c) as exercises.

Section 3.5   Exercises

1. Let $f : \mathbf{R} \to \mathbf{R}$ be defined by $f(x) = 2x - 1$. Let $A = \{x : -3 \le x \le 5\}$, and let $B = \{x : 0 \le x \le 2\}$. Find $f(A)$, $f^{-1}(B)$, $f(\mathbf{R}^+)$, and $f^{-1}(\mathbf{R}^+)$.

2. Let $f : \mathbf{R} \to \mathbf{R}$ be defined by $f(x) = x^2 + 1$. Let $A = \{x : -1 \le x \le 4\}$. Find $f(A)$ and $f^{-1}(A)$.

3. Define $f : \mathbf{R} \to \mathbf{R}$ by $f(x) = \lfloor x \rfloor$ and let n be an integer. Find $f^{-1}(\{n\})$. Why is $f^{-1}(n)$ not defined? (Note that the first expression has braces whereas the second does not.)

4. Let $f : \mathbf{Z} \to \mathbf{Z}$ be defined by $f(m) = m + |m|$. Let $S = \{2, -1, 1\}$. Find $f(S)$. Find $f^{-1}(\{0\})$. Find $f(\mathbf{Z})$.

5. Let E^* be the set of all words over $E = \{a, b\}$ and let L be the length function on E^*. Let A be the language of all words containing 3 or fewer a's and 4 or fewer b's. Find $L(A)$. How many elements are there in $L^{-1}(\{4\})$? in $L^{-1}(\{n\})$? Let $X_n = \{1, 2, 3, \ldots, n\}$. Find a formula for the number of elements in $L^{-1}(X_n)$.

6. Give an example to show that, in general, $f(f^{-1}(A)) \ne A$. Under what conditions would equality hold, however?

Bridge to Computer Science

3.6 A Look at Coding Theory (optional)

In digital technology, symbol sets such as the set of letters in the English alphabet or the set of digits of base-ten arithmetic are represented by strings of 0's and 1's. Suppose that S is such a set of symbols, and let B represent the set $\{0, 1\}$. Fix a length m and agree that all the symbols of S will be represented by strings of length m. Mathematically, the representation of the symbols of S by strings of 0's and 1's of length m is simply an injective function f from S into $B^m = B \times B \times \cdots \times B$. Thus if x is a symbol in S, then $f(x)$ is its representation in terms of 0's and 1's. [We identify a string of the form . . . 010001 . . . with the m-tuple $(. . . , 0, 1, 0, 0, 0, 1, . . .)$.] The injectivity of f reflects the fact that we cannot allow two different symbols to be represented by the same string.

Example 1. Suppose that S is the set of the 26 letters of the English alphabet. To represent all 26 letters by strings of 0's and 1's, we must take m to be at least 5. (The set B^4 has only 16 strings; B^5 has 32.) One injective function from S into B^5 follows. (The ith letter of the alphabet is represented by the binary representation of the number i.)

a	00001
b	00010
c	00011
.	
.	
.	
y	11001
z	11010

Example 2. The following is a representation of the integers from 0 to 15 by elements of B^4.

0	0000
1	0001
2	0011
3	0010
4	0110
5	0111
6	0101
7	0100
8	1100

9	1101
10	1111
11	1110
12	1010
13	1011
14	1001
15	1000

(If we allowed longer strings, we could represent larger integers according to the same rule: Upon reaching the mth power of 2, place a 1 in $(m + 1)$th position from the right and then repeat the preceding list in the remaining m positions in reverse order.) This representation of the integers is known as a Gray Code. It has the property that the representation of any integer n differs from the representation of $n + 1$ in exactly 1 position.

Now let $x = (x_1, x_2, \ldots, x_m)$ be an element of B^m. We shall call x a **message** and shall call the term x_j the "jth bit of x." If we were to transmit the message x over some electronic medium, there would be some chance of a transmission error; that is, a 0 might be received as a 1, or vice versa. Without extra information, the receiver could not detect the presence of an error in the received code. One way to provide for error detection is to add a "parity bit" to the message to be transmitted, as follows. To the message $x = (x_1, x_2, x_3, \ldots, x_m)$ we add an extra bit y in such a way that the resulting code word $(x_1, x_2, \ldots, x_m, y)$ has an even number of 1's. Thus if $x = (0, 1, 1, 1)$ is the message to be transmitted, then the parity bit $y = 1$ is added at the end, and the resulting word $(0, 1, 1, 1, 1)$ is transmitted. If the receiver receives a word with an odd number of 1's, then he or she detects that an error has been transmitted. (If an even number of 1's are received, he or she will not detect the less likely possibility that two errors have been transmitted.) If the receiver decides that a message has been correctly transmitted, he or she drops the parity bit to decode the message.

We can represent the coding process mathematically as follows. Let B^m be the set of messages to be transmitted. The addition of the parity bit to a message $x = (x_1, x_2, \ldots, x_m)$ is represented by the function $f : B^m \to B^{m+1}$, where $f(x) = (x_1, x_2, \ldots, x_m, y)$ with $y = 1$ if the number of 1's in x is odd and $y = 0$ otherwise. The received encoded messages are elements of B^{m+1}. Error-free messages are elements of the image set $f(B^m)$. This is the subset of B^{m+1} of strings with an even number of 1's. The decoding process is represented by the function $g : B^{m+1} \to B^m$ given by $g(x_1, x_2, \ldots, x_m, x_{m+1}) = (x_1, x_2, \ldots, x_m)$. The composition $f \circ g$ is the identity function on B_m. This represents the fact that g correctly decodes messages transmitted without error.

We now wish to extend our notion of coding and investigate the possibility of error correction. To this end, we define an (m, n)-code where B^m is the set of messages to be transmitted and the encoded messages are elements of B^n.

> **DEFINITION.** An **(m, n)-code** consists of an encoding function $f : B^m \rightarrow B^n$ and a decoding function $g : B^n \rightarrow B^m$ where the composition $f \circ g$ is the identity function on B^m. The elements of the image set $f(B^m)$ are called code words.

The parity-bit code is an $(n, n + 1)$-code. It is a single-error-detecting code because a transmission error in exactly one bit results in a word that is not a code word in $f(B^{n+1})$. It is not error-correcting, however, because there is no way to know which bit was incorrectly transmitted, resulting in an odd parity word. The following triple-repetition code both detects and corrects all single errors.

Example 3. **Triple-Repetition Code.** This $(m, 3m)$-code is given by the encoding function $f : B^m \rightarrow B^{3m}$ defined as follows: $f(x_1, x_2, \ldots, x_m) = (x_1, x_1, x_1, x_2, x_2, x_2, \ldots, x_m, x_m, x_m)$ where each bit is repeated three times. The decoding function $g : B^{3m} \rightarrow B^m$ is defined as follows: $g(y_1, y_2, \ldots, y_{3m}) = (x_1, x_2, \ldots, x_m)$ where we let $x_j = 1$ if two or more of the bits $\{y_{3j-2}, y_{3j-1}, y_{3j}\}$ are equal to 1 and we let $x_j = 0$ otherwise. In other words, we set x_j equal to the majority symbol of the jth three-bit group. For example, if we let $m = 2$ and let $x = (1, 0)$ be the message to be transmitted, then the code for x is $f(x) = (1, 1, 1, 0, 0, 0)$. Now suppose a transmission error occurs in the second bit, resulting in the word $y = (1, 0, 1, 0, 0, 0)$. Applying the decoding function, we find that $g(y) = (1, 0)$, thus correcting the transmission error. In fact, this decoding function corrects all occurrences of single errors and certain multiple errors.

We say an (m, n)-code detects k errors if whenever there is a change of k or fewer bits in a code word, the resulting word is not a code word. The parity-bit encodement detects all single errors. The triple-repetition code detects both single and double errors, although it cannot distinguish between them. Even if $n > m$, not all codes are error-detecting. For instance, the $(2, 3)$-code that follows does not detect all single errors.

$$00 \rightarrow 000$$
$$01 \rightarrow 001$$
$$10 \rightarrow 100$$
$$11 \rightarrow 111$$

A change in the last bit of 000 results in 001, which is another code word. In order for a code to detect all occurrences of single errors, each code word must differ from all other code words in at least two places.

> **DEFINITION.** Let x and y be two code words in B^n. The **distance** $d(x, y)$ between x and y is the number of places in which x and y differ. The **weight** $w(x)$ of x is the number of 1's in x.

Theorem 1.   For a code to detect all occurrences of k or fewer errors, the distance between any two code words must be at least $k + 1$.

PROOF: If a code word x is changed in k or fewer places, the resulting word cannot be a code word because each code word is assumed to differ from x in at least $(k + 1)$ places.

Example 4.   Consider the $(3, 6)$-code defined as follows:

$$000 \rightarrow 000000$$
$$001 \rightarrow 001111$$
$$010 \rightarrow 010011$$
$$011 \rightarrow 011100$$
$$100 \rightarrow 100110$$
$$101 \rightarrow 101001$$
$$110 \rightarrow 110101$$
$$111 \rightarrow 111010$$

We can verify that the distance between any two code words is at least 3. This code therefore detects all double and single errors.

Now suppose that we have an (m, n)-code that let x be a code word. Let $S(x)$ be the subset of B^n containing x and all words obtainable from x by a change in one bit of x. We say that the decoding function $g : B^n \rightarrow B^m$ is single-error-correcting if, whenever y is a word in $S(x)$, we have that $g(y) = g(x)$.

Example 5.   Consider the $(1, 3)$-code given by triple repetition. Then $x = 000$ is a code word and $S(x) = \{000, 100, 010, 001\}$. Decoding any word y in $S(x)$ by the majority rule yields $g(y) = 0$, thus correcting any single error.

Generalizing, let x be a code word and let $S_k(x)$ be the set containing x and all words obtainable from x by a change in k or fewer bits. That is, $S_k(x)$ is the set of words in B^n that are a distance k or less from x. Let y be an element of $S_k(x)$. We say that a decoding function g corrects k errors if, for every code word x, $g(y) = g(x)$ for every y in $S_k(x)$.

Theorem 2.   If an (m, n)-code corrects k errors, then the distance between any two code words must be at least $2k + 1$.

The proof is similar to that of Theorem 1, and we leave it as an exercise.

Example 6. Consider the (2, 5) encoding function f defined as follows:

$$00 \to 00000$$
$$01 \to 01101$$
$$10 \to 10011$$
$$11 \to 11110$$

The distance between any two code words is at least 3 (and the code is thus double-error-detecting). If we define the simple decoding function $g : B^n \to B^m$ by $g(x_1, x_2, x_3, x_4, x_5) = (x_1, x_2)$, we find that g is *not* error-correcting. For example, a single error in the first bit of the code word 00000 results in the noncode word 10000, yet we have that $g(10000) = 10$ and g has not corrected this single error. However, we can construct a decoding function that will correct all single errors. Let $C = \{00000, 11101, 10011, 11110\}$ be the set of code words, and let $h : B^5 \to C$ be defined by letting $h(x)$ be the code word in C that is nearest to x. [Thus $h(10000) = 00000$.] Now apply g to obtain the new decoding function $g \circ h$ that corrects all single errors for f. The following theorem generalizes this procedure.

Theorem 3. Let f be the encoding function of an (m, n)-code, and suppose that the distance between any two code words in B^n is at least $2k + 1$. Then there is a decoding function that corrects all occurrences of k or fewer errors.

PROOF: Let $C = f(B^m)$ be the set of all code words. Let $h : B^n \to C$ be defined by letting $h(x)$ be the code word in C that is nearest to x. Let $g : C \to B^m$ be defined by $g(x) = y$ if $f(y) = x$. Then the composition $h \circ g$ is a k-error-correcting decoding function.

We shall return to our discussion of coding theory at the end of Chapter 8.

Section 3.6	**Exercises**

1. Encode the set $\{a, b, c, d, e, f, g, h\}$ by strings of 0's and 1's of length 5 so that any two code words differ in at least two places.

2. Extend the Gray Code given in Example 2 to the integers 0 through 32. (Use B^5.)

3. (a) Decode the following words according to the decoding rule of the (3, 9)-triple-repetition code: 111011100 and 010101010.

   (b) Show that if the word 000111000 were changed in the first two places, it would be decoded incorrectly.

   (c) Characterize the types of errors (beyond single errors) that can be corrected by the triple-repetition code.

4. Consider the (3, 6)-code given in Example 4.

   (a) Decode the words 001001 and 101101 according to the procedure given in Theorem 3.

   (b) Find the minimum distance between any two code words.

   (c) Determine how many errors this code can correct.

5. Can one design a (2, 4)-error-detecting code?

6. (a) Design a coding scheme for B^4 that detects all single errors.

   (b) Design a coding scheme for B^4 that corrects all single errors.

7. Prove Theorem 2.

key concepts

3.1 Functions	function $f : A \to B$ domain A codomain B function value $y = f(x)$ for $x \in A$ and $y \in B$ identity function id_A inductively defined function recurrence relation recursively defined function
3.2 Injective and Surjective Functions	injective function (one-to-one) surjective function (onto) bijective function (one-to-one correspondence)
3.3 Composition of Functions	composition $g \circ f$ f^n, f composed with itself n times
3.4 Inverse Functions	f^{-1}, the inverse of f: $\quad f^{-1} \circ f = \mathrm{id}_A$ $\quad f \circ f^{-1} = \mathrm{id}_B$
3.5 Functions and Set Operations	$f(S)$, the direct image of the set $\quad S \subset A : f(S) \subset B$ $f^{-1}(Q)$, the inverse image of the set $\quad Q \subset B : f^{-1}(Q) \subset A$

1. Define $f: \mathbf{Z} \to \mathbf{R}$ by $f(x) = (x - 1)/(x^2 + 2)$. Evaluate $f(x)$ where $x = 1, -3$, or $\frac{1}{2}$ where possible. Suppose a is an integer. Evaluate $f(a - 3)$.

2. Let $f: \mathbf{R}^+ \to \mathbf{R}$ be defined by $f(x) = (x + 5)$. Find $f(4), f(0), f(0.2)$, and $f(20)$ where possible. For what values of a is $f(a - 7)$ defined?

3. Define $f: \mathbf{R} \to \mathbf{R}$ by $f(x) = (3x^2) + 1$ if $x \geq 0$ and $f(x) = -x^2$ if $x < 0$. Evaluate $f(3), f(0)$, and $f(-7)$. For what values of b is $f(b + 3)$ positive?

4. Let $h: \mathbf{R} \to \mathbf{R}$ be defined by $h(x) = 1/(x^2 - 1)$. Find all the values of x for which $h(x) = 1, 6$, or 0.

5. Let $f: \mathbf{R} \times \mathbf{R} \to \mathbf{R}$ be defined by $f(x, y) = 3x - 2y + 5$. Evaluate $f(0, 0)$, $f(1, 2)$, and $f(2, 1)$. Describe geometrically the subset of $\mathbf{R} \times \mathbf{R}$ for which $f(x, y) = 0$.

6. A function $f: \mathbf{R} \to \mathbf{R}$ is said to be *linear* if for any constants a and b and any values of x and y, we have $f(ax + by) = af(x) + bf(y)$. Determine which of the following functions from $\mathbf{R}$ to $\mathbf{R}$ is linear.

   (a) $f(x) = 2x + 5$

   (b) $g(x) = -2x$

   (c) $h(x) = x^2$

   (d) $f(x) = mx + b$, where m and b are constants that are not 0

   (e) $f(x) = f_1(x) + f_2(x)$, where both f_1 and f_2 are linear functions

7. The following functions from $\mathbf{N}$ to $\mathbf{N}$ are defined inductively. Evaluate each for at least five values of n.

   (a) $f(1) = 1$ and for $n > 1, f(n) = f(n - 1) + n$

   (b) $f(1) = 2$ and for $n \geq 1, f(n + 1) = (n + 1)f(n)$

   (c) $f(1) = 3$ and for $n \geq 1, f(n + 1) = 3f(n) + 2$

   (d) $f(1) = 1, f(2) = 2$, and for $n > 2, f(n) = f(n - 1) + 2f(n - 2)$

8. For f defined as in part (c) of Exercise 7, prove by induction that $f(n) = 4(3^{n-1}) - 1$ for all n in $\mathbf{N}$.

9. Let a, b, and c be real numbers, $a \neq 1$. Define $f: \mathbf{N} \to \mathbf{R}$ inductively as follows: $f(1) = c$, and $f(n + 1) = af(n) + b$ for $n \geq 1$. Prove by induction that $f(n) = a^{n-1}c + b(a^n - 1)/(a - 1)$ for all values of n in $\mathbf{N}$.

10. Suppose that $f: \mathbf{W} \to \mathbf{W}$ is defined inductively as follows:

    (1) $f(0) = 3$ and $f(1) = 1$

    (2) $f(n) = 2f(n - 1) + 3f(n - 2)$ for $n > 1$

    (a) Compute the values of $f(n)$ for $n = 1$ through $n = 5$.

(b) Use induction to prove that for each value of n in $\mathbf{W}$, we have $f(n) = 3^n + 2(-1)^n$. (*Hint*: Use the Second Principle of Induction.)

11. Let $f : \mathbf{W} \to \mathbf{N}$ be defined by $f(0) = 1$, $f(1) = 1$, and for $n > 0$, $f(n) = f(n - 2) + f(n - 1)$. (This function defines the Fibonacci numbers as in Example 11 in Section 3.1.) Prove by induction that for all values of n in $\mathbf{W}$,

$$f(n) = \frac{1}{\sqrt{5}}\left[\left(\frac{1+\sqrt{5}}{2}\right)^{n+1} - \left(\frac{1-\sqrt{5}}{2}\right)^{n+1}\right]$$

Verify this formula directly for a few values of n.

12. Let $E = \{a\}$ and define $f : E^* \times E^* \to \mathbf{W}$ inductively as follows:

(1) If Λ is the empty word, then $f(\Lambda, \Lambda) = 0$.
(2) If x and y are in E^* and $f(x, y) = n$, then $f(xa, y) = n + 1$ and $f(x, ya) = n + 1$.

Evaluate $f(aaa, aa)$ and $f(aaa, \Lambda)$. Give an arithmetic interpretation of f.

13. Give an inductive definition of each of the following functions.

(a) $f : \mathbf{N} \to \mathbf{N}$ defined by $f(n) = \left(\frac{1}{2}\right)^n$
(b) $f : \mathbf{N} \to \mathbf{N}$ defined by $f(n) = n + 1$
(c) $f : \mathbf{N} \to \mathbf{W}$ defined by $f(n) = -3n$
(d) $f : \mathbf{N} \to \mathbf{N}$ defined by $f(n) = n(n + 1)/2$
(e) $f : \mathbf{N} \to \mathbf{N}$ defined by $f(n) = 2^n - 1$

14. Determine the injectivity and surjectivity of each of the following functions.

(a) $f : \mathbf{Z} \to \mathbf{Z}$ where $f(n) = 2n + 3$
(b) Let E be the set of even integers and let O be the set of odd integers. Define $f : E \to O$ by $f(x) = 2x + 1$.
(c) $f : \mathbf{R} \to \mathbf{Z}$ where $f(x) = \lfloor 2x \rfloor$
(d) $g : \mathbf{R}^+ \to \mathbf{R}^+$ where $g(x) = 1/x$
(e) $f : \mathbf{R} \to \mathbf{R}$ where $f(x) = (x + 3)/2$
(f) $h : \mathbf{R} \to \mathbf{R}$ where $h(x) = x^2 + 2$
(g) $f : \mathbf{N} \times \mathbf{N} \to \mathbf{Q}^+$ where $f(n, m) = n/m$ and $\mathbf{Q}^+$ is the set of positive rational numbers
(h) $g : \mathbf{R} \times \mathbf{R} \to \mathbf{R} \times \mathbf{R}$ where $g(x, y) = (x + y, 2x - 3y)$
(i) $g : \mathbf{Z} \times \mathbf{Z} \to \mathbf{Z} \times \mathbf{Z}$ where $g(n, m) = (m + n, 2m - 3n)$

15. Let n be a positive integer, let S be the set of integers $\{0, 1, 2, 3, \ldots, n\}$, and let X be a set containing exactly n elements. Define $g : \mathcal{P}(X) \to S$ by letting $g(A)$ equal the number of elements in the set A. Show that g is surjective. Characterize all sets for which g is injective. Find an injective function from X to $\mathcal{P}(X)$.

16. Find a function $f : \mathbf{Z} \to \mathbf{Z}$ that is:

(a) Injective but not surjective

(b) Surjective but not injective

(c) Neither surjective nor injective

17. Let f, g, and h be defined by the accompanying diagrams. Find whichever of $f \circ g$, $g \circ f$, $h \circ g$, $h \circ f$, and $f \circ h$ is defined.

Figure for Exercise 17

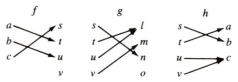

18. Let $f : \mathbf{R} \to \mathbf{R}$ and $g : \mathbf{R} \to \mathbf{R}$ be defined by $f(x) = x/(x^2 + 3)$ and $g(x) = x - 1$, respectively. Find the formulas for $f \circ g$, $g \circ f$, $f \circ f$, and $g \circ g$.

19. Let $f : \mathbf{R} \to \mathbf{R}$ be defined by $f(x) = x^3$ if $x > 0$ and by $f(x) = 3x + 1$ if $x \le 0$. Let $g : \mathbf{R} \to \mathbf{R}$ be defined by $g(x) = x^2 - 4$. Find $g \circ f$ and $f \circ g$.

20. Let $f : A \to B$ and $g : B \to C$ be functions, and suppose that $g \circ f$ is injective. Determine which of f or g must be injective, and prove your results.

21. Let $f : A \to B$ and $g : B \to C$ be functions and suppose that $g \circ f$ is surjective. Determine which of f or g must be surjective, and prove your results.

22. Let $g : A \to A$ be defined by the accompanying diagram.

Figure for Exercise 22

(a) Find g^n for $n = 0, 1, 2, 3,$ and 4.

(b) For each $n > 4$, characterize g^n in terms of the functions found in part (a), and prove your results by induction.

23. Let $f : \mathbf{R} \to \mathbf{R}$ be defined by $f(x) = ax + b$, where a and b are non-zero constants and $a \ne 1$. Prove that for all $n \ge 0$, we have $f^n(x) = a^n x + b(a^n - 1)/(a - 1)$.

24. Construct a bijection between the members of each of the following pairs of sets, and give the inverse of each.

(a) $\{1, 2, \ldots, 8\}$ and the power set of $\{a, b, c\}$

(b) The subset of $\mathbf{N} \times \mathbf{N}$ of all elements of the form (n, n) and the set $\mathbf{N}$

(c) The set $A = \{x : x \in \mathbf{R} \text{ and } x \ne 1\}$ and $B = \{x : x \in \mathbf{R} \text{ and } x \ne 2\}$

(d) $\mathbf{R}$ and the interval $\{x : -1 < x < 1\}$

25. Let a and b be real-valued constants such that $a < b$. Construct a bijection between the interval $\{x : 0 \le x \le 1\}$ and the interval $\{x : a \le x \le b\}$, and give its inverse.

26. Find the formula for f^{-1} for each of the following.

(a) $f : \mathbf{R} \to \mathbf{R}$ where $f(x) = (2x - 3)/5$

(b) $f : \mathbf{R} \times \mathbf{R} \to \mathbf{R} \times \mathbf{R}$ where $f(x, y) = (x + y, 2x - 3y)$

(c) $f : A \to B$ defined by $f(x) = (x + 3)^{1/2} + 2$ where $A = \{x : x > -3\}$ and $B = \{x : x > 2\}$

(d) $f : \mathbf{R} \to \mathbf{R}$ defined by $f(x) = 2x + 3$ if $x > 0$ and $f(x) = 3 - x^2$ if $x \le 0$

27. Let I be the interval $\{x : -2 < x < 3\}$ and J the interval $\{x : 1 < x < 4\}$. For each of the following functions from $\mathbf{R}$ to $\mathbf{R}$, find the sets specified.

$$f(I \cup J), \ f(I) \cup f(J), \ f(I \cap J), \ f(I) \cap f(J), \ f(I'), \ (f(I))'$$

(a) $f(x) = 3x + 2$ (b) $f(x) = x^2 + 1$ (c) $f(x) = \lfloor 2x \rfloor$

28. For the intervals I and J and the functions given in Exercise 27, find $f^{-1}(I \cup J), \ f^{-1}(I) \cup f^{-1}(J), \ f^{-1}(I \cap J), \ f^{-1}(I) \cap f^{-1}(J), \ f^{-1}(I')$, and $(f^{-1}(I))'$.

29. Let $f : \mathbf{R} \to \mathbf{R}$ be defined by $f(x) = x - \lfloor x \rfloor$. Find $f(R), f^{-1}(\{0\}), f^{-1}(\{\frac{1}{2}\})$, and $f^{-1}(\{1\})$.

30. Let $E = \{a, b, c\}$ and $S = \{0, 1\}$. Define the function $f : E^* \to S^*$ inductively as follows:

(1) $f(a) = 10, f(b) = 11$, and $f(c) = 01$

(2) If x and y are in E^* and $f(x) = s$ and $f(y) = t$, then $f(xy) = st$.

Find $f(abbc)$ and $f^{-1}(\{110110, 110011\})$. Let L be the subset of S^* defined by $\{y : y = 10z01 \text{ for } z \text{ in } S^*\}$, and characterize $f^{-1}(L)$. Characterize $(f(E^*))'$.

31. Show that a function $f : A \to B$ is surjective if and only if $f(A) = B$.

32. Let $f : A \to B$ be a function and let S be a subset of A. Prove that if f is injective, then $f^{-1}(f(S)) = S$.

33. Let $f : A \to B$ be a function and let T be a subset of B. Prove that if f is surjective, then $f(f^{-1}(T)) = T$.

34. Prove part (a) of Theorem 2 in Section 3.5.

35. Prove part (c) of Theorem 2 in Section 3.5.

36. Prove that if $f : A \to B$ is injective and S and T are subsets of A, then $f(S \cap T) = f(S) \cap f(T)$.

37. Suppose that $g : A \to B$ is surjective and that S is a subset of A. Either prove that $g(S') = (g(S))'$ or give a counterexample.

Chapter 4

COUNTING AND COUNTABILITY

outline

4.1 Counting Principles
4.2 Functions and Counting
4.3 Permutations and Combinations
4.4 Combinatorial Arguments
4.5 Infinite Sets and Countability
4.6 Bridge to Computer Science: An Algorithm Analyzed (optional)
Key Concepts
Exercises

Introduction

THE TASK of determining the number of elements in a finite set seems straightforward: just count 1, 2, 3, ... So let us try to find the number of elements in the set S of all words of length 7 that can be formed from the letters a, b, c, d, and e and that either begin with a or end with b. To "just count," we need a list of the words in S. Having the list, we need to know that it repeats no words and contains all the words in S. But to determine that our nonrepeating list is complete, we just might need to know the number of words in S before we start counting! In this chapter we present techniques for determining the size, or cardinality, of sets like S. The art of using such techniques is called combinatorics. (The cardinality of S is determined combinatorially and with little fuss in Example 14 in Section 4.1.)

The counting principles that we present in this chapter—when to add, when to multiply—are familiar and intuitive, but we begin by establishing their basis in the theory of functions. We do this so that we can extend our investigation of cardinality to infinite sets. Does the set of natural numbers $\mathbf{N} = \{1, 2, 3, \ldots\}$ have the same cardinality as the set of even numbers $E = \{2 \cdot 1, 2 \cdot 2, 2 \cdot 3, \ldots\}$? (The containment $E \subset \mathbf{N}$ is proper after all.) Or do all infinite sets have the same cardinality? (Aristotle believed this.) The key to the notion of "same cardinality" for infinite sets is the bijection or one-to-one correspondence: Two sets have the same cardinality if and only if there is a bijection between them. With this we shall find what would have surprised Aristotle—that not all infinite sets have the same cardinality. (But in fact, the sets $\mathbf{N}$ and E do!)

| Section 4.1 | Counting Principles |

We call the number of elements in a finite set A the **cardinality** of A, and we denote it by either card(A) or $|A|$. To determine that $|A| = n$, we count the elements of A. That is, we establish a one-to-one correspondence between the elements of A and the elements of the set $\{1, 2, 3, \ldots, n\}$. The following formal definition of cardinality reflects this very familiar procedure.

> **DEFINITION.** Let A be a nonempty finite set, and let n be a positive integer. Then card(A) $= n$ if and only if there is a bijection from the set A to the set $X_n = \{1, 2, 3, \ldots, n\}$. The cardinality of the empty set is 0.

The cardinality of the set $A = \{a, b, c\}$ is 3. One bijection (there are five others) from A to the set $X_3 = \{1, 2, 3\}$ is shown in Figure 4.1. The cardinality of the set $B = \{s, t, u\}$ is also 3, and there is a bijection from A to B. When two finite sets have the same cardinality, there is always a bijection between them.

Figure 4.1

Proposition 1. Let A and B be nonempty finite sets. Then $|A| = |B|$ if and only if there is a bijection from A to B.

PROOF: Suppose that both A and B have cardinality n. We shall find a bijection from A to B. We know we have bijections $f : A \rightarrow X_n$ and $g : B \rightarrow X_n$. The inverse of g, $g^{-1} : X_n \rightarrow B$, is also a bijection. We form the composition $g^{-1} \circ f$ to obtain the bijection $g^{-1} \circ f : A \rightarrow B$.

Now suppose that $|B| = n$ and that $f : A \rightarrow B$ is a bijection. We shall show that $|A| = n$. Suppose that $g : B \rightarrow X_n$ is a bijection. We form the composition $g \circ f$ to obtain the bijection $g \circ f : A \rightarrow X_n$. So A has cardinality n, and we conclude our proof.

Example 1. The set of integers $T = \{n : 0 \le n \le 15\}$ and the set S of strings of 0's and 1's of length 4 have the same cardinality. We can establish a bijection between these sets by corresponding each integer to the string that is its binary representation. (For example, 7 and 0111 correspond.) Because there are 16 integers between 0 and 15, there must be 16 strings of 0's and 1's of length 4.

The sets $A = \{a, b, c\}$ and $B = \{s, t\}$ are disjoint. The cardinality of $A \cup B$ is 5, the sum of the cardinalities of A and B, because no element of A is repeated in B. In general, the cardinality of two disjoint finite sets is the sum of their cardinalities. This fact is called the **addition principle** and is proved in the next theorem.

Theorem 1.   The Addition Principle. Let A and B be disjoint finite sets and suppose that $|A| = n$ and that $|B| = m$. Then $|A \cup B| = n + m$.

PROOF: If A is the empty set, then $A \cup B = B$ and $|A \cup B| = |B| = 0 + m$, as was to be proved. Similarly, the theorem is true if B is empty. Now suppose that both A and B are nonempty sets and that we have bijections $f : A \rightarrow X_n$ and $g : B \rightarrow X_m$. We now construct a bijection $h : A \cup B \rightarrow X_{n+m}$. First we rename the elements of A and B as follows: If x is an element of A and $f(x) = k$, rename x with a_k. If x is an element of B and $g(x) = j$, rename x with a_{n+j}. Thus $A = \{a_1, a_2, \ldots, a_n\}$ and $B = \{a_{n+1}, a_{n+2}, \ldots, a_{n+m}\}$. Because no element is a member of both A and B, each element has exactly one name.

Each of the names a_q, where $q = 1, 2, 3, \ldots, n + m$, is the name of exactly one member of $A \cup B$. The function $h : A \cup B \to X_{n+m}$ defined by $h(a_q) = q$ for $1 \leq q \leq n + m$ is thus a bijection, and we conclude our proof.

Example 2. Let $A = \{m, n, o\}$ and $B = \{s, t\}$. Define the bijection $f : A \to X_3$ by $f(m) = 2, f(n) = 1$, and $f(o) = 3$. Define the bijection $g : B \to S_2$ by $g(s) = 2$ and $g(t) = 1$. We now name the elements of $A \cup B = \{m, n, o, s, t\}$ according to the procedure given in the proof of Theorem 1.

$$m = a_2 \qquad s = a_5$$
$$n = a_1 \qquad t = a_4$$
$$o = a_3$$

(Because s is an element of B, we add 3 to the value of $g(s)$ to obtain its name and proceed similarly for t.) The bijection h prescribed in the proof of Theorem 1 is thus as shown in Figure 4.2.

Figure 4.2

Suppose that we have a collection of sets. We say that the sets of the collection are **mutually disjoint** if the intersection of each pair of sets taken from the collection is empty. The sets of the collection $A_1 = \{a, b, c\}$, $A_2 = \{s, t\}$, and $A_3 = \{x, y, z\}$ are mutually disjoint. No element appears in any two sets. The cardinality of $A_1 \cup A_2 \cup A_3$ is 8, the sum of the cardinalities of each of the sets in the collection. The addition principle extends to finite collections of mutually disjoint finite sets, and we have the following corollary.

Corollary to Theorem 1. Suppose that $A_1, A_2, \ldots, A_n$ is a finite collection of mutually disjoint sets and that $|A_1| = m_1, |A_2| = m_2, \ldots,$ and $|A_n| = m_n$. Then $|A_1 \cup A_2 \cup \cdots \cup A_n| = m_1 + m_2 + \cdots + m_n$.

We leave the inductive proof of the corollary as an exercise.

We noted in Chapter 2 that if A is a finite set containing n elements, then its power set $\mathcal{P}(A)$ contains 2^n elements. We now prove this fact, using induction and the addition principle.

Theorem 2. If A is a finite set containing n elements, then its power set $\mathcal{P}(A)$ contains 2^n elements.

PROOF: Suppose that A is the empty set so that $n = 0$. Then $\mathcal{P}(A) = \{\emptyset\}$ and contains exactly one element. Because $2^0 = 1$, the theorem is true for $n = 0$. Now assume that if the cardinality of a set B is equal to n, then $|\mathcal{P}(B)| = 2^n$. (This is the induction hypothesis.) Let A be a set containing $(n + 1)$ elements. We will show that $\mathcal{P}(A)$ contains 2^{n+1} elements. Let y be an element of A and let $B = A\backslash\{y\}$. The set B contains all the elements of A except y. Thus $|B| = n$ and $|\mathcal{P}(B)| = 2^n$ by the induction hypothesis. Now $\mathcal{P}(A)$ is the union of two disjoint collections of subsets of A: the collection H, which contains all of the subsets of A that contain the element y, and the collection K, which contains all the subsets that do not. The set K is precisely $\mathcal{P}(B)$ and thus contains 2^n elements. Now let S be a member of H so that S is a subset of A that contains y. Let $T = S\backslash\{y\}$. Then $S = \{y\} \cup T$ and T is a member of K. We have a one-to-one correspondence from K to H given by $T \rightarrow T \cup \{y\}$ for each T in H. Thus the cardinality of H is the same as that of K—namely 2^n. By the addition principle we have $|\mathcal{P}(A)| = |H| + |K| = 2^n + 2^n = 2^{n+1}$, as was to be proved.

Example 3. We wish to determine how many committees are formable from a set of 5 people. We assume that no committee is empty and that any person can serve on any and all committees. The set of possible committees is thus in one-to-one correspondence with the nonempty subsets of the set of 5 people. There are $2^5 - 1 = 31$ possible committees formable. (We subtract 1 to omit the empty committee.)

Proposition 2. Suppose that A is a subset of a finite universal set U. Then $|A'| = |U| - |A|$.

PROOF: We know that A and A' are disjoint and that $A \cup A' = U$. Thus, by the addition principle, we have that $|A| + |A'| = |U|$. Solving for $|A'|$ we see that $|A'| = |U| - |A|$.

Example 4. Suppose we wish to determine the number of integers between 1 and 100 that are not divisible by 3. We know there are $\lfloor 100/3 \rfloor = 33$ integers that are divisible by 3. Applying Proposition 2, we see that there must be $100 - 33 = 67$ integers that are not divisible by 3.

Proposition 3. Suppose that A and B are are finite sets. Then $|A\backslash B| = |A| - |A \cap B|$.

PROOF: We know that $A\backslash B$ and $A \cap B$ are disjoint and that $A = (A\backslash B) \cup (A \cap B)$. So $|A| = |A\backslash B| + |A \cap B|$. Solving for $|A\backslash B|$, we see that $|A\backslash B| = |A| - |A \cap B|$.

Example 4 (continued). Now suppose we wish to determine the number of integers between 1 and 100 that are divisible by 3 but not by 5. The number of integers divisible by both 3 and 5 is $\lfloor 100/15 \rfloor = 6$. Thus the number divisible by 3 but not by 5 is $33 - 6 = 27$.

If we let $A = \{a, b, c, d\}$ and $B = \{a, d, s, t, v\}$, then $|A| = 4$ and $|B| = 5$, but $|A \cup B| = 7$. Simply adding the cardinalities of the sets A and B would count the elements in the intersection $A \cap B$ twice. To obtain the correct answer 7, we evaluate $|A| + (|B| - |A \cap B|)$. This way we do not count twice the elements in B that are also in A. We summarize our reasoning in the following theorem.

Theorem 3.   The Inclusion–Exclusion Principle. Let A and B be sets, and suppose that $|A| = m$, $|B| = n$, and $|A \cap B| = r$. Then $|A \cup B| = |A| + |B| - |A \cap B|$ or $|A \cup B| = m + n - r$.

PROOF: We can write $A \cup B$ as the disjoint union of three sets: $A \cup B = (A \backslash B) \cup (B \backslash A) \cup (A \cap B)$. By Proposition 3 we know that $|A \backslash B| = m - r$ and $|B \backslash A| = n - r$. By the addition principle we have that $|A \cup B| = |A \backslash B| + |B \backslash A| + |A \cap B| = (m - r) + (n - r) + r$. Simplifying, we obtain $|A \cup B| = m + n - r$, as was to be proved.

Example 5. Suppose that A and B are sets and that $|A| = 100$, $|B| = 150$, and $|A \cup B| = 200$. We use the inclusion–exclusion principle to determine that $|A \cap B| = 50$ by solving for $|A \cap B|$ in the equation $|A \cup B| = |A| + |B| - |A \cap B|$—that is, by solving for $|A \cap B|$ in $200 = 100 + 150 - |A \cap B|$.

Example 6. In a survey, 200 people are asked whether they read Magazine A or Magazine B. Suppose it is found that 90 read A, 120 read B, and 40 read both. Because $170 = 90 + 120 - 40$, we can use the inclusion–exclusion principle to conclude that 170 people read at least one of the magazines and hence 30 read neither.

Example 7. We wish to determine how many integers between 1 and 1235 are divisible by either 2 or 3. There are $\lfloor 1235/2 \rfloor = 617$ integers divisible by 2, and there are $\lfloor 1235/3 \rfloor = 411$ integers divisible by 3. There are $\lfloor 1235/6 \rfloor = 205$ integers divisible by both 2 and 3. Applying the inclusion–exclusion principle, we see that there are $617 + 411 - 205 = 823$ integers divisible by either 2 or 3.

We can extend the inclusion–exclusion principle to collections of more than two sets. For instance, suppose that A, B, and C are finite sets. The sum of their cardinalities, $|A| + |B| + |C|$, counts the elements in $A \cap B$ twice: first when they appear in A and then when they appear in B. The elements of $A \cap C$ and $B \cap C$ are also counted twice. The elements of $A \cap B \cap C$ are counted three times.

However, the sum $|A| + |B| + |C| - |A \cap B| - |A \cap C| - |B \cap C|$ counts (exactly once) only the elements of $A \cup B \cup C$ that are not in $A \cap B \cap C$. Thus we add $|A \cap B \cap C|$ to obtain the inclusion–exclusion principle for three sets:

$$|A \cup B \cup C| = |A| + |B| + |C| - |A \cap B| - |A \cap C| - |B \cap C| + |A \cap B \cap C|$$

Example 8. Suppose that 100 people are surveyed and it is found that 30 read *Time*, 45 read *Newsweek*, and 60 read *Mad*. It is also found that 10 read both *Time* and *Newsweek*, 20 read both *Newsweek* and *Mad*, 15 read both *Time* and *Mad*, and 5 read all three. We wish to determine how many of the people surveyed read none of these magazines and how many read exactly two of them. We let T denote the readers of *Time*, N the readers of *Newsweek*, and M the readers of *Mad*. Then, by the inclusion–exclusion principle, we have

$$|T \cup N \cup M| = 30 + 45 + 60 - 10 - 20 - 15 + 5 = 95$$

Thus 95 people read at least one of the magazines, and so 5 read none of them. To find the number who read exactly two of the magazines, we calculate

$$|T \cap N| + |T \cap M| + |N \cap M| - 3|T \cap N \cap M| = 10 + 15 + 20 - 15 = 30$$

Example 9. We use the inclusion–exclusion principle to determine how many integers between 1 and 1000 are divisible by 2, 3, or 5. The number divisible by 2 is $\lfloor 1000/2 \rfloor = 500$. The number divisible by 3 is $\lfloor 1000/3 \rfloor = 333$, and the number divisible by 5 is $\lfloor 1000/5 \rfloor = 200$. Now the number divisible by both 2 and 3 is $\lfloor 1000/6 \rfloor = 166$, that divisible by both 2 and 5 is $\lfloor 1000/10 \rfloor = 100$, and that divisible by both 3 and 5 is $\lfloor 1000/15 \rfloor = 66$. The number of integers divisible by 2, 3, and 5 is $\lfloor 1000/30 \rfloor = 33$. By the inclusion–exclusion principle, we find that the number of integers between 1 and 1000 that are divisible by 2, 3, or 5 is

$$500 + 333 + 200 - 166 - 66 - 100 + 33 = 734$$

The general inclusion–exclusion principle for a collection of n finite sets is given by the following theorem.

Theorem 4. $|A_1 \cup A_2 \cup \cdots \cup A_n| = |A_1| + |A_2| + \cdots + |A_n| - |A_1 \cap A_2| - |A_1 \cap A_3| - \cdots - |A_{n-1} \cap A_n| + |A_1 \cap A_2 \cap A_3| + \cdots + (-1)^{n+1} |A_1 \cap A_2 \cap \cdots \cap A_n|$

(First we add (include) the cardinalities of each of the sets. Next we subtract (exclude) the cardinalities of the intersections of each of the possible pairs of sets. Then we add (include) the cardinalities of the intersections of each of the possible selections of three sets. We continue until we reach the intersection of all the sets, changing from inclusion to exclusion (or vice versa) every time we increment the number of sets we intersect.)

The proof of Theorem 4 is based on induction and the distributive laws. We sketch it below.

PROOF (sketch): We have proved the theorem directly for $n = 2$. (There is nothing to prove for $n = 1$.) Suppose the theorem is true for a collection of n sets, and suppose we wish to determine $|A_1 \cup A_2 \cup \cdots \cup A_n \cup A_{n+1}|$. By the inclusion–exclusion principle for two sets, we have

$$|(A_1 \cup A_2 \cup \cdots \cup A_n) \cup A_{n+1}| = |A_1 \cup A_2 \cup \cdots \cup A_n| + |A_{n+1}|$$
$$- |A_{n+1} \cap (A_1 \cup \cdots \cup A_n)|$$

Distributing in the last term, we obtain

$$A_{n+1} \cap (A_1 \cup A_2 \cup \cdots \cup A_n) = (A_{n+1} \cap A_1) \cup (A_{n+1} \cap A_2) \cup \cdots$$
$$\cup (A_{n+1} \cap A_n)$$

Let $B_k = A_{n+1} \cap A_k$. We can now use our induction hypothesis and apply the inclusion–exclusion principle to determine the cardinality of $B_1 \cup B_2 \cup \cdots \cup B_n$ and the cardinality of $A_1 \cup A_2 \cup \cdots \cup A_n$. Substituting back into the original expression, we obtain the stated result. (We will ask the reader to carry out the details for a collection of three sets in the exercises for this chapter.)

Now we use the addition principle to prove a counting principle about the cartesian product of two finite sets.

Theorem 5. The Multiplication Principle. Suppose that A and B are two nonempty finite sets and that $|A| = m$ and $|B| = n$. Then $|A \times B| = m \cdot n$.

PROOF: Suppose that $A = \{a_1, a_2, \ldots, a_m\}$ and that $B = \{b_1, b_2, \ldots, b_n\}$. We can write $A \times B$ as the union of the following family of m disjoint sets:

$$S_1 = \{(a_1, b_1), (a_1, b_2), (a_1, b_3), \ldots, (a_1, b_n)\}$$
$$S_2 = \{(a_2, b_1), (a_2, b_2), (a_2, b_3), \ldots, (a_2, b_n)\}$$
$$\vdots$$
$$S_m = \{(a_m, b_1), (a_m, b_2), (a_m, b_3), \ldots, (a_m, b_n)\}$$

Now each of the sets in this family contains exactly n elements, and there are m sets in the family. Thus by the addition principle, there are $n + n + \cdots + n$ (m times) elements in $S_1 \cup S_2 \cup \cdots \cup S_m$. That is, there are $m \cdot n$ elements in $A \times B$, as was to be proved.

Corollary to Theorem 5. If $A_1, A_2, \ldots, A_n$ is a family of nonempty finite sets such that $|A_1| = m_1, |A_2| = m_2, \ldots, |A_n| = m_n$, then $|A_1 \times A_2 \times \cdots \times A_n| = m_1 \cdot m_2 \cdot \cdots \cdot m_n$.

Example 10. Let $X_{100} = \{1, 2, \ldots, 100\}$ and $X_{20} = \{1, \ldots, 20\}$; then $|X_{100} \times X_{20}| = 2000$.

Example 11. We use the multiplication principle to determine how many two-symbol labels can be formed if the first symbol can be any letter of the English alphabet E and the second can be any digit in the set of digits $D = \{0, 1, 2, \ldots, 9\}$. (Such labels might be $a3$ and $f7$ but not $3n$ or $g22$.) Each label can be associated with exactly one of the elements in $E \times D$, and conversely. For instance, we can associate $d5$ with $(d, 5)$. Thus, by the multiplication principle, there are $26 \cdot 10 = 260$ possible labels.

Example 12. We use the corollary to Theorem 5 to determine how many different six-place license plates are possible if the first three places must be letters and the next three must be digits. The answer is $26 \cdot 26 \cdot 26 \cdot 10 \cdot 10 \cdot 10 = 17{,}576{,}000$. If no letter or digit can be repeated, the answer is $26 \cdot 25 \cdot 24 \cdot 10 \cdot 9 \cdot 8 = 11{,}232{,}000$. (After the first letter is chosen, there are only 25 possibilities for the second letter and 24 for the third. After the first digit is chosen, there are only 9 possibilities for the second digit and 8 for the third.)

Example 13. A code word containing only the symbols 0 and 1 can be 3, 4, or 5 symbols in length. We wish to determine how many such code words are possible. Using the multiplication principle, we find that there are 2^3 words of length 3, 2^4 words of length 4, and 2^5 words of length 5. Because these cases are disjoint, we use the addition principle to determine that there are $2^3 + 2^4 + 2^5 = 56$ possible code words.

Example 14. We wish to determine how many words of length 7 can be formed from the letters $\{a, b, c, d, e\}$ if the first letter must be a or the last letter b. There are 5^6 words that begin with a, there are 5^6 words that end with b, and there are 5^5 words that both begin with a and end with b. By the inclusion–exclusion principle, there are $5^6 + 5^6 - 5^5 = 28{,}125$ words that begin with a or end with b.

Section 4.1   Exercises

1. Let $A = \{a, b, c, d, e\}$, let $B = \{d, e, f, g\}$, and let $C = \{g, h, i\}$. Find the cardinalities of each of the following sets: $A \cup C$, $A \times B$, $A \cup B$, $\mathscr{P}(A)$, $\mathscr{P}(A) \cup \mathscr{P}(C)$, and $\mathscr{P}(A) \cup \mathscr{P}(B)$.

2. Let $A = \{a, b, c, d\}$ and $B = \{s, t, u\}$. Let $f : A \to X_4$ and $g : B \to X_3$ be

defined as shown in the accompanying diagrams. Find $h : A \cup B \rightarrow X_7$ as prescribed in Theorem 1.

Figure for Exercise 2

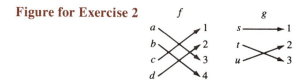

3. Let $A = \{a, b, c, d\}$ and $B = \{s, t, u\}$. Find a one-to-one correspondence between $\mathcal{P}(B)$ and the collection of subsets of A that do not contain the element c.

4. A man has 3 hats, 2 ties, and 6 shirts. How many different outfits can he assemble?

5. Two programs will be run simultaneously. The first requires 1200 bytes of memory and the second requires 900 bytes. If 200 bytes can be shared, what is the total number of bytes that must be allotted in order to run both programs?

6. Suppose that 100 people are surveyed and it is found that 70 like oranges, 50 like kiwis, and 10 like neither. How many like both? How many like oranges but not kiwis?

7. How many different words of length 5 can be made up from the letters q, m, y, t, s, h, and o? What if no letter can appear twice? What if no two adjacent letters can be the same?

8. Suppose the lead-off player in a Scrabble game draws 7 letters with no repeats. How many different 5-letter words (including illegal words) can he start the game with? How many different words of length 3 can he start with? How many words in all can he start with?

9. A coin is tossed 5 times, and the sequence of heads and tails is noted. How many different sequences are possible? How many sequences are possible that either begin or end with heads?

10. How many ways can you roll a red die and a green die? What if the red die always comes up even and the green die always comes up odd? What if the sum is always 7?

11. How many integers between 1 and 567 are divisible by either 3 or 5?

12. How many integers between 999 and 9999 either begin or end with 3?

13. How many integers between 1 and 567 are divisible by 3, 5, or 7?

14. How many integers between 1 and 1000 are divisible by 2, 3, 5, or 7?

15. How many words of length 7 over the alphabet $\{a, b, c\}$ have the letter b appearing in the first, fourth, or seventh position?

| **Functions and Counting**

In this section we apply the counting principles of Section 4.1 to the study of functions between finite sets.

Suppose that A is the set $\{a_1, a_2, \ldots, a_m\}$ and that B is the set $\{b_1, b_2, \ldots, b_n\}$ so that $|A| = m$ and $|B| = n$. To determine a function $f : A \to B$, we choose a value b_j in B for each a_k in A. Thus for each a_k there are n possible choices for $f(a_k)$. Because there are m elements a_k in A, and n choices for each $f(a_k)$, we apply the multiplication principle to find that there are $n \cdot n \cdot n \cdot \cdots \cdot n$ (m times) functions—that is, n^m possible functions from A to B. We summarize our reasoning in the following proposition.

Proposition 1. Suppose that A and B are finite nonempty sets and that $|A| = m$ and $|B| = n$. There are n^m different functions possible from A to B.

Example 1. Let $A = \{a, b, c\}$ and $B = \{s, t\}$. There are $2^3 = 8$ functions from A to B and $3^2 = 9$ functions from B to A.

Example 2.   Logical Functions. An n-argument logical function is a function of the form $Q = f(P_1, P_2, \ldots, P_n)$, where each of the variables $P_1, P_2, \ldots, P_n$ and Q can assume only the value 0 or the value 1. For example, the function $f(P, R) = \sim(P \wedge R)$ is a 2-argument or binary logical function. Let $B = \{0, 1\}$. Because each variable P_k in an n-argument logical function f can assume only the value 0 or the value 1, the domain of f is in one-to-one correspondence with the cartesian product of B with itself n times—that is, $B \times B \times \cdots \times B$ (n times). The domain of f thus has 2^n elements. Since $Q = f(P_1, P_2, \ldots, P_n)$ can assume only the value 0 or the value 1, the codomain of f contains only 2 values. By Proposition 1 there are exactly 2^{2^n} different n-argument logical functions. So, for example, there are $2^{2^2} = 16$ possible binary logical functions and $2^{2^3} = 256$ possible 3-argument or ternary logical functions.

Suppose now that $|B| \geq |A|$. We wish to determine the number of injective functions from A to B. Again we assume that $A = \{a_1, a_2, \ldots, a_m\}$ and $B = \{b_1, b_2, \ldots, b_n\}$. To construct an injective function f from A to B, we have n choices for the value of $f(a_1)$. We have $(n - 1)$ choices for the value of $f(a_2)$, namely any element of B except $f(a_1)$. For $f(a_3)$ we have $(n - 2)$ choices and so forth, until finally we have $n - (m - 1) = n - m + 1$ choices for $f(a_m)$. By the multiplication principle we can construct $n(n - 1)(n - 2) \cdots (n - m + 1) = n!/(n - m)!$ different injective functions from A to B. We summarize our reasoning in the following proposition.

Proposition 2. Suppose A and B are nonempty finite sets and that $|A| = m$ and $|B| = n$. Suppose also that $n \geq m$. Then the number of injective functions from A to B is $n!/(n - m)!$.

Corollary to Proposition 2. Suppose that $|A| = n$. Then there are $n!$ different bijections of the form $f: A \rightarrow A$.

Example 3. Suppose that $A = \{a, b, c\}$ and $B = \{s, t, u, v, w\}$. Then there are $5!/2! = 60$ different injective functions from A to B. There are $3!/0! = 6$ different bijective functions from A to A.

Example 4. We wish to determine how many different ways we can form a code for the digits in the set $D = \{0, 1, 2, 3, 4, 5, 6, 7, 8, 9\}$ using words of length 4 over the symbol set $E = \{a, b, c, d, e\}$. Call this set of words B. Now there are $5^4 = 625$ different words in the set B. We wish to assign each digit exactly one of these words, and no two digits should be assigned the same word. That is, we want an injective function from the set D into the set B. By Proposition 2 there are $625!/615!$ ways to do this.

Suppose that A and B are finite sets and suppose that $|A| > |B|$. Then there are no bijections from A to B. Let $f: A \rightarrow B$. Because f is not injective, we can find distinct elements s and t in A such that $f(s) = f(t)$. Thus we have the "pigeon-hole principle."

The Pigeon-Hole Principle. Suppose that $|A| > |B|$ and that $f: A \rightarrow B$. There are elements s and t in A, $s \neq t$, such that $f(s) = f(t)$.

Example 5. Suppose that 11 pigeons wish to roost in 10 pigeon holes. Then either some pair of pigeons must share a roost or one pigeon will remain without a roost. These pigeons cannot establish an injective function between themselves and the pigeon holes.

Example 6. Suppose we have a bag containing 5 apples and 5 pears. How large a sample must be chosen to ensure that at least one pear is chosen? The answer is 6. Because $6 > 5$, we cannot establish a one-to-one correspondence between our choices and the set of apples in the bag.

Example 7. If 13 pennies are distributed among 6 children, then at least 1 child receives at least 3 pennies. First we give each child 1 penny. Then 1 of the 6 children must receive at least 2 of the 7 remaining pennies.

> **DEFINITION.** Let A be a nonempty set. A bijection $f : A \to A$ is called a **permutation** of A.

Suppose that $|A| = n$. We know from the corollary to Proposition 2 that there are $n!$ different permutations of A. Thus if $A = \{a, b, c, d\}$, there are $4! = 24$ permutations of A.

In standard English usage, the word "permutation" means a rearrangement of the elements of a set given in a specific order. For example $AMHT$ is a permutation of the symbols of the word $MATH$. This meaning of the word "permutation" is consistent with our definition. If we think of the letters of the word $MATH$ as given in the order M-A-T-H, then the order of the letters given by the rearrangement $AMHT$ defines the bijection from the set $\{M, A, T, H\}$ to itself that is shown in Figure 4.3.

The letter M goes to the first letter appearing in the rearrangement, A goes to the second, and so forth. Conversely, every bijection gives us a rearrangement. For example, the bijection given by Figure 4.4 defines the rearrangement $TAMH$. We take the first letter of the rearrangement to be the letter assigned to M, the second letter of the rearrangement to be the letter assigned to A, and so forth. Thus there are $4!$ different arrangements of the letters of the word $MATH$. In general, there are $n!$ permutations of the elements of a set of cardinality n.

Figure 4.3   **Figure 4.4**

Example 8. There are $6! = 720$ different ways in which 6 people can form a queue. There are $5! = 120$ ways in which 6 people can form a circle. (Here we consider only the relative positions of the 6 people in the circle.) So if we fix 1 person at "twelve o'clock," there are $5!$ different ways for the others to line up around him.

Example 9. There are $6!$ permutations or arrangements of the letters in the word $MISTER$. However, there are only $6!/3! = 120$ ways to rearrange the letters of the word $CHEESE$. To see this, we number the E's as E_1, E_2, and E_3. There are $3!$ orders in which we can put these E's. When counting $6!$ ways in which to arrange the letters of $CHEESE$, we count C-H-E_1-E_2-S-E_3 as different from C-H-E_3-E_1-S-E_2. Thus a count of $6!$ would count each arrangement of the letters in $CHEESE$ $3!$ times, once for each arrangement of the E's. To ignore the arrangements of the E's, we divide $6!$ by $3!$ to obtain the correct answer of 120.

Example 10. There are $11!/4!4!2!$ ways to arrange the letters in the word *MISSISSIPPI*.

Example 11. We wish to determine how many ways there are in which to arrange a sequence of 6 identical white balls and 5 identical red balls (11 balls in all). If we labeled each of the balls so that we could distinguish between balls of the same color, then there would be $11!$ different possible arrangements. However, in a given sequence of red and white balls, we wish to disregard the $5!$ different arrangements of the red balls and the $6!$ different arrangements of the white balls. Thus there are $11!/5!6! = 462$ different sequences of red and white balls possible.

Section 4.2	**Exercises**

1. Let A and B be sets and suppose that $|A| = 3$ and $|B| = 4$. Find the number of:

   (a) Functions from A to B

   (b) Injective functions from A to B

   (c) Functions from B to A

   (d) Injective functions from B to A

   (e) Functions from A to $\mathcal{P}(A)$

   (f) Injective functions from A to $\mathcal{P}(A)$

2. Let $A = \{a_1, a_2, \ldots, a_m\}$ and let $B = \{b_1, b_2, \ldots, b_n\}$. Suppose that $A \cap B = \{b_1, b_2, \ldots, b_r\}$, where $r \le n$ and $r \le m$. Find a bijection from $A \cup B$ to X_{m+n-r}.

3. For each of the following sets called A, find a bijection from A to X_n for the appropriate value of n.

   (a) Let A be the set of words over $\{a, b\}$ of length 3.

   (b) Let A be the set of integers between 0 and 1000 that are divisible by 2.

   (c) Let A be the set of integers between 0 and 100 that are divisible by either 2 or 3.

   (d) Let A be the set of all permutations of the set $\{1, 2, 3, 4\}$.

4. Suppose we wish to represent the letters of the alphabet and the digits from 0 to 9 by strings of 0's and 1's of length 8. How many ways can this be done? What if each of the strings must have an even number of 1's?

5. In how many ways can we assign a number between 1 and 100 to each of 50

people? What if each person must be assigned a different number? What if no two numbers can have a difference of 1 and each person is assigned a different number?

6. We wish to represent the numbers from 0 to 10,000 by strings of 0's and 1's of length n. How large must n be?

7. Suppose we wish to distribute leaflets to a crowd of 2000 people. How many leaflets must we distribute if we wish at least 1 person to receive at least 4 leaflets?

8. Suppose we have a jar that contains 20 pennies and 5 nickels. How many coins must we choose in order to guarantee that we have at least 22 cents? How many coins must we choose in order to guarantee that we can give someone exactly 11 cents in change?

9. In how many ways can the letters in the word *PASCAL* be arranged? In how many ways can the letters in the word *LETTERS* be rearranged?

10. In how many different ways can 4 identical red books, 6 identical green books, and 3 identical blue books be arranged on a shelf?

11. In how many ways can 10 children form a line? a circle?

12. How many ways can we fill 3 different positions from a pool of 10 job candidates? How many ways can we fill 3 identical positions from a pool of 10 job candidates?

| Section 4.3 | **Permutations and Combinations** |

We can solve both of the following problems using the multiplication principle.

(i) Determine how many different words of length 5 can be formed from the letters of the set $\{a, f, g, e, t, k, l\}$ if no letter can be repeated. (The answer is $7 \cdot 6 \cdot 5 \cdot 4 \cdot 3 = 2520$.)

(ii) Determine in how many ways first and second prizes can be awarded in a contest with 10 competitors. (The answer is $10 \cdot 9 = 90$.)

In both of the preceding situations, the counting procedure involves choosing all subsets of the correct size from the source set and counting each possible arrangement or permutation of each of these subsets. More generally, we wish to know in how many ways we can choose and arrange subsets of cardinality r from a set of cardinality n, where $n \geq r$. We know there are n choices for the first position in an arrangement, $(n - 1)$ choices for the second position, and so on until finally we have $n - (r - 1) = n - r + 1$ choices for the rth position. By the multiplication principle we know that there are thus $n(n - 1)(n - 2) \cdots$ $(n - r + 1) = n!/(n - r)!$ ways in which we can choose and arrange r elements from a set of n elements. We extend the use of the word "permutation" to this more general situation.

DEFINITION. Suppose that $0 \leq r \leq n$. The number of permutations of r elements chosen from n elements is denoted by $P(n, r)$. The number $P(n, r) = n(n - 1)(n - 2) \cdots (n - r + 1) = n!/(n - r)!$.

If we use the notation of the definition, the answer to problem (i) is $P(7, 5)$, and the answer to problem (ii) is $P(10, 2)$. We note that the number of ways in which it is possible to arrange all n objects chosen from a set of size n is $P(n, n)$ $= n!$. This agrees with our previous use of the word "permutation."

Example 1.

(a) The number of ways in which it is possible to choose a president, vice president, and secretary from a slate of 15 candidates is $P(15, 3)$ $= 15 \cdot 14 \cdot 13 = 2730$.

(b) The number of 5-letter words (meaningful and otherwise) over the English alphabet in which no letter is repeated is $P(26, 5) = 26!/21!$ $= 7,893,600$. We simply choose and arrange a set of 5 letters from 26 possibilities.

(c) The number of ways in which a winner and a runner-up can finish in a race with 100 people is $P(100, 2) = 100 \cdot 99 = 9900$.

(d) The number of injective functions from a set with r elements to a set with n elements is $P(n, r)$. (We saw the reasoning behind this in Section 4.2.)

Now suppose we wish to count the number of ways in which we can select a committee of 4 people from a set of 10 people. Here we wish to count the number of subsets of cardinality 4 but disregard their arrangements. The number $P(10, 4)$ counts each of the subsets of cardinality 4 in each of its 4! arrangements. Thus we divide by 4! to find that $P(10, 4)/4!$ different committees (or subsets) can be chosen from the original set of 10. More generally, if $n \geq r \geq 0$, the number of ways in which it is possible to choose a subset of cardinality r from a set containing n elements is $P(n, r)/r!$. We divide $P(n, r)$ by $r!$ to count each subset of cardinality r exactly once.

DEFINITION. For any pair of integers n and r such that $0 \leq r \leq n$, we define the number $C(n, r)$ to be $P(n, r)/r! = n!/[(n - r)!r!]$. We call $C(n, r)$ the number of **combinations of r objects chosen from n,** or more simply, "n choose r."

Thus $C(n, r)$ is the number of ways in which we can choose a subset of size r from a set of size n. (The numbers $C(n, r)$ are also called the "binomial coefficients." We shall observe their use in the binomial theorem later in this chapter.) The number $C(n, r)$ is often denoted by $\binom{n}{r}$. We shall use both notations frequently.

Example 2.

   (a) The number of ways in which we can form a committee of 4 people from among 10 people is $C(10, 4) = 10!/6!4! = 210$.

   (b) The number of ways in which a child can choose 2 flavors for a double-scoop ice cream cone from Mr. Sticky's 25 delicious flavors is $\binom{25}{2}$ $= 25 \cdot 24/2 = 300$.

   (c) The number of ways in which a basketball team (5 players) can be chosen from a field of 12 hopefuls is $\binom{12}{5}$.

   (d) There are $C(52, 13)$ different 13-card bridge hands that can be dealt out of a standard 52-card deck. There are $C(52, 5)$ different poker hands.

In Examples 3 through 5, we emphasize the difference between the use of $P(n, r)$ and that of $C(n, r)$. It is useful to remember that $P(n, r) = r!C(n, r)$: The number $P(n, r)$ counts each of the $r!$ arrangements of each of the $C(n, r)$ subsets of size r chosen from a source set of size n.

Example 3. Suppose a librarian has 10 distinct math books, and suppose that he wishes to display 4 on a shelf. He can choose the set of 4 books to be displayed in $C(10, 4) = 210$ ways. Now if the librarian decides that 2 different arrangements of the *same* set of 4 books will be counted as 2 *different* displays, then there are $P(10, 4) = 5040$ possible ways in which he can display his 4 math books. Each set of 4 books can be arranged in 4! different ways, and $P(10, 4) = 4!C(10, 4)$.

Example 4. There are $C(26, 7)$ ways in which a Scrabble player can choose a set of 7 distinct letters. However, there are $P(26, 7)$ ways in which he can form a 7-letter word with no repeated letters, because the same set of 7 letters can be arranged into 7! different words. $P(26, 7) = 7!C(26, 7)$.

Example 5. There are $C(7, 3)$ ways to form a 3-person committee from a set of 7 people willing to serve. However, there are $P(7, 3)$ ways to choose a president, vice president, and bursar for a club from a slate of 7 candidates. Each set of 3 candidates can form 3! different administrations.

In the examples that follow, we combine the use of $P(n, r)$ and $C(n, r)$ with the other counting principles.

Example 6. Suppose a man owns 7 shirts, 3 pairs of pants, and 5 ties. We wish to determine how many ways he can pack a travel bag with 4 shirts, 2 pairs of pants, and 3 ties. There are $\binom{7}{4}$ ways to choose the shirts, $\binom{3}{2}$ ways to choose the pants, and $\binom{5}{3}$ ways to choose the ties. Using the multiplication principle, we find that there are $\binom{7}{4}\binom{3}{2}\binom{5}{3} = 35 \cdot 3 \cdot 10 = 1050$ different ways in which the travel bag can be packed.

Example 7. We wish to determine how many 5-symbol codes can be formed if the first 2 symbols are letters and the next 3 are digits but no symbol is repeated. There are $P(26, 2)$ ways we can choose and arrange the first 2 symbols and $P(10, 3)$ ways we can choose and arrange the next 3. By the multiplication principle, there are thus $P(26, 2) \cdot P(10, 3) = 650 \cdot 720$ codes that can be so formed.

Example 8. We wish to determine how many ways 3 committees can be formed from 20 people if the first has 4 members, the second has 6, the third has 5, and no person can serve on more than 1 committee. There are $\binom{20}{4}$ ways to form the first committee. From the 16 people left, there are $\binom{16}{6}$ ways to form the second committee. From the remaining 10 people, there are $\binom{10}{5}$ ways to form the third committee. Thus there are $\binom{20}{4}\binom{16}{6}\binom{10}{5}$ ways to form 3 such committees.

Example 9. Suppose we have an urn that contains 6 numbered white balls and 4 numbered red balls. (We number the balls so that we can distinguish between any 2 white balls and similarly between any 2 red balls.) There are $C(10, 5)$ different ways to draw a sample of 5 balls from the urn. We wish to determine how many of these samples contain exactly 3 white balls (and hence exactly 2 red balls). There are $C(6, 3)$ ways to choose 3 white balls from among the 6 in the urn and $C(4, 2)$ ways to choose 2 red balls from among the 4 in the urn. Thus by the multiplication principle there are $C(6, 3) \cdot C(4, 2) = 20 \cdot 6 = 120$ different ways to draw a sample of size 5 that contains exactly 3 white and 2 red balls.

    Now we wish to determine how many samples of size 5 that contain at least 3 white balls can be drawn from our urn. The samples we wish to count can contain 3, 4, or 5 white balls. These cases are disjoint, and so we compute each case separately and add our results. (We use the addition principle.) There are $C(6, 3) \cdot C(4, 2) = 120$ samples with 3 white balls. There are $C(6, 4) \cdot C(4, 1) = 15 \cdot 4 = 60$ samples with 4 white balls. There are $C(6, 5) \cdot C(4, 0) = 6 \cdot 1 = 6$ ways to choose a sample with 5 white balls. Adding, we find that there are thus $120 + 60 + 6 = 186$ samples of size 5 that contain at least 3 white balls.

Example 10. A fair coin is tossed 5 times. There are various sequences of heads and tails that might occur, such as HHTTH or TTTTH. In all, there are $2^5 = 32$ different ways to toss a fair coin 5 times. We wish to determine how many different ways heads can occur exactly 3 times. To determine a sequence with exactly 3 heads, we choose the positions of the 3 H's in the sequence that describes the toss. The other 2 positions must have T's. There are $C(5, 3)$ ways to choose 3 positions from 5 and hence $C(5, 3) = 10$ ways to obtain exactly 3 heads. To determine how many sequences have at most 4 heads, we use the addition principle. We separately compute the number of ways it is possible to obtain exactly 0, 1, 2, 3, or 4 heads and then add our results. The answer is $C(5, 0) + C(5, 1) + C(5, 2) + C(5, 3) + C(5, 4) = 31$. Alternatively, there is only $C(5, 5) = 1$ way to obtain more than 4 heads. So there are $32 - 1 = 31$ ways to obtain at most 4 heads.

The problems solved in the preceding two examples are frequently discussed combinatorial problems, because they provide models for so many different situations. We examine two such situations.

Example 9a. We wish to determine how many different committees of 5 can be formed from 6 men and 4 women on which exactly 3 men and 2 women serve. The answer is $C(6, 3) \cdot C(4, 2)$. (Here the men are the white balls and the women are the red balls of Example 9.)

Example 10a.   Block Walking. We wish to determine how many different routes there are from point A to point B in Figure 4.5 if we allow travel only directly east and directly south. On any route the traveler must walk exactly 5 blocks, of which exactly 3 must be in the south direction. The various routes can be represented by such sequences as SSEES and EESSS. There are exactly $C(5, 3)$ ways to choose the positions of the S's and hence exactly $C(5, 3) = 10$ ways to travel from point A to point B. (Here we simply replace the H by S and the T by E in Example 10.)

Figure 4.5

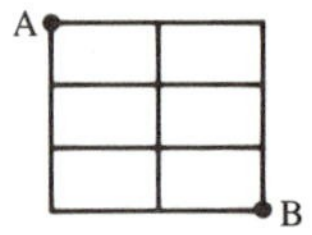

Example 11. We wish to determine how many strings of length 7 can be formed from the letters $\{a, s, t, v, u\}$ in which the letter s appears exactly twice. First we choose the positions for the 2 s's. There are $C(7, 2)$ ways in which to place the s's. The remaining 5 positions can be filled in with any of the remaining 4 letters. There are 4^5 ways to do this. Applying the multiplication principle, we find that there are $4^5 C(7, 2) = 21,504$ ways to form such words.

Example 12. A die is rolled 5 times and the sequence of faces is noted. We wish to determine the number of different sequences in which the face "2" occurs exactly 3 times or the face "5" occurs exactly twice. There are $C(5, 3)5^2$ ways in which the face "2" can appear exactly 3 times. To see this, note that we must first choose the 3 positions in the sequence for the face "2." There are $C(5, 3)$ ways to do this. Then we can fill in each of the 2 remaining positions with any of the 5 remaining faces. There are 5^2 ways to do this. Similarly, there are $C(5, 2)5^3$ ways in which the face "5" can appear exactly twice. There are $C(5, 3)$ ways in which "2" can appear 3 times and "5" twice. (We choose the positions for "2," and the remaining positions must all be filled in with "5.") We apply the inclusion–exclusion principle to conclude that there are $C(5, 3)5^2 + C(5, 2)5^3 - C(5, 3)$ sequences in which "2" appears 3 times or "5" appears twice.

Example 13. We wish to determine in how many ways it is possible to distribute 7 identical (indistinguishable) balls among 3 distinguishable children. To do this, let us represent the balls by little o's and the division of the balls among the children by slashes. The distribution represented by

$$oo/oooo/o$$

indicates that the first child receives 2 balls, the second 4, and the third just 1. Now suppose the first and second children receive no balls and the third receives all 7. We represent this by

$$//ooooooo$$

Thus any distribution can be represented by a string of 9 symbols, 2 of which are slashes and the rest o's; and conversely, every such sequence determines a unique distribution. Just as in the coin-tossing and block-walking problems, we find that our solution is $\binom{9}{2}$, because we need only choose the two positions for the slashes to determine a distribution.

Example 14. We wish to determine how many non-negative integer solutions there are to the equation $x + y + z + w = 10$. If we represent the number 10 by a sequence of 10 o's, we obtain a solution to our equation by placing 3 slashes in the sequence. Thus the sequence

$$oo/ooooo/oo/o$$

gives us the solution $x = 2$, $y = 5$, $z = 2$, and $w = 1$. Any solution can be represented by such a sequence. Thus the number of solutions to the equation is $\binom{13}{3}$.

We use Examples 13 and 14 as models to formulate the following proposition.

Proposition 1. The number of ways in which it is possible to distribute n identical items into r (distinguishable) boxes is $\binom{n + r - 1}{r - 1}$.

PROOF: We can represent n items by n circles and the distribution of n items into r distinguishable boxes by a sequence of n circles and $(r - 1)$ slashes. There are $\binom{n + r - 1}{r - 1}$ ways in which to do this.

Example 15. Suppose we have an urn containing 6 identical red balls, 6 identical green balls, and 6 identical white balls. We shall choose a sample of 8 balls, but every time one is selected it is replaced by an identical ball. We wish to determine how many different samples can be chosen. To do this, let us label 3 boxes with the colors red, white, and green. Every time we choose a ball, let us place it in the box labeled with its color. A sample may be represented again by a sequence of o's and slashes. For instance, oo/ooo/ooo represents a sample with 2 red balls, 3 white balls, and 3 green balls. Because every sample can be uniquely represented this way, we find that there are $\binom{10}{2}$ possible samples.

We can use the model of Example 15 to reformulate Proposition 1 as follows:

Proposition 1a. The number of ways in which we can choose n items from among r types with replacement is $\binom{n + r - 1}{r - 1}$.

Here we consider the items of a given type indistinguishable and the supply of each type inexhaustible, because each item chosen is replaced by an identical item.

Section 4.3   **Exercises**

1. Evaluate $P(6, 4)$, $C(6, 4)$, $P(7, 1)$, $C(7, 1)$, $P(8, 6)$, $C(8, 6)$, and $C(8, 2)$.

2. Show that:
   (a) $C(n, 0) = 1$ (b) $C(n, 1) = n$ (c) $C(n, n - 1) = n$

3. In how many ways can a president and vice president be chosen from a slate of 30 candidates?

4. In how many ways can 5 children out of a class of 20 line up for a picture?

5. In how many ways can a groundskeeper fly 3 differently colored flags on 3 flag poles if there are 8 colors available to her and if she flys 1 flag on each pole?

6. In how many ways can a steering committee of 5 be chosen from a group of 50 activists?

7. In how many ways can a team of 9 players be chosen from among 30 hopefuls?

8. In how many ways can a hungry student choose 3 toppings for his pizza from list of 10 delicious possibilities?

9. A librarian wants to place 3 groups of books on 3 shelves: 6 books on the first shelf, 3 on the second, and 5 on the third. In how many ways can this be done if the librarian has 20 books from which to choose? (This librarian does not care how each set of books is arranged.)

10. In how many ways can 25 students be assigned to 3 homerooms if the first room can admit 10 students, the second 8 students, and the third 7 students?

11. A die is rolled 6 times and the sequence of faces is noted. How many sequences are possible? In how many sequences does the face "5" appear exactly 3 times? In how many sequences does the face "5" appear an even number of times?

12. Let $E = \{a, b, c, d, e\}$ be an alphabet. How many words of length 5 are there in which the letter c appears at least 3 times? In how many words of length 8 does the letter c appear 3 times and the letter b twice? In how many words of length 5 does the subsequence abc appear?

13. How many 7-card hands can be drawn from a standard 52-card deck of cards? How many of these hands have exactly 4 hearts and 3 clubs?

14. In how many ways can a committee of 5 be formed from 10 club members if one of Mr. Smith or Mrs. Smith must serve, but not both?

15. In how many ways can 10 identical pennies be distributed into 5 different cups? What if the cups are also identical?

16. In how many ways can 7 balls be selected from 10 identical red ones and 10 identical white ones?

17. How many nonnegative integer solutions are there to the equation $x + y + z = 15$?

| Section 4.4 | Combinatorial Arguments |

Suppose that A is a set of cardinality n and that r is an integer such that $0 \leq r \leq n$. Each subset B of A of cardinality r determines a unique subset of cardinality $n - r$, namely $A \backslash B = A \cap B'$. Conversely, every subset C of cardinality $n - r$ determines a unique subset of cardinality r—namely $A \backslash C = A \cap C'$. Thus there is a one-to-one correspondence between the subsets of A of cardinality

r and the subsets of A of cardinality $n - r$. Because there are $C(n, r)$ subsets of cardinality r, there must be $C(n, r)$ subsets of cardinality $n - r$. But we know that there are $C(n, n - r)$ subsets of cardinality $n - r$. Thus we must conclude that $C(n, r) = C(n, n - r)$.

The preceding paragraph is a **combinatorial proof** for the identity

(1) $$C(n, r) = C(n, n - r)$$

(We shall also give an algebraic proof.) Many identities involving combinations and permutations can be reasoned both algebraically and combinatorially. In a combinatorial proof, we generally find a combinatorial question (that is, a counting problem) that can be answered by both sides of the given identity. We will give some examples here. But first we prove identity (1) algebraically.

Proposition 1. $C(n, r) = C(n, n - r)$

ALGEBRAIC PROOF: We have that $C(n, n - r) = \dfrac{n!}{[n - (n - r)]!(n - r)!}$.
Simplifying, we find that $C(n, n - r) = \dfrac{n!}{r!(n - r)!}$, which in turn is equal to $C(n, r)$, as was to be shown.

Proposition 2. $C(n, k) = C(n - 1, k) + C(n - 1, k - 1)$ for $n > k > 0$

COMBINATORIAL PROOF: Let A be a set of cardinality n and let k be an integer such that $0 < k < n$. There are $C(n, k)$ different subsets of A of cardinality k. Let y be an element of A. Every subset of A either includes y or does not. There are $C(n - 1, k)$ different subsets of A of cardinality k that do not include y. (We form such a subset from the $(n - 1)$ elements of A that are not equal to y.) There are $C(n - 1, k - 1)$ ways to choose a subset of A of cardinality k that includes y. (To form such a subset, we must choose $(k - 1)$ elements in addition to y from the $(n - 1)$ elements in A that are not equal to y.) Adding, we find that there are $C(n - 1, k) + C(n - 1, k - 1)$ ways in which to choose a subset of cardinality k from A. Thus $C(n, k) = C(n - 1, k - 1) + C(n - 1, k)$.

ALGEBRAIC PROOF: By definition, we have $C(n - 1, k) + C(n - 1, k - 1) = \dfrac{(n - 1)!}{[(n - 1) - k]!k!} + \dfrac{(n - 1)!}{[(n - 1) - (k - 1)]!(k - 1)!}$.
Simplifying, finding a common denominator, and adding, we find that the sum is equal to $\dfrac{(n - k)(n - 1)! + (n - 1)!k}{(n - k)!k!}$. Using the distributive law, we find that the latter sum is in turn equal to $\dfrac{n!}{(n - k)!k!}$, as was to be proved.

Often combinatorial proofs provide straightforward proofs for identities that are somewhat intractable algebraically. But even where the algebra is very simple, the combinatorial alternative may offer insight into the meaning and use of a given identity. The following combinatorial proof for the binomial theorem sheds light on why the coefficients of the expansion of $(x + y)^n$ should arise from combinatorial considerations.

Theorem 1. Binomial Theorem. For any integer $n \geq 0$, we have

$$(x + y)^n = \binom{n}{0}x^n y^0 + \binom{n}{1}x^{n-1}y^1 + \binom{n}{2}x^{n-2}y^2 + \cdots + \binom{n}{n}x^0 y^n$$

That is,

$$(x + y)^n = \sum_{r=0}^{n} \binom{n}{r}x^{n-r}y^r$$

COMBINATORIAL PROOF: After we multiply but before we collect terms, each term in the product $(x + y)^n = (x + y)(x + y) \cdots (x + y)$ is of the form $a_1 a_2 \cdots a_n$, where each a_k is either an x or a y from the kth factor. To determine the coefficient of $x^{n-r}y^r$, we must determine how many of these sequences of x's and y's have exactly r y's and $(n - r)$ x's. As in the coin-tossing problem, there are exactly $\binom{n}{r}$ different ways in which the sequence can have exactly r y's and $(n - r)$ x's. Thus, after collecting terms, we find that the coefficient of $x^{n-r}y^r$ is $\binom{n}{r}$, as was to be proved.

INDUCTIVE PROOF: Let $n = 1$. Then $\binom{1}{0} = \binom{1}{1} = 1$. Because $(x + y)^1 = x + y$, the theorem is certainly true for $n = 1$. Now assume that the theorem is true for the integer n. We shall show that the coefficient of $x^{(n+1)-r}y^r$ is $\binom{n+1}{r}$ in the expansion of the product $(x + y)^{n+1}$. By the induction hypothesis, we have that $(x + y)^{n+1} = (x + y)(x + y)^n = (x + y)\left[\binom{n}{0}x^n y^0 + \cdots + \binom{n}{n}x^0 y^n\right]$. When multiplying, we obtain a term of the form $x^{(n+1)-r}y^r$ by either multiplying $x^{n-r}y^r$ by x or multiplying $x^{(n+1)-r}y^{r-1}$ by y. Thus the coefficient of $x^{(n+1)-r}y^r$ is the sum $\binom{n}{r} + \binom{n}{r-1}$. By Proposition 2 this sum is equal to $\binom{n+1}{r}$, as was to be proved.

Example 1.

(a) $(x + y)^3 = x^3 + 3x^2 y + 3xy^2 + y^3$, which is in turn equal to $\binom{3}{0}x^3 + \binom{3}{1}x^2 y + \binom{3}{2}xy^2 + \binom{3}{3}y^3$, as asserted by the binomial theorem.

(b) To expand $(x - 2)^5$ using the binomial theorem, we set y equal to -2 in $(x + y)^5$. We obtain

$$(x - 2)^5 = x^5 + \binom{5}{1}x^4(-2) + \binom{5}{2}x^3(-2)^2$$
$$+ \binom{5}{3}x^2(-2)^3 + \binom{5}{4}x(-2)^4 + (-2)^5$$

Simplifying, we obtain

$$(x - 2)^5 = x^5 - 10x^4 + 40x^3 - 80x^2 + 80x - 32$$

Example 2. Setting both x and y equal to 1 in $(x + y)^n$ and applying the binomial theorem, we find that

(2)
$$2^n = \binom{n}{0} + \binom{n}{1} + \binom{n}{2} + \cdots + \binom{n}{n}$$

We can obtain identity (2) by combinatorial reasoning as well. We know that if A is a set of cardinality n, then the cardinality of $\mathcal{P}(A)$ is 2^n; that is, the set A has 2^n subsets. The number $C(n, r)$ gives us the number of subsets of A of cardinality r. Summing $\binom{n}{0} + \binom{n}{1} + \cdots + \binom{n}{n}$, we again obtain the number of all possible subsets of A, thereby showing that identity (2) holds.

Section 4.4 | **Exercises**

---

1. Use the binomial theorem to expand each of the following.
   (a) $(x + y)^6$ (b) $(x - y)^5$ (c) $(x - 3)^4$ (d) $(x + 3y)^4$

2. Give a combinatorial argument to show that $C(n, 1) = C(n, n - 1)$.

3. Give a combinatorial argument to prove that

$$C(n, m) \cdot C(m, k) = C(n, k) \cdot C(n - k, m - k)$$

4. Verify the following identity for a few values of n. Give a combinatorial argument to prove it.

$$\binom{n}{0}\binom{n}{n} + \binom{n}{1}\binom{n}{n-1} + \binom{n}{2}\binom{n}{n-2} + \cdots + \binom{n}{n}\binom{n}{0} = \binom{2n}{n}$$

5. Let A be a set and suppose that $|A| = n$. Let S be the set of words over $\{0, 1\}$ of length n.

   (a) Find $|S|$.

   (b) Find a bijection from S to $\mathcal{P}(A)$.

   (This exercise gives another combinatorial proof for the cardinality of $\mathcal{P}(A)$.)

Section 4.5   ## Infinite Sets and Countability

The reader is familiar with many different infinite sets, such as the set of natural numbers, the set of real numbers, and the set of all words over the alphabet $\{a, b\}$. In this section we discuss the cardinality of infinite sets. We discuss criteria for showing which sets have the same cardinality as the natural numbers—these sets are called **countably infinite**—but we also show that not all infinite sets have the same cardinality.

> **DEFINITION.** A nonempty set A is an **infinite** set if there is no bijection between $X_n = \{1, 2, 3, \ldots, n\}$ and A for any positive integer n.

This definition merely says that we shall call a nonempty set infinite if it is not finite.

The following proposition shows that the set of natural numbers is indeed infinite under the preceding definition.

Proposition 1. The set $\mathbf{N}$ of natural numbers is infinite.

PROOF: Suppose that n is a positive integer and that $f : X_n \to \mathbf{N}$ is an injective function. (We shall show that f cannot be surjective.) For $k = 1, 2, \ldots, n$ let $x_k = f(k)$. Thus the image set $f(X_n) = \{x_1, x_2, \ldots, x_n\}$ is a finite set and we can find its largest element. Denote the largest element of the set $f(X_n)$ by x. The value $x + 1$ is thus a natural number larger than x, but it is not an element of $f(X_n)$. The function f is not bijective because it is not surjective. Since there is no bijection from X_n to $\mathbf{N}$ for any value of n, the set $\mathbf{N}$ satisfies our definition of an infinite set.

In Theorem 1 in Section 4.1, we proved that two finite sets A and B have the same cardinality if and only if there exists a bijection between them. We use this notion of a one-to-one correspondence to define when two infinite sets have the same cardinality.

> **DEFINITION.** Two sets A and B have the **same cardinality** if and only if there exists a bijection between them.

Example 1. The sets $\mathbf{N}$ and $\mathbf{Z}$ have the same cardinality. The following function $f : \mathbf{N} \to \mathbf{Z}$ is a bijection between them.

$$f(n) = n/2 \text{ if } n \text{ is even}$$
$$f(n) = (1 - n)/2 \text{ if } n \text{ is odd}$$

The function f places the even natural numbers in one-to-one correspondence with the positive integers, and it places the odd natural numbers in one-to-one correspondence with the integers that are less than or equal to 0.

Example 2. Let E denote the set of even numbers, E^+ the set of even natural numbers, and O the set of odd numbers. The bijection $f : \mathbf{N} \to E^+$ defined by $f(n) = 2n$ shows that $\mathbf{N}$ and E^+ have the same cardinality. The bijection $g : E \to O$ defined by $g(n) = n - 1$ shows that E and O have the same cardinality.

Example 3. This example shows that any two intervals of real numbers, no matter what their length, have the same cardinality. For any pair of real numbers a and b where $a \neq b$, let $[a, b]$ denote the interval $\{x : a \leq x \leq b\}$. Then the function $f : [a, b] \to [s, t]$, defined as follows, is a bijection.

$$f(x) = \left(\frac{t - s}{b - a} \right)(x - a) + s$$

Thus the intervals $[2, 4]$ and $[1, 100]$ have the same cardinality. The function $f : [2, 4] \to [1, 100]$ given by $f(x) = (99/2)(x - 2) + 1$ establishes a one-to-one correspondence between these sets.

Examples 1, 2, and 3 point out an essential difference between finite and infinite sets: An infinite set can be put in one-to-one correspondence with a proper subset of itself, whereas a finite set cannot.

In the proposition that follows, we show that if B is any infinite set, there is an injective function from the natural numbers into B. For two finite sets A and C we know that if there is an injective function from A to C, then the cardinality of A is less than or equal to that of C. It is in this sense that the proposition that follows shows that among all infinite sets, the set of natural numbers has the smallest cardinality.

Proposition 2. Let B be an infinite set. There is an injective function from **N** into B.

PROOF: We construct an injective function f from **N** into B. First we choose an element of B and call it b_1. Let $f(1) = b_1$. The set $B_2 = B\backslash\{b_1\}$ is not empty because B is infinite. Choose an element b_2 of B_2 and let $f(2) = b_2$. We continue inductively. Suppose we have found distinct elements $\{b_1, b_2, \ldots, b_n\}$ and have set $f(1) = b_1, f(2) = b_2, \ldots$, and $f(n) = b_n$. The set $B_{n+1} = B\backslash\{b_1, \ldots, b_n\}$ is not empty. So we can choose an element b_{n+1} from B_{n+1} and set $f(n + 1) = b_{n+1}$. Thus for each value of n in **N** we determine a unique value for $f(n)$ and so construct an injective function from **N** to B.

> **DEFINITION.** A set A is called **countably infinite** if it has the same cardinality as the natural numbers **N**—that is, if there is a bijection between **N** and A. A set is called **countable** if it is either finite or countably infinite.

The bijections given in Examples 1 and 2 show that the set **Z** of all integers and the set E^+ of even natural numbers are both countably infinite sets.

Example 4.

(a) The set **W** of whole numbers is countably infinite. The function $f: \mathbf{W} \rightarrow \mathbf{N}$ defined by $f(x) = x + 1$ is a bijection from **W** to **N**.

(b) Let H be the language over $\{a, b\}$ of all words containing no b's. The language H is a countably infinite set. Let $L(x)$ denote the length of the word x. The function $f: H \rightarrow \mathbf{N}$ defined by $f(x) = L(x) + 1$ is a bijection. (We need to add 1 because the empty word is in H.)

(c) Let S be the language over $\{a, b\}$ defined as follows:

(1) The word a is in S.

(2) If x is in S and x ends with a, then xbb is in S. If x is in S and x ends with b, then xa is in S.

(3) Only words so formed are in S.

We shall show that S is countable. Recall that if x is a word, we denote the word $xxx \cdots xx$ (that is, x concatenated with itself n times) by x^n. The word x^0 is the empty word. The words in $S = \{a, abb, abba, abbabb, \ldots\}$ are of the form $(abb)^k$ for some $k \geq 1$ if they end with b. They are of the form $a(bba)^k$ for some $k \geq 0$ if they end with a. The following is a bijection f from **N** to S:

$$f(n) = (abb)^{n/2} \text{ if } n \text{ is even}$$
$$f(n) = a(bba)^{(n-1)/2} \text{ if } n \text{ is odd}$$

Suppose that B is a countably infinite set and that $f : \mathbf{N} \to B$ is a bijection. If $x = f(k)$, call it b_k. In so doing, we have labeled each element of B with a unique integer and can thus speak of the first element of B (namely b_1), the second element of B (namely b_2), and so on. In this way B inherits the order structure of the natural numbers.

Example 5. Under the instructions of the preceding paragraph and with the bijection given in Example 1, we can impose on the set of integers $\mathbf{Z}$ a new order that is different from the usual order $\ldots, -2, -1, 0, 1, 2, 3, \ldots$. The new order is $0, 1, -1, 2, -2, 3, -3, \ldots$. (We have taken the elements $f(1), f(2), \ldots$ in sequence.)

The propositions and their corollaries that follow in this section provide criteria by which we can determine when a set is countably infinite without necessarily demonstrating an explicit bijection between that set and $\mathbf{N}$.

Proposition 3. Suppose A is a set and B is a countable set. If there is a bijection between A and B, then A is also countable.

We leave the proof as an exercise.

Example 6. Let S be the set of all integer powers of 2. The function $g : \mathbf{Z} \to S$ defined by $g(n) = 2^n$ is a bijection. Because $\mathbf{Z}$ is countably infinite, Proposition 3 guarantees that S is also countably infinite.

Let $P = \{2, 3, 5, 7, \ldots\}$ be the set of prime numbers. (It is a fact of number theory that there is an infinite number of primes.) The set P is a countably infinite subset of $\mathbf{N}$. The following diagram illustrates how to construct a bijection $g : \mathbf{N} \to P$. (We have simply assigned the natural numbers to the prime numbers in the order in which they appear.)

$$
\begin{array}{cccccccc}
0 & 1 & 2 & 3 & 4 & 5 & 6 & 7 \ldots \\
  &   & \uparrow & \uparrow &   & \uparrow &   & \uparrow \\
  &   & 1 & 2 &   & 3 &   & 4
\end{array}
$$

In general, any subset of a countable set is also countable, as the following proposition asserts.

Proposition 4. Suppose that B is a subset of a countably infinite set A. Then B is countable.

PROOF: If B is finite, there is nothing more to prove. So suppose that B is an infinite subset of the countably infinite set A. Let $f : \mathbf{N} \to A$ be a bijection.

We now construct a bijection $g : \mathbf{N} \to B$. Let i_1 be the smallest value of i for which $f(i)$ is also in B, and let $g(1) = f(i_1)$. (We have assigned 1 to the first element of A that is also in B.) Now let i_2 be the smallest value of i such that $f(i)$ is in $B \backslash \{g(1)\}$, and let $g(2) = f(i_2)$. (We have assigned 2 to the next element of A that is also in B.) We proceed inductively. Suppose that we have defined $g(1), g(2), g(3), \ldots, g(n-1)$. Let i_n be the smallest value of i for which $f(i)$ is an element of $B \backslash \{g(0), g(1), \ldots, g(n-1)\}$, and let $g(n) = f(i_n)$. We thus have a bijection from $\mathbf{N}$ onto the subset B, showing that B must also be countable.

Proposition 4 gives another argument for the fact that the set E^+ of even natural numbers is countable: It must be countable simply because it is a subset of a countable set. Similarly, the set of all multiples of 10 must be countable because it is also a subset of $\mathbf{Z}$, and $\mathbf{Z}$ is a countable set.

We know that the set of even integers E is countably infinite and that the set O of odd integers is countably infinite. Any countably infinite set can be put in one-to-one correspondence with either of these sets. The union $E \cup O = \mathbf{Z}$ is also countably infinite. In general, the union of any two countable sets is itself countable.

Proposition 5. Let A and B be disjoint countable sets. Then their union $A \cup B$ is also countable.

PROOF: If A and B are both finite, their union is finite and hence countable. We leave the case where A is finite and B countably infinite as an exercise. We now assume that both A and B are countably infinite and disjoint. Suppose $A = \{a_1, a_2, \ldots\}$ and $B = \{b_1, b_2, \ldots\}$. We define a bijection $f : A \cup B \to \mathbf{N}$ as follows: $f(a_i) = 2i - 1$ and $f(b_i) = 2i$. That is, we put the elements of A in one-to-one correspondence with the odd natural numbers, and we put the elements of B in one-to-one correspondence with the even natural numbers.

Corollary 1. (by induction). Let $A_1, A_2, \ldots, A_n$ be a finite collection of mutually disjoint countable sets. Then their union $A_1 \cup A_2 \cup \cdots \cup A_n$ is also countable.

Corollary 2. Let A and B be countable sets (not necessarily disjoint). The union $A \cup B$ is countable.

PROOF: We can write $A \cup B$ as the union of three mutually disjoint countable sets: $A \cup B = (A \backslash B) \cup (B \backslash A) \cup (A \cap B)$.

Example 7. Let Q be the set of all words over $\{a, b\}$ that are either strings of all a's, strings of all b's, or empty. We know from Example 4 that the set S_a of all words that contain no b's is countable. Similarly, the set S_b of all words that contain no a's is countable. The set $Q = S_a \cup S_b$, the union of two countable sets. By Corollary 2, Q is countable.

Suppose that $A = \{a_1, a_2, \ldots\}$ and $B = \{b_1, b_2, \ldots\}$ are both countably infinite sets. We can display the elements of the cartesian product $A \times B$ in an infinite array as follows:

When we follow the arrows in the diagram and label the elements of $A \times B$ in the order in which they appear, we establish a one-to-one correspondence between $\mathbf{N}$ and $A \times B$. We conclude that $A \times B$ is also countably infinite. The proposition that follows proves this fact by giving a formula for the counting procedure illustrated in the preceding array.

Proposition 6. Let $A = \{a_1, a_2, \ldots\}$ and $B = \{b_1, b_2, \ldots\}$ be countable sets. Then $A \times B$ is also countable.

PROOF: (We prove only the case where both A and B are countably infinite.) We construct a bijection $f: \mathbf{N} \rightarrow A \times B$ inductively as follows:

(1) Let $f(1) = (a_1, b_1)$.
(2) Suppose that $f(q) = (a_m, b_n)$.
    If $m \neq 1$, let $f(q + 1) = (a_{m-1}, b_{n+1})$.
    If $m = 1$, let $f(g + 1) = (a_{n+1}, b_1)$.

We evaluate f explicitly for a few values of q:

$$f(1) = (a_1, b_1), \quad f(2) = (a_2, b_1), \quad f(3) = (a_1, b_2),$$
$$f(4) = (a_3, b_1), \quad f(5) = (a_2, b_2), \ldots$$

We display the values of f as follows:

$f(1)$ $f(3)$ $f(6)$ $f(10)$ $\ldots$
$f(2)$ $f(5)$ $f(9)$ $\ldots$
$f(4)$ $f(8)$ $\ldots$
$f(7)$ $\ldots$

Corollary to Proposition 6. Let $A_1, A_2, \ldots, A_n$ be a finite collection of countably infinite sets. Then $A_1 \times A_2 \times \cdots \times A_n$ is also countably infinite.

The corollary follows by induction.
That the sets $\mathbf{N} \times \mathbf{N}$, $E \times O$, and $\mathbf{Z} \times \mathbf{Z} \times \mathbf{Z} \times \mathbf{Z}$ are all countably infinite follows from the corollary.

Example 8. Let $\mathbf{Q}$ denote the set of rational numbers. The set $\mathbf{Q}$ is countably infinite. To show this, we first note that the set $\mathbf{Z} \times \mathbf{N}$ is countably infinite. Now each rational number can be written in exactly one way as a ratio s/t, where s is an integer, t is an integer strictly greater than 0, and s and t share no common factors. (Observing these restrictions, we must rewrite the rational number $-(6/9)$ as $(-2)/3$. Thus $s = -2$ and $t = 3$.) We put $\mathbf{Q}$ in one-to-one correspondence with a subset of $\mathbf{Z} \times \mathbf{N}$ by associating s/t with (s, t). Because an infinite subset of a countably infinite set must also be countably infinite, we conclude that $\mathbf{Q}$ is countably infinite.

Example 9. Let S be the language over the alphabet $\{a, b\}$ defined as follows:

    (1) The empty word Λ is in S.

    (2) If x is in S, then ax and xb are in S.

    (3) Only words so formed are in S.

We can characterize S as the set of words in which a string of a's (perhaps empty) is followed by a string of b's (perhaps empty). Some words in S are $aaabb$, $bbbb$, aa, ab, and $aaaaab$. The set S is countably infinite. We note that the set of whole numbers $\mathbf{W}$ is countably infinite, and we define a bijection $f : \mathbf{W} \times \mathbf{W} \to S$ by letting $f(n, m) = a^n b^m$, the word with n a's followed by m b's. [So $f(2, 3) = aabbb$ and $f(0, 0) = \Lambda$.] Because $\mathbf{W} \times \mathbf{W}$ is countably infinite, so is S.

Example 10. The set of polynomials of the form $a_0 + a_1x + a_2x^2 + \cdots + \cdots + a_nx^n$ with integer coefficients is countably infinite. We form a one-to-one correspondence between the countable set $\mathbf{Z} \times \mathbf{Z} \times \cdots \times \mathbf{Z}$ $(n + 1$ times) and this set of polynomials by associating each $(n + 1)$-tuple of integers $(a_0, a_1, \ldots, a_n)$ with the polynomial $a_0 + a_1x + a_2x^2 + \cdots + a_nx^n$. Thus if $n = 4$, the 4-tuple $(2, -3, 0, 7)$ is associated with the polynomial $p(x) = 2 - 3x + 0x^2 + 7x^3$.

Proposition 7. The union of a countable collection of countable sets is countable.

    PROOF: (We prove only the case where we have a countably infinite collection of disjoint countably infinite sets.) Suppose that $A_1, A_2, \ldots$ is a countably infinite collection of disjoint countably infinite sets. Suppose that we denote the elements of the set A_k by $\{a(k, 1), a(k, 2), a(k, 3), \ldots, a(k, j), \ldots\}$. Then we can form a one-to-one correspondence between the union $\overset{\infty}{\underset{k=1}{\cup}} A_k$ and the countably infinite set $\mathbf{N} \times \mathbf{N}$ by associating each pair (k, j) with the element $a(k, j)$. That is, we associate (k, j) with the jth element of A_k.

Example 11. Let $E = \{a, b\}$. The set E^* is countably infinite. We can write E^* as the disjoint union of the following countably infinite collection of countable (actually finite) sets as follows: For each integer $k \geq 0$, let E_k be the set of all words over $\{a, b\}$ of length k. Each E_k is a finite set of words, and every word in E^* is an element of E_k for some $k \geq 0$. Thus $E^* = \bigcup\limits_{k=0}^{\infty} E_k$. We apply Proposition 7 to conclude that E^* is countably infinite. Because any language L over $\{a, b\}$ is a subset of E^*, it too must be countable.

Thus far we have examined only infinite sets with the same cardinality as the set of natural numbers. The examples that follow exhibit infinite sets with cardinalities different from that of **N**. Such infinite sets are called **uncountable**. In light of the remarks preceding Proposition 2 of this section, we may view uncountable sets as being of larger cardinality than **N**.

Example 12. The power set of **N** is not countable. We shall denote a subset A of **N** by a sequence $(a_1, a_2, a_3, \ldots)$, where $a_k = 1$ if the subscript k is an element of A and $a_k = 0$ if k is not. Thus the finite set $\{2, 4, 5\}$ is represented by the sequence $(0, 1, 0, 1, 1, 0, 0, 0, \ldots)$ where the second, fourth, and fifth entries in the sequence are 1's and all the remaining entries are 0. The even natural numbers are represented by the sequence $(0, 1, 0, 1, 0, \ldots)$. Now suppose that we have a countably infinite collection of subsets of **N**—namely $\{S_1, S_2, S_3, \ldots\}$. (We shall construct a subset X of **N** that is not included in this collection.) Let $[a(k, 1), a(k, 2), a(k, 3), \ldots]$ be the sequence that denotes S_k. Now we construct a new subset X of **N** by the following rule: The integer k is a member of X if and only if $a(k, k) = 0$—that is, if and only if the integer k is not a member of S_k. (So if the sequence denoting S_3 happens to be $(1, 1, 0, 0, \ldots)$, then the number 3 is not in S_3 and so 3 must be in X by our rule. If the sequence representing S_2 happens to be $(0, 1, 0, 1, 0, \ldots)$, then the number 2 is in S_2 and therefore 2 is not in X.) Now the sequence denoting X differs in at least one place from each of the sequences denoting each of the S_k's. Namely, it differs in the kth place. Thus the set X is different from each of the S_k's and is not a member of the collection $\{S_1, S_2, \ldots\}$. Our construction of X did not depend on any particular property of the collection $\{S_k\}$. Thus any countable collection of subsets of **N** excludes some subset of **N**. Because $\mathcal{P}(\mathbf{N})$ does not exclude any subset of **N**, we conclude that it cannot be a countable collection of sets. The set $\mathcal{P}(\mathbf{N})$ is uncountable.

We can display the elements of the sets in the collection $\{S_1, S_2, S_3, \ldots\}$ of Example 12 in an infinite array as follows:

$a(1, 1)$ $a(1, 2)$ $a(1, 3)$ $\ldots$
$a(2, 1)$ $a(2, 2)$ $a(2, 3)$ $\ldots$
$a(3, 1)$ $\ldots$
$\vdots$

We constructed the elements of X by considering only the diagonal elements of this array. Because of this, our argument to show that $\mathcal{P}(\mathbf{N})$ is not countable is called a **diagonalization argument.** Such arguments occur frequently in mathematics. We now give a very similar diagonalization argument to show that the set of real numbers between 0 and 1 is uncountable.

Example 13. Let $I = \{x : 0 \leq x < 1\}$. Each element x in I has a unique decimal representation $x = 0.a_1a_2a_3a_4 \ldots$ where each a_k is a digit between 0 and 9. (We shall not allow a number to be represented with a non-terminating sequence of 9's.) Now suppose that $S = \{x_1, x_2, x_3, \ldots\}$ is a countably infinite subset of I and that the decimal representation of x_k is $0.a(k, 1)a(k, 2)a(k, 3) \ldots$, where each $a(k, j)$ is a digit between 0 and 9. Let y be the element of I with the following decimal representation: $y = 0.b_1b_2b_3 \ldots$, where $b_k = 1$ if $a(k, k) \neq 1$ and $b_k = 2$ if $a(k, k) = 1$. Now y differs in at least one place from each of the elements in the set S—namely, y differs in the kth place from x_k. So y is not equal to any of the x_k and is not a member of S. Our construction of y did not depend on any particular property of S. Thus every countable subset of I excludes some element of I. Thus I itself cannot be countable.

Proposition 8. If a set B contains an uncountable subset A, then B is also uncountable.

> PROOF: The contrapositive of Proposition 8 says that if B is countable, then any subset of B is also countable. We already know that this is true.

Example 14. The set **R** of all real numbers is uncountable because it contains the uncountable subset $\{x : 0 \leq x < 1\}$.

Section 4.5	**Exercises**

1. Show that the members of each of the following pairs of sets have the same cardinality by finding a bijection between them.
   (a) $A = \{x : x \text{ is real and } 0 < x < 3\}$ and $B = \{x : x \text{ is real and } 1 < x < 7\}$
   (b) $\mathbf{N} \times \mathbf{N}$ and $\mathbf{Z} \times \mathbf{Z}$
   (c) The set A of all integer multiples of 2 and the set B of all integer multiples of 3
   (d) $S = \{x : x \text{ is real and } x > 1\}$ and $T = \{x : x \text{ is real and } -1 > x\}$

2. Find a bijection between **N** and each of the following sets.
   (a) The set of positive odd integers
   (b) $\{1, 6, 11, 16, \ldots\}$

(c) The set of all multiples of 3

(d) {2, 4, 8, 16, . . .}

3. Show that each of the following languages over {a, b} is countable by explicitly finding a bijection to the natural numbers.

(a) The language S defined as follows:

(1) Λ is an element of S.

(2) If x is a word in S, then axb is a word in S.

(3) Only words so formed are in S.

(b) The language Q defined as follows:

(1) b is a word in Q.

(2) If x is a word in Q, then xab is a word in Q.

(3) Only words so formed are in Q.

(c) The language A defined as follows:

(1) Λ is in Q.

(2) If x is a word in Q, then so is aaxbbb.

(3) Only words so formed are in Q.

4. Show that each of the following sets is countable. (Use any of the propositions or examples in this section.)

(a) $S = \{x : 0 < x < 1 \text{ and } x \text{ has a finite decimal expansion}\}$

(b) The set of 2×2 matrices with integer entries

(c) The set of all polynomials $p(x) = a_0 + a_1 x + \cdots + a_n x^n$, where $n \geq 0$ and each a_k is rational

5. Identify the following sets as countable or uncountable.

(a) The interval $\{x : -1 < x < 1\}$

(b) The set of all words over {a, b, c}

(c) The x-y plane

(d) The set of finite sequences of integers

(e) The set of all sequences of integers

6. Prove Proposition 3.

Bridge to Computer Science

4.6 An Algorithm Analyzed (optional)

In Section 4.4 we looked at a few combinatorial identities. A complete list of all currently known combinatorial identities would be very long and almost impossible to keep up to date. (Knuth's book *The Art of Computer Programming,* 2d ed. (Reading, Mass.: Addison Wesley, 1973), Volume 1, contains an interesting survey of the various types of combinatorial identities.) Mathematicians have known about and manipulated the binomial coefficients for many centuries. High school students use Pascal's triangle (Figure 4.6), published in 1653, to compute $C(n, r)$.

Figure 4.6. Pascal's Triangle

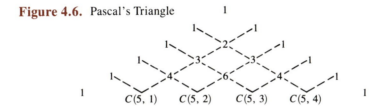

To use the triangle, we first note that for any non-negative value of n, $C(n, 0)$ $= C(n, n) = 1$. Otherwise, to compute $C(n, r)$ we go to the $(n + 1)$th row of the triangle and find the $(r + 1)$th entry in that row. We note that, if we form a triangle with $C(n, r)$ and the two values above it, then $C(n, r)$ is the sum of these two values. So to generate the $(n + 1)$th row of Pascal's triangle, we start with 1 on the far left and then, under each pair of terms already appearing in the triangle, we enter their sum. We end our new row with 1. The triangle is based on the identity

$$(1) \qquad C(n, r) = C(n - 1, r - 1) + C(n - 1, r), \quad 0 < r < n$$

Of more current interest, the same identity gives us a recursive procedure for computing the binomial coefficients. We define the function $C : \mathbf{W} \times \mathbf{W} \to \mathbf{N}$ recursively as follows:

(i) For any value of n, $C(n, 0) = 1$ and $C(n, n) = 1$.

(ii) For $0 < r < n$, $C(n, r) = C(n - 1, r - 1) + C(n - 1, r)$.

From step (i) we have $C(0, 0) = 1$, $C(1, 0) = 1$, and $C(1, 1) = 1$. From step (ii) we have $C(2, 1) = C(1, 0) + C(1, 1) = 2$, and so forth. We can take advantage of the recursive capabilities of the language Pascal and use the following program to generate the binomial coefficients.

```
function C(n,r:integer);integer;
begin
```

```
   if (r=0) or (n=r) then C:=1
   else C:=C(n-1,r-1) + C(n-1,r)
end;
```

The advantage that our recursive procedure has over the direct computation of $C(n, r)$ by the formula $C(n, r) = n!/[r! \, (n - r)!]$ is that it avoids the use of the factorial function. The number $n!$ is very large even for rather small values of n. For instance $6! = 720$, whereas $C(6, 3) = 20$. If the largest integer available on a computer is 32,768, then the largest value of n for which $n!$ can be computed is 7. (Of course, we are making no accommodations for "large integers.")

For a fixed value of n, the largest binomial coefficient of the form $C(n, r)$ is $C(n, \lfloor n/2 \rfloor)$. We know that $C(n, 0) + C(n, 1) + C(n, 2) + \cdots + C(n, n) = 2^n$ and so $C(n, \lfloor n/2 \rfloor) \leq 2^n$. For very large values of n, $n!$ is very much larger than 2^n. (See Table 4.1.) In the limit as n goes to infinity, the quotient $2^n/n!$ approaches 0. In our recursive procedure, we never use a number larger than $C(n, r)$ to compute $C(n, r)$. So our simple procedure can compute $C(17, 9) = 24,318$ without exceeding 32,768 anywhere in the routine.

Table 4.1

n	2^n	$n!$
1	2	1
2	4	2
3	8	6
4	16	24
5	32	120
6	64	720
7	128	5,040
8	256	40,320

In the foregoing discussion, we used our knowledge of combinatorics to write a routine for computing $C(n, r)$. Now we employ our counting techniques to analyze an algorithm. The following simple algorithm, a version of which appears in Volume 1 of *The Art of Computer Programming*, 2d ed., finds the maximum of a sequence of n numbers, $(x_1, x_2, \ldots, x_n)$. We shall assume that all the numbers are distinct.

Algorithm Max

(This algorithm finds the maximum value of a sequence of n numbers, $x_1, x_2, \ldots, x_n$, and returns it as m.)

    (1) (Initialize.) Set $k = n - 1$, and $m = x_n$.

    (2) (All tested?) If $k = 0$, the algorithm terminates.

    (3) (Compare.) If $x_k \leq m$, go to step 5.

    (4) Set $m = x_k$. (Now m is the current maximum.)

(5) (Decrease k.) Decrease k by 1 and return to step 2.

The algorithm checks the values of the sequence in reverse order, starting with x_n. After i iterations of steps 2 through 5, m is the maximum of the values x_{n-i} through x_n. Step 5 is executed on the next iteration only if $m < x_{n-i-1}$. To get an index for the average length of time the algorithm takes to terminate when it is applied to a sequence of length n, we shall determine the average number of times step 4 is executed. We call this number $Av(n)$. In what follows we shall develop a recursive formula for $Av(n)$.

Let $S = \{a_1, a_2, \ldots, a_n\}$ be a set of n distinct numbers, and suppose that max is the largest element in this set. Let P be the set of all $n!$ different sequences that can be formed from the numbers in S. Let $s = (x_1, x_2, \ldots, x_n)$ be a particular sequence in P. Suppose that max occurs in the $(n - i)$th position of s; that is, suppose $x_{n-i} = $ max. The key to finding a recursive formula for $Av(n)$ is the fact that the number of times step 4 is executed when it is applied to the sequence s is exactly one more than the number of times step 4 is executed when the algorithm is applied to the subsequence $(x_{n-i+1}, x_{n-i+2}, \ldots, x_n)$ that has length i. (See Figure 4.7.)

Figure 4.7

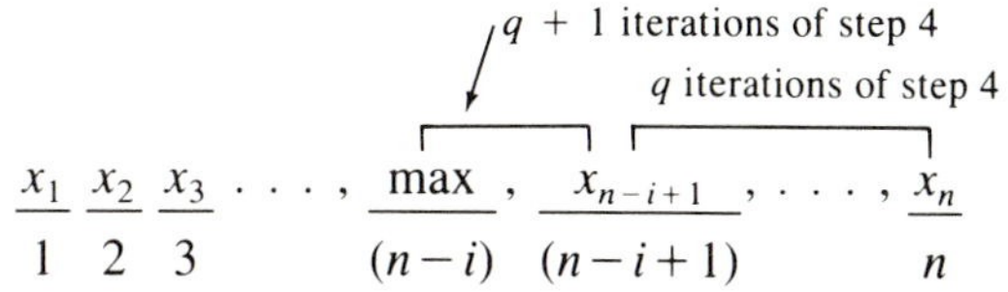

Let $\text{Tot}(i)$ be the total number of times step 4 is executed when the algorithm is applied to all the $i!$ permutations of a set of i distinct numbers. Let P_i be the subset of P of all sequences in which max occurs in the $(n - i)$th position. For $1 \leq i \leq n - 1$, there are $C(n - 1, i)$ different sets of numbers that can follow max in the i positions numbered $(n - i + 1)$ through n. Each permutation of each of these sets occurs $(n - i - 1)!$ times in P_i, once for each of the $(n - i - 1)$ sequences that can precede max in positions 1 through $(n - i - 1)$. The number of times step 4 is executed when the algorithm is applied to each of the $(n - 1)!$ different sequences in P_i is thus

$$(2) \qquad (n - i - 1)!\, C(n - 1, i)\text{Tot}(i) + (n - 1)!$$

Remember that the number of times step 4 is applied to a sequence s in P_i is one more than the number of times it is applied to a sequence of i values following max and that there are $(n - 1)!$ sequences in P_i. Hence the added term $(n - 1)!$ in expression (2). If $i = 0$ and s is in P_0, then max is in the nth position of the sequence, and step 4 is not executed. Because P can be expressed as the disjoint union $P = P_0 \cup P_2 \cup \cdots \cup P_{n-1}$, we sum our results for $i = 1$ through $n - 1$ to find

$$\text{Tot}(n) = \sum_{i=1}^{n-1} [(n - i - 1)!\, C(n - 1, i)\text{Tot}(i) + (n - 1)!]$$

The expression for Tot(n) is recursive. We compute Tot(1) directly. If we set $n = 1$ in step 1 of the algorithm, then $k = 0$ and the algorithm halts in step 2. So step 4 is never executed and Tot(1) = 0. We note that the sum of $(n - 1)!$ taken $(n - 1)$ times is $(n - 1)[(n - 1)!]$ and that $(n - i - 1)! C(n - 1, i) = (n - 1)!/i!$. We can thus simplify our formula somewhat to obtain

$$\text{Tot}(n) = (n - 1)[(n - 1)!] + \sum_{i=1}^{n-1} [(n - 1)!/i!]\text{Tot}(i)$$

Finally, we have our formula for Av(n):

$$\text{Av}(n) = \text{Tot}(n)/n!$$

Table 4.2 gives the values of Tot(n) and Av(n) explicitly for a few values of n.

Table 4.2

n	Tot(n)	Av(n)
1	0	0
2	1	1/2 (0.500)
3	5	5/6 (~0.833)
4	26	26/24 (~1.083)
5	154	154/120 (~1.283)
6	1044	1044/720 (1.450)
7	8028	8028/5040 (~1.593)

In Table 4.3 we verify the value of Tot(3) directly by listing the 6 permutations of the set {1, 2, 3} and counting how many times step 4 is executed on each of these permutations.

Table 4.3

Sequence	Number of Times Step 4 Is Executed		
1, 2, 3	0		
2, 1, 3	0		
1, 3, 2	1		
2, 3, 1	1		
3, 1, 2	1		
3, 2, 1	2	Tot(3) = 5	Av(3) = 5/6

In his book, Knuth also finds a formula for Av(n). But the techniques he employs are very different and more advanced. His results for Av(n) are the same as ours, but his techniques enable him to obtain even more statistical information about the algorithm. The subject of combinatorics is deep, rich, and varied, but even the basic techniques developed in this chapter can carry us quite far.

Exercises

1. Generate the next two rows of Pascal's triangle (see Figure 4.1).

2. Discuss some of the problems that arise in our analysis if we do not assume that the values in a sequence are distinct.

3. Show that $Av(n) = \dfrac{n-1}{n} + \dfrac{1}{n}\displaystyle\sum_{i=1}^{n-1} Av(i)$ for $n > 1$ and $Av(1) = 0$.

4. Develop a formula for the average number of times step 5 is executed on a sequence of length n in the following algorithm, which sorts a sequence $(x_1, x_2, \dots , x_n)$ of n distinct values into ascending order.

Algorithm Sort

     (1) Let $j = n$.

     (2) If $j - 1 = 0$, stop; the sequence is now in ascending order.

     (3) Set $p = j$ and $k = j - 1$.

     (4) If $x_p > x_k$, go to step 6.

     (5) Set $p = k$.

     (6) Decrease the value of k by 1.

     (7) If $k = 0$, go to step 8; otherwise, go to step 4.

     (8) Interchange to values of x_p and x_j.

     (9) Decrease the value of j by 1 and go to step 2.

Hint: Steps 2 through 7 find the maximum value of the sequence $(x_1, x_2, \dots , x_j)$ in the same way as the algorithm discussed in the text.

key concepts

4.1 Counting Principles	cardinality of a finite set A: $\text{card}(A)$ or $|A|$ addition principle: $|A \cup B| = |A| + |B|$ for A and B disjoint $|\mathcal{P}(A)| = 2^{|A|}$ inclusion–exclusion principle: $|A \cup B| = |A| + |B| - |A \cap B|$ multiplication principle: $|A \times B| = |A| \cdot |B|$
4.2 Functions and Counting	number of functions from A to B: $|B|^{|A|}$ number of injective functions from A to B: $n!/(n - m)!$ if $|A| = m$, $|B| = n$, and $m \leq n$ pigeon-hole principle permutation of A: a bijection from A to A
4.3 Permutations and Combinations	permutation: a rearrangement of a set $P(n, m) = n!/(n - m)!$: the number of permutations of m elements chosen from n $C(n, m) = n!/[m! (n - m)!]$: the number of combinations of m elements
4.4 Combinatorial Arguments	combinatorial argument binomial theorem infinite set same cardinality for infinite sets
4.5 Infinite Sets and Countability	countably infinite set countable set: If A and B are countable, then so are $A \cup B$, $A \times B$, and $S \subset A$. uncountable set: $\mathcal{P}(\mathbf{N})$, $\mathbf{R}$ diagonalization argument

| Chapter 4 | Exercises |

1. For each of the following sets A, find a bijection $f : A \to S_n$ for the appropriate value of n.

   (a) Let A be the set of even integers between 0 and 10 inclusively.

   (b) Let A be the set of vowels in the word *SQUEAKING*.

   (c) Let A be the set of integers from 1 to 100 that are divisible by either 3 or 4.

   (d) Let q be a positive integer, and let $A = \{x : x$ is an odd integer and $0 \le x \le q\}$.

2. Suppose $A = \{2, 4, 6, 8\}$ and $f : A \to X_4$ is defined by $f(x) = (10 - x)/2$. Suppose that $B = \{1, 3, 5, 7\}$ and that $g : B \to X_4$ is defined by $g(x) = (x + 1)/2$. Find the bijection $h : A \cup B \to X_B$ given by the procedure described in Theorem 2 in Section 4.1.

3. Find a bijection between the power set of X_n and the subsets of S_{n+1} that do not contain 1.

4. Suppose that A and B are subsets of a universal set U.

   (a) Find $|A \cap B|$ if $|A| = 100$, $|B| = 70$, and $|A \cup B| = 130$.

   (b) Find $|A \cup B|$ if $|A| = 120$, $|B| = 90$, and $|A \cap B| = 35$.

   (c) Find $|A'|$ if $|U| = 240$, $|A \cup B| = 100$, $|B| = 30$, and $|A \cap B| = 10$.

5. A school of 200 people is surveyed, and it is found that 70 students take gym, 100 take Latin, and 20 take both. How many students take neither? How many take exactly one of gym or Latin?

6. Let q be an integer. How many numbers between 0 and q (inclusively) are divisible by either 2 or 3?

7. Suppose that A, B, and C are subsets of a universal set U. Suppose that $|U| = 200$, $|A \cap B \cap C| = 20$, $|A \cap B| = 30$, $|A \cap C| = 25$, $|B \cap C| = 35$, $|A| = 40$, $|B| = 60$, and $|C| = 70$. Find each of the following.

   (a) $|A \cup B \cup C|$

   (b) $|A' \cap B|$

   (c) $|A' \cap B' \cap C'|$

8. Find the number of integers between 1 and 500 that are not divisible by 2, 3, or 7.

9. Complete the proof of the inclusion–exclusion principle given for Theorem 5 in Section 4.1 for three sets A, B, and C.

10. A man has 6 ties, 10 shirts, and 5 pairs of pants. How many different outfits can he put together?

11. A die is cast 5 times and the sequence of faces is noted. How many different sequences are possible? How many different sequences are possible in which no face appears more than once? How many different sequences are possible in which no face appears twice in a row?

12. How many 4-digit numbers are there in which either the first or the last digit is a 5? How many 4-digit numbers are there that either begin or end with 5 or 3?

13. How many different words of length 5 or less can be formed from the symbols in the set $\{r, a, d, g, i, l\}$? What if no letter can be repeated in any given word?

14. How many different functions are there from S_5 to S_8? How many of these are injective? For how many injective functions does $f(1) = 1$?

15. Let S be the set of all words over the symbols $\{a, b, c\}$ of length 3 or less. How many different functions are there from S to the set $\{0, 1\}$? How many different permutations of S are there?

16. How large must n be in order to ensure that there are at least 10^6 different n-argument logical functions?

17. We wish to form a code for the digits $\{0, 1, 2, 3, 4, 5, 6, 7, 8, 9\}$ using words over the symbols $\{a, b\}$ of length 5. In how many ways can this be done? In how many ways can this be done if the first symbol in the word must be an a?

18. Suppose that $A = S \cup T$, where S and T are disjoint, and that $|S| = m$ and $|T| = n$. Find the number of functions $f : A \rightarrow A$ such that:

(a) $f(S) \subset S$ and $f(T) \subset T$

(b) $f(S) \subset S$ and $f(T) \subset T$ and f is injective

(c) $f(x) = x$ for each x in S and f is injective

19. (a) In how many ways can 8 people line up for a picture?

(b) In how many ways can 4 couples line up for a picture if spouses must be next to each other?

(c) In how many ways can 4 couples form a circle if spouses must face each other?

20. How many different words can be formed by rearranging the letters in the word *SMILE*? in the word *QUEUE*? in the word *COMMITTEE*?

21. How many words of length 7 are there in which there are exactly 2 vowels? (The set of vowels is $\{a, e, i, o, u\}$.) How many of these words have no letter repeated?

22. (a) In how many ways can a child select 3 books from among 10 different books?

(b) In how many ways can a president, vice president, and secretary be selected from among 10 candidates?

(c) In how many ways can each of 10 students sign up for one of 3 different math courses?

23. How many 9-man baseball squads can be selected from among 23 men? How many different line-ups can be formed?

24. In how many ways can 3 committees be formed from 20 people if the first has 6 people, the second has 4 people, and the third has 5 people. (No person can serve on more than 1 committee.)

25. (a) How many words of length 7 can be formed from the letters $\{a, b, d, e, f, r\}$ if the letter a must appear exactly 3 times or the letter b appear exactly 4 times?

   (b) How many words of length 5 can be formed from the letters $\{a, b, e, d, f, r\}$ if the sequence ab must appear?

26. A fair coin is tossed 7 times and the sequence of heads and tails is noted.

   (a) How many different sequences are possible?

   (b) How many different sequences are possible in which heads occurs exactly 5 times?

   (c) How many different sequences are possible in which at most 6 heads occur?

   (d) How many different sequences are possible in which either 3 or 5 heads occur?

27. Twelve numbered balls are placed in an urn. Four are white and 8 are red. Five balls are selected from the urn.

   (a) How many different selections are possible?

   (b) How many selections are possible that contain exactly 2 white balls?

   (c) How many selections are possible that contain at least 2 white balls?

28. In the accompanying diagram, how many different direct routes are there from point A to point B? How many of these routes pass through point C?

Figure for Exercise 28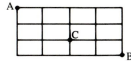

29. How many words of length 3 are there over the alphabet $\{a, b, c, d, e\}$ that have at least 1 letter appearing at least twice in a row? What about words of length 4?

30. How many 7-card hands in which at most 2 suits appear can be drawn from a standard 52-card deck?

31. In how many different ways can 4 pairs of people be chosen from among 30 people? (First, try solving the simpler problem of choosing 2 pairs of people from among 7 people.)

32. In how many ways can 10 identical pencils be distributed among 5 students?

33. In how many ways can 12 crayons be chosen from an unlimited supply of red, green, purple, blue, and yellow ones?

34. A groundskeeper wishes to fly 1 flag on each of 5 flag poles. He can choose from an unlimited supply of red flags, white flags, green flags, and blue flags. (Flags of the same color are indistinguishable.) In how many ways can he do this? What if at most 2 flags can be of the same color?

35. Find the number of non-negative integer solutions to the equation $x + y + z + w = 20$. Find the number of positive integer solutions.

36. A jar contains 20 pennies, 10 nickels, and 30 dimes. How many coins must one choose to guarantee that at least 15 are of the same denomination?

37. (a) Use the binomial theorem to prove the following identity:

$$C(n, 0) - C(n, 1) + C(n, 2) - \cdots + (-1)^n C(n, n) = 0$$

(b) Prove the same identity combinatorially.

38. Prove the following identity combinatorially:

$$C(m, 0) \cdot C(n, r) + C(m, 1) \cdot C(n, r - 1) + \cdots$$
$$+ C(m, r) \cdot C(n, 0) = C(m + n, r)$$

39. Give a combinatorial argument for the following identity:

$$C(2n, 2) = 2C(n, 2) + n^2$$

40. Prove that the members of each of the following pairs of sets have the same cardinality by finding a bijection between them.

(a) The set of natural numbers and the set of odd integers

(b) $\{x : x < 1\}$ and $\{x : x > 0\}$

(c) The intervals $[-2, 3]$ and $[5, 6]$

41. Prove that each of the following sets is countably infinite by finding a bijection to the natural numbers.

(a) $\{x : x$ is a natural number divisible by 2 or 3$\}$

(b) $\{x : x$ is real and $x - \lfloor x \rfloor = 0$ or $0.5\}$

(c) The set of words over $\{a, b\}$ in which a and b alternate

42. Prove that the following sets are countably infinite. Use any of the results or propositions from Section 4.5.

(a) The set of all points in $\mathbf{R} \times \mathbf{R}$ with rational coordinates

(b) The set of all functions of the form $y = mx + b$, where m and b are integers

(c) The language Q over the symbols $\{a, b\}$ defined inductively as follows:

    (1) The empty word is in Q.

    (2) If x is in Q, then so are *aaax* and *xbb*.

    (3) Only words so defined are in Q.

(d) The language S over the symbols $\{a, b\}$ defined inductively as follows:

    (1) The symbol a is in S.

    (2) If x is in S, then so are *xa* and *xbb*.

    (3) Only words so formed are in S.

43. Prove that the set of all finite subsets of **N** is countable.

44. Let E be the alphabet $\{a, b\}$, and let Q be the set of all languages over Q. Decide whether Q is countable and justify your answer.

45. Suppose that A is a finite set and that B is countably infinite. Prove that $A \cup B$ is countable by finding a bijection from $A \cup B$ to N.

46. Show that each of the following sets is uncountable.

(a) The set of real numbers greater than 0

(b) The x-y plane

(c) The set of all sequences of rational numbers

47. Which of the following sets are uncountable?

(a) The set of all finite sequences of real numbers

(b) The set of all finite sequences of rational numbers

(c) The set of all functions of the form $f : \mathbf{N} \to \mathbf{N}$

48. Let A be an uncountable set. Prove that $A \times A$ is uncountable.

Chapter 5

RELATIONS

outline

5.1 Relations
5.2 Composition of Relations
5.3 Equivalence Relations
5.4 Equivalence Classes
5.5 Order Relations
5.6 Bridge to Computer Science: Two's-Complement Arithmetic (optional)
    Key Concepts
    Exercises

Introduction

CONSIDER the following sets:

$A = \{7, 3, 1, 2, 4, 8, 5, 10, 9, 6\}$
$B = \{$cat, canary, frog, sparrow, salamander, dog, cow$\}$
$C = \{$Henry IV, Charles II, Elizabeth I, Richard I, James II$\}$

Perhaps the reader has already challenged chaos by relisting the elements in set A in ascending order and classifying the animals in set B as avian, amphibian, or mammalian—while perhaps leaving it to the historically acute among us to place that English royalty in set C upon its family tree. We seek such patterns in a given set because we cannot access information from or about a set until the structure of the relations that exist among its elements is established. In this chapter we characterize such relations in terms of set theory, and we study the properties of the most useful and familiar relations. The set of integers provides us with our prototypes: the partial-order relation and the equivalence relation.

Because we can compare any pair of integers with respect to size, we can sort any finite set of integers in ascending order. Sets that admit similar comparisons are called partially ordered sets, or "posets." For example, the dictionary is a poset under alphabetical order, and a family tree establishes a partial order on a set of relatives. To sort large sets (names, numbers, and the like) into a predetermined order occupies most of today's available computer time. Thus the means of doing so efficiently are of major interest. (We look at the sorting problem in Chapter 6.)

When we classified the animals in set B, we regarded the frog and the salamander as the same. When we classify integers as to parity, either even or odd, we regard a pair of integers as the same if they have the same remainder after division by 2. Such relations are called equivalence relations. In this chapter we shall study the modular relation, a generalization of the parity relation, in which we classify the integers according to their remainders after division by a fixed integer m. This relation has proved to be a powerful tool in the investigation of the nature of prime numbers, the classical concern of number theory. But it also has contemporary utility as the basis of computer arithmetic and as the key forming certain relatively unbreakable codes. We shall touch on these topics in the bridge section of this chapter and again in Chapter 8.

| Section 5.1 | Relations |

Many natural and familiar relationships exist between pairs of integers. For instance:

(i) We write $m \leq n$ for a given ordered pair of integers (m, n) if $n - m$ is a nonnegative integer.

(ii) We say that a pair of integers (m, n) has a common divisor if m and n share a common factor other than $+1$ or -1.

(iii) We say that m divides n for a given ordered pair of integers (m, n) if we can find an integer q such that $n = qm$.

In each of the preceding situations we can decide when an ordered pair is a related pair. For example, in situation (i), the pair $(3, 5)$ is a related pair because $5 - 3 = 2$, and 2 is a nonnegative integer. But $(5, 3)$ is not a related ordered pair because $3 - 5$ is negative. In situation (ii), $(9, 12)$ is a related pair since 9 and 12 share a common factor of 3. In situation (iii), $(3, 6)$ is a related pair since $6 = 2 \cdot 3$ and $q = 2$ is an integer, whereas $(6, 3)$ is not a related pair since $3 = (1/2) \cdot 6$ and $q = 1/2$ is not an integer. In each of these cases the set of related ordered pairs forms a subset of the cartesian product $\mathbf{Z} \times \mathbf{Z}$. We use the cartesian product to formalize our notion of a relation and extend it to arbitrary sets.

> **DEFINITION.** Let A be a set. A subset R of the cartesian product $A \times A$ is called a **relation** on A. If a pair (x, y) is in R, we write $x R y$ and say that x is related to y. Otherwise we write $x \not R y$.

Thus if R_i is the relation described in situation (i), we could write $2 R_i 3$ because $(2, 3)$ is an ordered pair in R_i. (However, we shall retain the notation $m \leq n$ for a pair in this familiar relation.) Similarly, if R_{iii} is the relation described in situation (iii), we write $2 R_{iii} 6$ for the related pair $(2, 6)$. Often we shall write $x \mid y$ for ordered pairs (x, y) in this relation.

A relation can be quite arbitrary. For example, let $A = \{a, b, c, d, e\}$ and let $R = \{(a, b), (b, c), (c, e), (a, d), (d, a), (d, d)\}$. In this relation we have $a R b$ and $d R a$ but $c \not R d$. We can represent a relation on a finite set S with a diagram called a **directed graph** or **digraph**. To do this, we represent each element in S with a labeled dot called a **node** or a **vertex**. If (s, t) is a related pair, we connect the two nodes labeled s and t with an arrow from s to t. (This arrow is called a **directed edge**.) The digraph that represents the relation R on the set A we have described is shown in Figure 5.1. Because both (a, d) and (d, a) are elements of the relation R,

Figure 5.1

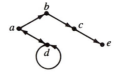

there are two directed edges between the nodes *a* and *d*: one from *a* to *d* and another from *d* to *a*. Sometimes, for the sake of neatness, we represent these with one line and two arrowheads. The directed edge from *d* to itself is called a **self-loop**. The pictorial representation of a relation is not unique. Its shape varies with the original distribution of nodes on the page. But any two digraphs representing the same relation have the same set of nodes and directed edges.

Example 1. Let $S = \{1, 2, 3, 4, 5\}$ and let $R = \{(1, 1), (1, 2), (2, 3), (1, 3),$ $(1, 4), (4, 5), (5, 1), (1, 5), (4, 1)\}$. A digraph representing this relation is shown in Figure 5.2.

Example 2. Let $A = \{a, b, c, d\}$ and let $R = \{(a, a), (a, b), (b, b), (b, a), (c, b),$ $(c, c), (b, c), (d, a), (d, d), (a, d)\}$. Figure 5.3 shows a digraph for this relation.

Figure 5.2   **Figure 5.3**

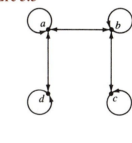

Example 3. Let $S = \{2, 3, 4, 6, 8, 9, 12\}$ and let R be the relation on S defined by $x\,R\,y$ if x divides y. A digraph for this relation is shown in Figure 5.4.

Figure 5.4

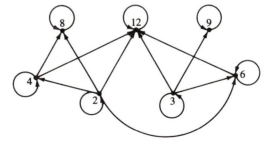

Every function of the form $f : A \rightarrow A$ defines a relation R_f on the set A as follows: $x\,R_f\,y$ if and only if $y = f(x)$. For example, let $A = \{a, b, c, d, e, f\}$ and let f be the function defined by the diagram shown in Figure 5.5. The digraph for R_f is shown in Figure 5.6.

Figure 5.5

f

a — a
b — b
c — c
d — d
e ⟶ e
f ⟶ f

Figure 5.6

If $f : \mathbf{R} \to \mathbf{R}$ is a function from the set of real numbers to itself, then the set of ordered pairs in the relation R_f is the graph of the function $y = f(x)$ in the x-y plane. Thus if $f(x) = x^2 + 1$, the relation R_f is the parabola sketched in Figure 5.7.

Figure 5.7

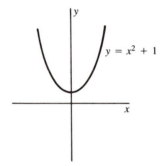

$y = x^2 + 1$

Conversely, let R be a relation on a set A. If there is exactly one ordered pair of the form (a, y) in R for each element a of the set A, then we can use R to define a function $f_R : A \to A$ as follows: let $f_R(a) = y$, where y is the second coordinate of the unique pair (a, y) that appears in R. For example, consider the relation R on the set $A = \{a, b, c, d, e\}$ defined by the digraph shown in Figure 5.8. There is exactly one out-going arrow from each node. Thus each element of A appears exactly once as the first coordinate of a pair in the set of ordered pairs defining R. The function f_R defined by R is given in Figure 5.9.

Figure 5.8

Figure 5.9 f_R

a — a
b — b
c — c
d — d
e — e

In the next few paragraphs we define certain properties that are common to many important relations. We shall use these properties later to classify our relations.

Suppose that R is a relation on a set A. We say that R is a **symmetric** relation if whenever we have $x R y$, we also have $y R x$. The digraph shown in Figure 5.10 defines a relation that is symmetric: if there is a directed edge from x to y, there is also one from y to x. The relation R_{ii} described in situation (ii) is a symmetric relation because, if m and n share a common factor, then both (m, n) and (n, m) must be related pairs. The $\leq$ relation defined on the set of integers $\mathbf{Z}$ is not symmetric. Although we know that $2 \leq 3$, it is not the case that $3 \leq 2$.

Figure 5.10

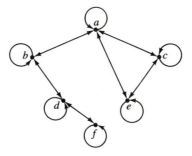

We say that a relation R on a set A is **reflexive** if for each x in A, we have $x R x$. The relation defined by the digraph shown in Figure 5.10 is reflexive. The relation $\leq$ defined on $\mathbf{Z}$ is reflexive because, for every integer n, we have $n \leq n$. The relation defined by $<$ is not reflexive because, for instance, $1 \not< 1$.

We say that a relation R on a set A is **transitive** if whenever we have $x R y$ and $y R z$, we also have $x R z$. The relation defined by the digraph shown in Figure 5.11 is transitive. That given in Figure 5.10 is not, because $f R d$ and $d R b$ but $f \not{R} b$. The relations $<$ and $\leq$ defined on $\mathbf{Z}$ are both transitive.

Figure 5.11

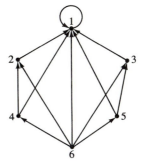

We say a relation R on a set A is **irreflexive** if for each x in A, we have that $x \not{R} x$. The relation $<$ on the integers is an irreflexive relation. The relation shown in Figure 5.11 is neither reflexive nor irreflexive.

We say that a relation R on a set A is **antisymmetric** if for any pair of elements x and y, whenever we have both $x R y$ and $y R x$, then $x = y$. Both the relations $<$ and $\leq$ are antisymmetric relations on the integers. The relation shown in Figure 5.11 is antisymmetric.

A relation R on a finite set $A = \{a_1, a_2, \ldots, a_n\}$ can be represented by an $n \times n$ matrix M_R of 0's and 1's. To do this, we let the entry in the ith row and jth column of M_R equal 1 if $a_i \, R \, a_j$, and we let it equal 0 if $a_i \, \not{R} \, a_j$.

Example 4. The matrix that represents the relation given in Example 1 is:

$$
\begin{array}{c|ccccc}
 & 1 & 2 & 3 & 4 & 5 \\
\hline
1 & 1 & 1 & 1 & 1 & 1 \\
2 & 0 & 0 & 1 & 0 & 0 \\
3 & 0 & 0 & 0 & 0 & 0 \\
4 & 1 & 0 & 0 & 0 & 1 \\
5 & 1 & 0 & 0 & 0 & 0 \\
\end{array}
$$

Figure 5.12

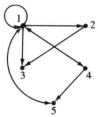

A digraph for this relation is again shown in Figure 5.12.

Example 5. The matrix for the relation of Example 2 is:

$$
\begin{array}{c|cccc}
 & a & b & c & d \\
\hline
a & 1 & 1 & 0 & 1 \\
b & 1 & 1 & 1 & 0 \\
c & 0 & 1 & 1 & 0 \\
d & 1 & 0 & 0 & 1 \\
\end{array}
$$

Figure 5.13

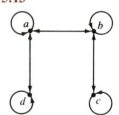

A digraph for this relation is again shown in Figure 5.13.

The matrix of a reflexive relation always has a 1 in each of the diagonal positions. In the matrix for a symmetric relation, the (i, j)th entry always equals the (j, i)th entry.

Every $n \times n$ matrix M of 0's and 1's defines a relation R_M on a set $\{a_1, a_2, \ldots, a_n\}$ as follows. We let $a_k \, R_M \, a_j$ if and only if the (k, j)th entry of M is equal to 1.

Example 6. Let M be the matrix

$$
\begin{pmatrix}
0 & 1 & 1 \\
1 & 0 & 1 \\
1 & 0 & 0
\end{pmatrix}
$$

The digraph of the relation that M defines on $\{a_1, a_2, a_3\}$ is shown in Figure 5.14.

Figure 5.14

Matrices of 0's and 1's provide a convenient data structure for representing a relation on a finite set on the computer. Matrix multiplication and addition are also readily implemented on the computer. In the next section, we shall see how we can use these matrix operations to gain more information about relations.

Section 5.1 **Exercises**

1. Draw a digraph for each of the following relations.

   (a) Let $A = \{a, b, c, d\}$ and let $R = \{(a, b), (b, d), (a, d), (d, a), (d, b), (b, a), (c, c)\}$.

   (b) Let $A = \{1, 2, 3, 4, 5, 6, 7, 8\}$ and let $x\,R\,y$ whenever y is divisible by x.

   (c) Let $A = \{1, 2, 3, 4, 5, 6, 7, 8\}$ and let $x\,R\,y$ whenever x and y share no common factor other than 1.

2. Determine which of the relations given in Exercise 1 are reflexive, which are symmetric, which are transitive, which are antisymmetric, and which are irreflexive.

3. Let $A = \{a, b, c\}$ and let $\mathscr{P}(A)$ be the power set of A. Define the relation R on $\mathscr{P}(A)$ as follows: if S and T are subsets of A, then $S\,R\,T$ if $S \subset T$. Draw a digraph for this relation. Determine its reflexivity, transitivity, symmetry, and antisymmetry.

4. Let R be the relation on $\mathbf{N}$ defined by $x\,R\,y$ if x and y share a common factor other than 1. Determine the reflexivity and transitivity of R.

5. Draw the digraph of the relation R_f defined by the following function.

   Figure for Exercise 5

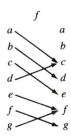

6. Determine which of the following relations defines a function.

Figure for Exercise 6

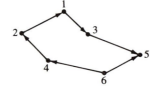

7. Find the matrix that represents each of the relations given in Exercise 1.

8. Draw the digraph of the relation defined on $\{a_1, a_2, a_3, a_4\}$ by the matrix

	a_1	a_2	a_3	a_4
a_1	0	1	1	1
a_2	0	0	1	1
a_3	1	1	0	0
a_4	1	0	0	0

9. Characterize the matrix of an irreflexive relation. What about that of a relation that is neither reflexive nor irreflexive?

Section 5.2   Composition of Relations

Suppose that R is a relation on a set A. We define the **composition** of R with itself, written $R \circ R$, to be the relation on A defined as follows: $x(R \circ R)y$ if there is an element z in A such that $x R z$ and $z R y$.

Example 1. Let $A = \{a, b, c, d\}$ and let $R = \{(a, b), (b, c), (c, a), (d, d), (d, a)\}$. A digraph for this relation is shown in Figure 5.15. The relation $R \circ R$ is given by $\{(a, c), (b, a), (c, b), (d, d), (d, a), (d, b)\}$. A digraph for $R \circ R$ is shown in Figure 5.16.

Figure 5.15

Figure 5.16

If we look at a digraph for a relation R on a set A, we see that two elements x and y are related under $R \circ R$ if there is a directed path from x to y that passes through exactly one intermediary node. (See Figure 5.17. The solid arrows represent R and the broken arrow represents $R \circ R$.)

Figure 5.17

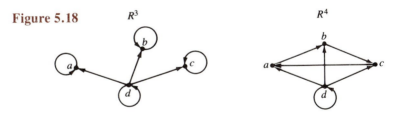

Example 2. If we let $R_<$ represent the relation on **Z** defined by $x\ R_<\ y$ if $x < y$, then we have $2\ R_<\ 3$ and $3\ R_<\ 4$ and so $2(R_< \circ R_<)4$. But we do not have $2(R_< \circ R_<)3$. However, if $R_\le$ represents the relation defined on **Z** by $x\ R_\le\ y$ if $x \le y$, we have $2(R_\le \circ R_\le)3$ because $2\ R_\le\ 2$ and $2\ R_\le\ 3$.

Let R be a relation on a set A. We define R^n, the relation R composed with itself n times, inductively as follows:

(1) For $n = 0$ we let $x\ R^0\ y$ if and only if $x = y$.

(2) Suppose that we have defined R^n. Then $x\ R^{n+1}\ y$ if and only if we can find an element z such that $x\ R^n\ z$ and $z\ R\ y$.

Using this inductive definition, we have $R^1 = R$ and $R^2 = R \circ R$.

Example 3. Consider the relation R on the set A given in Example 1. Digraphs for the relations R^3 and R^4 are shown in Figure 5.18.

Figure 5.18

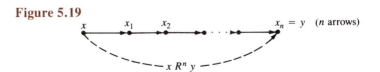

In terms of a digraph for a relation R on a finite set A, we have $x\ R^n\ y$ if there is a directed path of length n from x to y—that is, if we can follow exactly n arrows in going from x to y (see Figure 5.19).

Figure 5.19

Suppose that R is a relation on a finite set and that M_R is the matrix that represents R. Then the matrix that represents $R \circ R$ is $M_R \times M_R$ when we use the following addition and multiplication tables for the symbols 0 and 1:

$$
\begin{array}{c|cc}
\cdot & 0 & 1 \\
\hline
0 & 0 & 0 \\
1 & 0 & 1
\end{array}
\qquad\qquad
\begin{array}{c|cc}
+ & 0 & 1 \\
\hline
0 & 0 & 1 \\
1 & 1 & 1
\end{array}
$$

If we think of 0 and 1 as denoting the truth values false and true, respectively, then multiplication $\cdot$ is the "and" operation (conjunction) and addition $+$ is the "or" operation (disjunction). Conjunction is called **Boolean multiplication,** and disjunction is called **Boolean addition.** If we let $a_{m,n}$ denote the entry in the mth row, nth column of M_R, we find $c_{i,j}$, the entry in the ith row, jth column of $M_R \times M_R$, as follows:

$$
c_{i,j} = a_{i,1} \cdot a_{1,j} + a_{i,2} \cdot a_{2,j} + \cdots + a_{i,n} \cdot a_{n,j}
$$

Example 4. Let $A = \{a, b, c\}$ and let $R = \{(a, b), (b, c), (c, a)\}$. Then $R \circ R = \{(a, c), (b, a), (c, b)\}$. Digraphs for R and $R \circ R$ are shown in Figure 5.20.

Figure 5.20

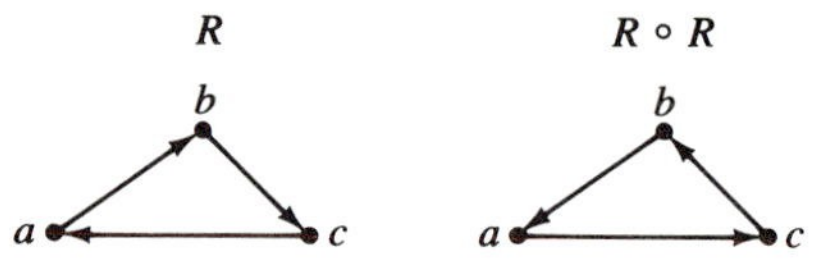

The matrix representing R is M_R:

$$
\begin{array}{c|ccc}
 & a & b & c \\
\hline
a & 0 & 1 & 0 \\
b & 0 & 0 & 1 \\
c & 1 & 0 & 0
\end{array}
$$

We find $M_R \times M_R$ by using the addition and multiplication operations just defined. For instance, to find the entry in the second row (labeled b), first column (labeled a) of the matrix $M_R \times M_R$, we compute

$$
0 \cdot 0 + 0 \cdot 0 + 1 \cdot 1 = 0 + 0 + 1 = 1
$$

Similarly, the entry in the first row, second column is

$$
0 \cdot 1 + 1 \cdot 0 + 0 \cdot 0 = 0 + 0 + 0 = 0
$$

Completing the computations, we find that $M_R \times M_R$ is

$$
\begin{array}{c|ccc}
 & a & b & c \\
\hline
a & 0 & 0 & 1 \\
b & 1 & 0 & 0 \\
c & 0 & 1 & 0
\end{array}
$$

The relation defined by $M_R \times M_R$ is exactly $R \circ R$. So $M_R \times M_R = M_{R \circ R}$.

Let us see why the relation $R \circ R$ is represented by the matrix $M_R \times M_R$. We have

$$c_{j,k} = a_{j,1} \cdot a_{1,k} + a_{j,2} \cdot a_{2,k} + \cdots + a_{j,n} \cdot a_{n,k}$$

Now $c_{j,k} = 1$ if and only if one of the terms on the right of the form $a_{j,m} \cdot a_{m,k}$ is equal to 1. Further, $a_{j,m} \cdot a_{m,k} = 1$ if and only if both $a_{j,m}$ and $a_{m,k}$ are equal to 1. Both $a_{j,m}$ and $a_{m,k}$ are equal to 1 if and only if we can find a node a_m such that $a_j R a_m$ and $a_m R a_k$—that is, if and only if $a_j R \circ R a_k$.

Proceeding inductively, we find that the matrix defining the relation R^n is M_R^n, the product of M_R with itself n times. (We leave the proof as an exercise.)

Example 4 (continued). The digraph for R^3 is shown in Figure 5.21.

Figure 5.21

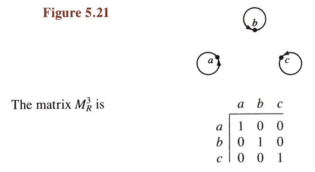

The matrix M_R^3 is

	a	b	c
a	1	0	0
b	0	1	0
c	0	0	1

Note: A brief introduction to matrices and matrix operations can be found in the appendix.

Section 5.2   **Exercises**

---

1. Let $A = \{a, b, c, d, e\}$ and let $R = \{(a, b), (a, a), (a, c), (c, c), (b, d), (b, b) (d, d)\}$. Draw a digraph for R. Find the ordered pairs in $R \circ R$ and draw a digraph for $R \circ R$.

2. Let $A = \{a, b, c, d\}$ and let $R = \{(a, b), (b, c), (c, a), (a, d), (d, a)\}$. Find digraphs for R^k for $0 \leq k \leq 6$. What if $k > 6$?

3. Let $A = \{a, b, c, d, e\}$ and let $f : A \to A$ be defined by the accompanying diagram. Let R_f be the relation defined by f. Find the digraph for $R_f \circ R_f$ and the digraph for $R_{f \circ f}$, and compare them.

Figure for Exercise 3

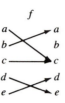

4. Let f be a function on a set A and let R_f be the relation defined by f. Prove that $R_{f \circ f} = R_f \circ R_f$.

5. Let R be the relation on the set $\{a, b, c, d, e, f\}$ defined by the accompanying digraph. Find the digraphs for R^2 and R^3. What can you conclude?

Figure for Exercise 5

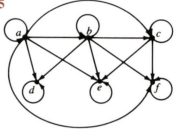

6. Let R_M be the relation on the set $\{a_1, a_2, a_3, a_4\}$ defined by the following matrix M:

$$\begin{pmatrix} 0 & 0 & 0 & 1 \\ 1 & 0 & 0 & 0 \\ 1 & 0 & 0 & 0 \\ 0 & 1 & 0 & 1 \end{pmatrix}$$

(a) Draw the digraph of the relation R_M defined by M.

(b) Find $M \times M$.

(c) Find the digraph for $R_M \circ R_M$.

(d) Find the matrix for $R_M \circ R_M$ and compare it with $M \times M$.

7. Let M and R_M be as in Example 4.

(a) Find M^n for all $n > 0$.

(b) Find the digraphs for all R_M^n for all $n > 0$.

8. Let R be a relation on a set A. Define the relation R^T (called the **transitive closure of R**) as follows: $a \, R^T \, b$ if $a \, R^n \, b$ for some $n > 0$. Find the transitive closure of each of the following relations.

(a) $A = \{a, b, c, d, e\}$ and $R = \{(a, b), (b, c), (c, d), (d, a), (a, e)\}$

(b) The relation given in Exercise 5

(c) The relation on $\{a, b, c, d, e, f, g\}$ defined by the following diagram:

Figure for Exercise 8(c)

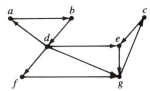

9. Prove that R^T is transitive. (See Exercise 8.)

| Section 5.3 | Equivalence Relations |

A relation that is reflexive, symmetric, and transitive is called an **equivalence relation.** For example, the relation R defined on a set A by $x\ R\ y$ if and only if x equals y is an equivalence relation. As the word "equals" suggests, this trivial relation serves as the model for more general equivalence relations.

Example 1. Let $A = \{a, b, c, d, e\}$. The following relation R is reflexive, symmetric, and transitive and hence an equivalence relation: $R = \{(a, a), (a, b), (b, a), (b, b), (b, c), (c, b), (c, c), (a, c), (c, a), (d, d), (d, e), (e, d), (e, e)\}$. Its digraph is shown in Figure 5.22.

Figure 5.22

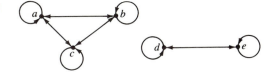

The relation defined on the integers by the relation $\leq$ is not an equivalence relation because it fails to be symmetric.

Example 2. Let E be the alphabet $\{a, b\}$. The relation R defined on E^* by $x\ R\ y$ if and only if x and y are words of the same length is an equivalence relation. So $aa\ R\ ba$ and $bbb\ R\ aaa$. The empty word is related only to itself: $\Lambda\ R\ \Lambda$.

Example 3. The relation R defined on the power set of the natural numbers by $A\ R\ B$ if and only if A and B are subsets of $\mathbf{N}$ having the same cardinality is an equivalence relation. The sets $\{1, 3, 5\}$ and $\{13, 35, 67\}$ are related. The set of odd natural numbers and the set of even natural numbers are related because both are countably infinite. In fact, any two infinite subsets of $\mathbf{N}$ are related because all such sets are countably infinite.

Example 4. We define the relation R_5 on the integers $\mathbf{Z}$ as follows: $m\ R_5\ n$ if and only if $m - n$ is divisible by 5—that is, if and only if we can find an integer q such that $m - n = 5q$. For example, we have $18\ R_5\ 3$ since $18 - 3 = 15$, and 15 is divisible by 5. Similarly, we have $-2\ R_5\ 8$ since $-2 - 8 = -10$, and $-10 = -2 \cdot 5$. All the numbers within each of the following sets are related:

$$E(0) = \{\ldots, -10, -5, 0, 5, 10, 15, \ldots\}$$
$$E(1) = \{\ldots, -9, -4, 1, 6, 11, \ldots\}$$
$$E(2) = \{\ldots, -8, -3, 2, 7, 12, \ldots\}$$
$$E(3) = \{\ldots, -7, -2, 3, 8, 13, \ldots\}$$
$$E(4) = \{\ldots, -6, -1, 4, 9, 14, \ldots\}$$

We now prove that R_5 is an equivalence relation.

(a) (Reflexivity) Given any integer n, we have that $n - n = 0$, and 0 is divisible by 5. Thus $n\ R_5\ n$ and R_5 is reflexive.

(b) (Symmetry) Suppose that $m\ R_5\ n$ so that $m - n$ is divisible by 5. Its negative, $-(m - n)$, is also divisible by 5. Since $-(m - n) = n - m$, we have $n\ R_5\ m$. Thus R_5 is symmetric.

(c) (Transitivity) Suppose that $m\ R_5\ n$ and that $n\ R_5\ s$. We know that $m - n = 5k$ and that $n - s = 5j$ for some integers j and k. Thus $(m - n) + (n - s) = (m - s) = 5(k + j)$. Since $m - s$ is also divisible by 5, we have that $m\ R_5\ s$. Thus R_5 is transitive.

We now generalize the R_5 relation. Let m be any positive integer. We define R_m on $\mathbf{Z}$ as follows: $x\ R_m\ y$ if and only if $x - y$ is divisible by m. That is, $x\ R_m\ y$ if and only if there is an integer q such that $x - y = qm$. If x and y are integers and $x\ R_m\ y$, we write $x \equiv y \pmod{m}$ and we say that x is **congruent** to y (modulo m). The integer m is called the **modulus** of the relation. So, for example, we have $4 \equiv 10 \pmod{3}$ since $4 - 10$ is divisible by 3. Also $1 \equiv -1 \pmod{2}$ and $23 \equiv 42 \pmod{19}$. To prove that the mod m relation is an equivalence relation, we need only replace 5 by m in proof that R_5 is an equivalence relation. (We leave the details as an exercise.)

The mod m relation is the basis for modular arithmetic, which we shall develop more fully in Chapter 8. Modular arithmetic is an old and important tool in the study of number theory, but it has many current applications as well. At the end of this chapter we shall see an application to computer arithmetic. In Chapter 8 we shall see how congruences are used for cryptic coding.

Example 5.

(a) To solve the congruence

$$(1) \qquad\qquad x + 1 \equiv 3 \quad (\text{mod } 5)$$

we look for the set S of all integers for which expression (1) holds. The solution set is $S = \{\ldots, -3, 2, 7, 12, \ldots\}$. To see this, we note that $x + 1 \equiv 3$ (mod 5) if and only if $(x + 1) - 3 = 5q$ for some integer q. Simplifying, we find that $x - 2 = 5q$, which is to say that $x \equiv 2$ (mod 5). Thus S is the set of all integers that are congruent to 2 (modulo 5).

(b) The solution set to the congruence $2x \equiv 1$ (mod 4) is empty because $2x - 1$ is always odd and hence never a multiple of 4.

(c) To solve the congruence

$$(2) \qquad\qquad 2x \equiv 4 \quad (\text{mod } 5)$$

we find all integer values of x such that $2x - 4 = 5q$ for some integer q. Because $2x - 4$ is even and 5 is odd, we know that q must be even and so $q = 2k$ for some integer k. Thus we have $2(x - 2) = 2 \cdot 5k$ or $x - 2 = 5k$. We find that x is a solution to our congruence when $x \equiv 2$ (mod 5). The solution set is $\{\ldots, -3, 2, 7, \ldots\}$.

(d) To solve the congruence

$$(3) \qquad\qquad 2x \equiv 4 \quad (\text{mod } 8)$$

we must find all integer values of x such that $2x - 4 = 8q$ for some integer value of q, or $x - 2 = 4q$ for some integer value of q. Thus the solution set is $\{\ldots, -2, 2, 6, 10, 14, \ldots\}$, the set of all integers congruent to 2 (modulo 4). (Note that had we simply canceled 2 from both sides of the original congruence, we would have missed the solutions $x = 6, 14, 22, \ldots$.)

Example 6. Let A be a finite set and let $f : A \to A$ be a permutation of A—that is, a bijection from A to itself. The function f defines an equivalence relation $\mathcal{R}_f$ on A as follows: $x \, \mathcal{R}_f \, y$ if we can find $n > 0$ such that $y = f^n(x)$. (Recall that f^n is the function f composed with itself n times.) For example, let $S = \{r, s, t, u, v, w, z\}$ and let g be the permutation of A defined by the diagram shown in Figure 5.23.

Figure 5.23

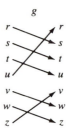

Then $r \, \mathcal{R}_g \, t$ since we have $t = g^2(r)$. Also $s \, \mathcal{R}_g \, s$ since $s = g^4(s)$. The reader will find that all the elements of the subset $\{r, s, t, u\}$ are related to each other and that all the elements of the subset $\{v, w, z\}$ are related to each other. The digraph is shown in Figure 5.24. The relation $\mathcal{R}_f$ is the transitive closure of the relation R_f defined in Section 5.1. (For the definition of transitive closure, see Exercise 8 at the end of Section 5.2.)

Figure 5.24

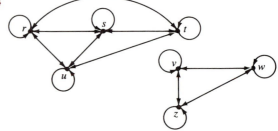

We now show that $\mathcal{R}_f$ establishes a symmetric and reflexive relation on A. (We leave it as an exercise to show that this relation is also transitive and is thus an equivalence relation.) Let a be an element of A and consider the sequence $\{a, f(a), f^2(a), f^3(a), \ldots\}$. (For S, g, and the element t as above, we would obtain the sequence $\{t, u, r, s, t, u, r, \ldots\}$.) The elements in the sequence must repeat because A is a finite set. The first element to make a second appearance is a itself. (Note that in the sequence $\{t, u, r, s, t, u, \ldots\}$ the element t is the first element in the sequence and the first element to be repeated.) To see that this must be the case, suppose that x is the first element to appear a second time. Suppose that it first appears as $x = f^j(a)$ and that it next appears as $f^k(a)$ so that $j < k$. Now if $j \neq 0$, we can write $x = f \circ f^{j-1}(a)$ and $x = f \circ f^{k-1}(a)$. Since the function f is injective, we have that $y = f^{j-1}(a) = f^{k-1}(a)$. But this says that the element y occurs in the sequence before the element x and is repeated before the second appearance of x. This contradicts our assumption that x is the first element to appear a second time. So it must be the case that $j = 0$ and that $x = a$. Let $f^m(a)$ be the second appearance of a. Since $a = f^m(a)$, we have proved that $a \, \mathcal{R}_f \, a$, and the relation is thus reflexive.

Now suppose that $a \, \mathcal{R}_f \, b$ and let q be the smallest integer such that $b = f^q(a)$. Then $q < m$. To show symmetry, we must show that $b \, \mathcal{R}_f \, a$ or that

$a = f^j(b)$ for some integer j. Let $j = m - q$. Then $f^j(b) = f^{m-q}(b) = f^{m-q}[f^q(a)]$ $= f^m(a) = a$. Thus we indeed have that $b \, \mathcal{R}_f \, a$ since $a = f^j(b)$, and we conclude that $\mathcal{R}_f$ is a symmetric relation.

Section 5.3	Exercises

1. Decide which of the following digraphs defines an equivalence relation.

(a)

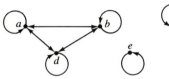

(b)

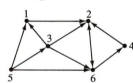

(c)

2. Determine which of the following defines an equivalence relation.

(a) The relation on the set of rational numbers defined by $x \, R \, y$ if $x - y$ is an integer

(b) The relation R on the x-y plane defined by $(x, y) \, R \, (s, t)$ whenever $x - s$ or $y - t$ is an integer

(c) $A = \{a, b, c\}$ and R is the relation on $\mathscr{P}(A)$ defined by $S \, R \, T$ if $S \cap T$ is not empty

3. Prove that the following defines an equivalence relation on the x-y plane: $(x, y) \, R \, (s, t)$ if $s - x$ and $t - y$ are both integers.

4. Prove that the mod m relation is an equivalence relation.

5. Find the solution sets of each of the following congruences.

(a) $x - 3 \equiv 2 \pmod{7}$ (b) $x + 5 \equiv 4 \pmod{9}$

(c) $2x \equiv 3 \pmod{6}$ (d) $2x \equiv 3 \pmod{5}$

(e) $3x \equiv 3 \pmod{6}$

6. Let A be a finite set and let $f : A \to A$ be a permutation of A. Prove that $\mathcal{R}_f$ is a transitive relation. (See Example 6 of this section.)

7. For each set A and permutation f that follow, draw the digraph for $\mathcal{R}_f$.

(a)   (b)

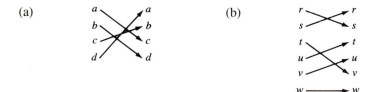

Section 5.4   Equivalence Classes

We saw in the preceding section that the mod 5 relation partitioned the set of integers into five disjoint subsets: $E(0)$, $E(1)$, $E(2)$, $E(3)$, and $E(4)$. We call these subsets the **equivalence classes** of the mod 5 relation. Every integer is in some equivalence class. Within each equivalence class, every pair of elements is related and no element in one equivalence class is related to an element in a different class. Every equivalence relation establishes a similar partition on its ambient set.

> **DEFINITION.** Let R be an equivalence relation on a set A. Let x be an element of A. The set $E(x) = \{y : y$ is an element of A and $y\,R\,x\}$ is called the **equivalence class of x**.

Example 1. Let R be the equivalence relation defined on the set $\{a, b, c, d, e, f\}$ by the digraph shown in Figure 5.25. The equivalence class of a is $E(a) = \{a, b, c\}$. The equivalence class of d is $E(d) = \{d, e\}$. $E(f) = \{f\}$. Note that $E(a) = E(b) = E(c)$ and that $E(d) = E(e)$.

Figure 5.25

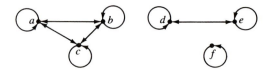

Example 2. Let $E = \{a, b\}$ and let R be the relation on E^* in which $x \mathrel{R} y$ when x and y are words of the same length. There is an infinite number of equivalence classes for this relation. Some of them are

$\{\Lambda\}$ $\{a, b\}$ $\{aa, ab, ba, bb\}$

$\{aaa, aab, abb, aba, baa, bab, bba, bbb\}$

Example 3. We list the equivalence classes of the elements 0, 1, 2, and 3 of the mod 3 relation defined on the integers:

$$E(0) = \{\ldots, -6, -3, 0, 3, 6, \ldots\}$$

$$E(1) = \{\ldots, -5, -2, 1, 4, 7, \ldots\}$$

$$E(2) = \{\ldots, -4, -1, 2, 5, 8, \ldots\}$$

$$E(3) = \{\ldots, -6, -3, 0, 3, 6, \ldots\}$$

We note that the sets $E(0)$ and $E(3)$ are identical, whereas $E(0)$ and $E(1)$ are disjoint. Any two equivalence classes of a given equivalence relation are either disjoint or identical, as the following proposition shows.

Proposition 1. Suppose that R is an equivalence relation on a set A, and let x and y be elements of A. Then $E(x)$ and $E(y)$ are either disjoint or equal.

PROOF: Suppose that $E(x)$ and $E(y)$ are not disjoint and that z is an element of both equivalence classes. We show that $E(x)$ is contained in $E(y)$. Let s be an element of $E(x)$. Then we have that $s \mathrel{R} x$. Since $x \mathrel{R} z$, we have $s \mathrel{R} z$ by transitivity. Furthermore, since $z \mathrel{R} y$, we have $s \mathrel{R} y$ by transitivity again. Thus s is an element of $E(y)$, and we have shown that $E(x)$ is contained in $E(y)$. A similar argument shows that $E(y)$ is contained in $E(x)$. Thus $E(x)$ is equal to $E(y)$.

Example 4. Let A be a finite set and let f be a permutation of A. The equivalence classes of $\mathcal{R}_f$ are called the **orbits** of the permutation f. For example, let $A = \{a, b, c, d, e, f, g\}$ and let $f : A \rightarrow A$ be as shown in Figure 5.26. The orbits of $\mathcal{R}_f$ are $\{a, b, c\}$, $\{d, e\}$, and $\{f, g\}$. The orbits of a permutation f enable us to

Figure 5.26

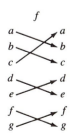

write f in a form called the **cycle decomposition** of f. To do this, we first select an orbit E of f and an element x of E. We then form the sequence or "cycle" $(x, f(x), f^2(x), \ldots, f^n(x))$, stopping when $f^{n+1}(x) = x$. (We note that each element of E appears exactly once in the cycle.) This sequence is called a **cycle** of f. If y is an element of a cycle, then $f(y)$ is the next element to the right unless y is the last element of the cycle. In this case, $f(y)$ is the first element of the cycle. We do this for each orbit of f to obtain the cycle decomposition of f. Thus a cycle decomposition of f given above is $(a, b, c)(d, e)(f, g)$. The cycle decomposition of a permutation is not unique. Another cycle decomposition of f is $(b, c, a)(g, f)(e, d)$.

Example 5. Let $A = \{a, b, c, d, e, f, g\}$ and let g be the permutation defined by the diagram shown in Figure 5.27. The cycle decomposition of g is $(a, c, d, f, b, g)(e)$.

Figure 5.27

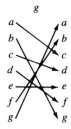

Example 6. If h is a permutation of the set $S = \{s, t, u, v, w, x, y, z\}$ with cycle decomposition given by $(s, t, w, u)(v, x)(y, z)$, then h is also given by the diagram shown in Figure 5.28.

Figure 5.28

To determine an equivalence class E of an equivalence relation R given on a set A, we need only identify one element x of E. That element is called a **representative** of E. A subset of A containing exactly one representative of each of the equivalence classes of R is called a **complete set of representatives of R.**

Example 7. A complete set of representatives of the mod 5 relation on the integers is the set $S = \{0, 1, 2, 3, 4\}$. Another complete set of representatives is the set $Q = \{5, 11, 17, 24, 13\}$. Each of the sets S and Q contains exactly one element from each equivalence class of the mod 5 relation. The usual complete set of representatives of the mod m relation is the set $\{0, 1, \ldots, m - 1\}$. Here we identify each equivalence class by the smallest positive remainder that each of its elements has after division by m.

Example 8. Let R be the equivalence relation defined on the x-y plane by $(x, y)\ R$ (s, t) if both $x - s$ and $y - t$ are integers. A complete set of representatives is the set $\{(x, y) : 0 \le x < 1 \text{ and } 0 \le y < 1\}$. To see this, note that for every real number z, we have $0 \le (z - \lfloor z \rfloor) < 1$ and that $(x, y)\ R\ (x - \lfloor x \rfloor, y - \lfloor y \rfloor)$. The equivalence class of $(0, 0)$ is $\mathbf{Z} \times \mathbf{Z}$—that is, the set of all corners on the square grid shown in Figure 5.29. The equivalence class of $(1/2, 1/2)$ is the set of corners on the grid of broken lines.

Figure 5.29

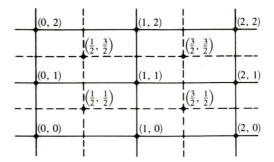

Section 5.4   Exercises

1. List the elements in the equivalence class of the integer 3 under the mod 7 relation. Do the same for -1 under the mod 9 relation.

2. How many equivalence classes are there under the mod 2 relation? In general, into how many equivalence classes does the mod m relation partition $\mathbf{Z}$?

3. Let $E = \{a_1, a_2 \ldots, a_n\}$ be an alphabet, and let R by the relation on E^* where two words are related if they have the same length. How many elements are there in each of the equivalence classes of R? Find a complete set of representatives of R.

4. Let R be an equivalence relation on a set A, and let $E(x)$ and $E(y)$ be equivalence classes of R. Prove that $E(x) = E(y)$ if and only if $x\ R\ y$.

5. Let R be the equivalence relation on the x-y plane given by $(x, y) R (s, t)$ if $x = s$. Find the equivalence classes of $(1, 2)$ and $\left(0, \frac{1}{2}\right)$ in the plane. Find a complete set of representatives of R.

6. Let R be the equivalence relation on the x-y plane given by $(x, y) R (s, t)$ if $x - s$ is an integer multiple of 2 and $y - t$ is an integer multiple of 3. Find the equivalence class of $(-1, 2)$ in the plane, and find a complete set of representatives of R.

7. Let f be the permutation of the set $\{a, b, c, d, e, f, g, h\}$ defined by the accompanying diagram. Find the orbits of f. Find a complete set of representatives of $\mathcal{R}_f$. Find the cycle decomposition of f.

Figure for Exercise 7

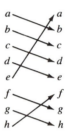

8. Let $A = \{a_1, a_2, \ldots, a_n\}$. Find the cycle decomposition of the identity function on A.

9. Draw the diagram of the permutation f on the set $\{a, b, c, d, e, f, g, h, i\}$ whose cycle decomposition is $(a, c, d)(e, g, i)(h)(b, f)$.

10. Find the cycle decomposition of $f \circ f$ for the permutation f given in Exercise 9.

Section 5.5   Order Relations

An equivalence relation on a set is a generalization of the relation $x = y$. In this section we discuss relations that are generalizations of the relation $x \leq y$ on the integers. Such relations are called **partial orders**.

> **DEFINITION.** A relation R on a set A is called a **partial order** if it is reflexive, transitive, and antisymmetric. The set A under the relation R is called a **partially ordered set** or a **poset**.

If R establishes a partial order on a set A, we often write $x \lesssim y$ whenever $x R y$. If $x \lesssim y$, we say that x precedes y.

Example 1. Let $A = \{a, b, c, d, e\}$ and $R = \{(a, a), (a, b), (b, b), (b, c),$ $(a, c), (c, c), (a, d), (d, d), (d, e), (e, e), (a, e)\}$. This relation is transitive, reflexive, and antisymmetric. Hence it is a partial order on A. A digraph for this relation is shown in Figure 5.30. Note that not all pairs are comparable. For instance, we have neither $b \lesssim d$ nor $d \lesssim b$. Hence the term "partial order."

Figure 5.30

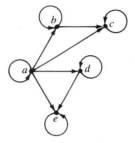

If a relation R is known to be a partial order, we can represent it with a simpler Hasse diagram rather than a full digraph. To do this, we omit the arrows that go from an element to itself because the relation is known to be reflexive. Also we know the relation is transitive, and so we omit the arrows implied by transitivity. That is, we omit the broken arrows in the following configuration:

Finally we omit the arrow heads under the convention that if $x R y$, then y appears above x on the page. The Hasse diagram for the partial order given in Example 1 is shown on the right in Figure 5.31.

Figure 5.31

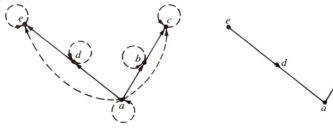

Example 2. Let S be a set and let $\mathscr{P}(S)$ be its power set. Then $\mathscr{P}(S)$ is a poset under the inclusion relation $A \subset B$, where A and B are subsets of S. Thus if $S = \{a, b, c\}$, the Hasse diagram for $\mathscr{P}(S)$ under this partial order is as shown in Figure 5.32.

Figure 5.32

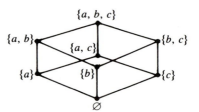

Example 3. Let $A = \{1, 2, 3, 4, 6, 8, 12, 24\}$. The relation $\preceq$ defined by $x \preceq y$ if and only if x divides y is a partial order on A. We usually write $x \mid y$ for $x \preceq y$ in this relation. The Hasse diagram for this relation is shown in Figure 5.33.

Figure 5.33

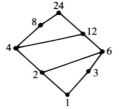

Example 4. Let $E = \{a, b\}$ and let $\preceq$ be the relation on E^* defined as follows: if x and y are words in E^*, then $x \preceq y$ if x is a prefix of y. That is, $x \preceq y$ if $y = xz$ for some word z in E^*. (We leave as an exercise to show that this relation is a partial order.) Thus we have $aa \preceq aaba$ but neither $aba \preceq bab$ nor $bab \preceq aba$. The Hasse diagram for this relation on words of length 3 or less is shown in Figure 5.34.

Figure 5.34   aaa   aab   aba   abb   baa   bab   bba   bbb

aa   ab   ba   bb

a   b

Λ

　　As Examples 1 through 4 show, it is often the case that if we choose a pair of elements x and y, then neither $x \preceq y$ nor $y \preceq x$ holds. Not all pairs of elements can be compared. However, if we look at the set of integers under the relation $\leq$ and choose any pair of elements x and y, it is always the case that either $x \leq y$ or $y \leq x$ holds. We say that this set is totally ordered under the relation $\leq$.

> **DEFINITION.** A partial order $\preceq$ on a set A is said to be a **total order** if for every pair of elements x and y, either $x \preceq y$ or $y \preceq x$ holds. The set A is said to be **totally ordered** under this relation.

The relation given in Example 3 is not a total order, but {1, 2, 4, 8} and {1, 3, 6, 12} are totally ordered subsets of *S*. (There are other totally ordered subsets of *S*.)

Example 5. The English alphabet taken in the usual order is a totally ordered set and the (finite) set of English words taken in alphabetical order is also totally ordered: we can always determine which of two words precedes the other in alphabetical order. We now describe **lexicographic order,** a generalization of alphabetical order. Suppose that *E* is a finite totally ordered alphabet and that *x* and *y* are words in *E**. Then *x* precedes *y* in lexicographic order if *any one* of the following conditions holds:

(1) *x* is the empty word.

(2) The first symbol of *x* precedes the first symbol of *y*.

(3) $x = ws$ and $y = wt$, where *w* is the longest common prefix of *x* and *y*, and *s* precedes *t* in lexicographic order.

Thus if we let *E* be the set {*a*, *b*, *c*, *t*, *o*, *e*, *i*} and order *E* by the Hasse diagram shown in Figure 5.35, then the following words are listed in lexicographic order: *tab, cob, cat, bit, bite*.

Figure 5.35

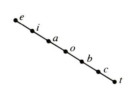

If we let $D = \{0, 1, 2, 3, 4, 5, 6, 7, 8, 9\}$ be the set of digits taken in the usual order and let D_3 be the set of words (or numbers) over *D* of length 3, then lexicographic order is identical to numerical order on this set.

> **DEFINITION.** Suppose that *A* is a partially ordered set under the relation $\lesssim$, and suppose that *t* is an element of *A* such that, for each element *x* of *A*, we have $x \lesssim t$. We call *t* the **greatest element** of *A*. Similarly, if *s* is an element of *A* such that $s \lesssim x$ for each *x* in *A*, we say that *s* is the **least element** of *A*.

Example 6. Consider the set $A = \{1, 2, 3, 4, 6, 8, 12, 24\}$ under the relation $x \mid y$ as given in Example 3. Then 24 is the greatest element and 1 is the least element of this poset. Let $B = \{2, 3, 4, 5, 6, 7, 8\}$ under the same relation. The poset *B* has neither a greatest nor a least element.

Example 7. The power set of {a, b, c} under the containment relation $\subset$ has the empty set as its least element and the set {a, b, c} as its greatest element.

Example 8. Let E^* and $\preceq$ be as given in Example 4. This set has no greatest element, but the empty word is its least element. The subset of E^* of all words of length 3 has neither a greatest element nor a least element.

Example 9. The set of natural numbers under the relation $\leq$ has no greatest element, but 0 is its least element. Every subset of **N** has a least element.

A subset S of the natural numbers may or may not have a greatest element, but it always has a least element under the $\leq$ relation. We call such a set a well-ordered set.

DEFINITION. A totally ordered set A is said to be a **well-ordered set** if every subset of A has a least element.

Although the set of natural numbers is well-ordered, the set of integers **Z** is not. The set **Z** itself does not have a least element. If we let T be the poset of all rational numbers $r \geq 0$ under the relation $\leq$, we find that this set is not well-ordered. The entire set T itself has 0 as a least element, but its subset $S = \{x : 0 < x < 1$ and x is rational} does not have a least element. To see this, let r be any element of the subset S. The element r cannot be the least element of S since the element $s = r/2$ is also in S, and $s < r$. Thus the subset S has no least element and T fails to be well-ordered.

Example 10. Suppose that E is a finite totally ordered set and that S is a *finite* language over E. Then S is well-ordered under lexicographic order. However, if $E = \{a, b\}$ with $a \preceq b$, the set E^* is not well-ordered. The infinite subset {b, ab, aab, $aaab$, . . .} has no least element.

Consider the partial order $\preceq$ defined on the set A by the Hasse diagram shown in Figure 5.36. It is not totally ordered because, for instance, neither $d \preceq e$ nor $e \preceq d$ holds. However, if S is any subset of A, we can find some x in A such that $s \preceq x$ for each s in S. For instance, if $S = \{h, i\}$ and $x = e$, then $h \preceq e$ and $i \preceq e$. If $S = \{d, e, f\}$, then we can let $x = a$. Such an element x is called an upper bound of S. Similarly, for each subset S we can find y in A such that $y \preceq s$ for each s in S. For instance, if $S = A$ and $y = j$, then $y \preceq s$ for each s in A. Such an element y is called a lower bound of S.

Figure 5.36

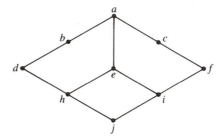

> **DEFINITION.** Let A be a poset under the relation $\leq$. Let S be a subset of A. An element x in A is called an **upper bound** of S if $s \leq x$ for each element s of S. An element y of A is called a **lower bound** of S if $y \leq s$ for each element s of S.

Upper and lower bounds of a set are not unique. In the preceding poset, both e and a are upper bounds of the set $\{h, i\}$.

Example 11. Let **N** be a poset under the relation $\leq$. Let $S = \{1, 2, 3, 4, 5\}$. Then any integer $n \geq 5$ is an upper bound of S. The only lower bound of S is 1. The subset of even natural numbers has no upper bound, but 2 and 1 are both lower bounds.

Example 12. Let A be a set and let $\mathcal{P}(A)$ be the power set of A under the inclusion relation. Then the empty set is a lower bound for any subset of $\mathcal{P}(A)$, and the set A is an upper bound for any subset of $\mathcal{P}(A)$.

Example 13. Let **N** be the set of natural numbers under the relation $x \, R \, y$ if y is divisible by x. Let $S = \{m_1, m_2, \ldots, m_n\}$ be a finite subset of **N**. The product $m_1 m_2 \cdots m_n$ is an upper bound of S. The integer 1 is a lower bound of any subset of **N**. For example, if $S = \{2, 3, 5\}$, then 30 is an upper bound of S and 1 is a lower bound. If $T = \{6, 9, 12\}$, then 648 is an upper bound of T (but so is 36), and both 1 and 3 are lower bounds.

In Example 11 we saw that any integer $n \geq 5$ is an upper bound of the set $\{1, 2, 3, 4, 5\}$. No integer $m < 5$ is an upper bound of this set, so that 5 is the smallest or least upper bound of $\{1, 2, 3, 4, 5\}$. Similarly, if we take the natural numbers under the division relation as in Example 13 and let $K = \{4, 6\}$, then 2 is a lower bound of K and there is no other lower bound q divisible by 2—that is, for which $2 \mid q$. Under this order, 2 is the greatest lower bound.

> **DEFINITION.** Let A be a poset under the relation $\leqslant$, and let S be a subset of A. An element z of A is called the **least upper bound** of S if z is an upper bound of S, and whenever x is any other upper bound of S we have $z \leqslant x$. We write $z = \mathrm{lub}\, S$. An element w is called the **greatest lower bound** of S if w is a lower bound of S, and whenever y is any other lower bound we have $y \leqslant w$. We write $w = \mathrm{glb}\, S$.

Example 14. Consider the natural numbers under $\leq$. The least upper bound of the set $\{4, 5\}$ is 5, and the greatest lower bound of this set is 4. No infinite subset of **N** has an upper bound, but every subset of **N** has a greatest lower bound—namely its least element.

Example 15. Consider the real numbers under the relation $\leq$. Let $S = \{x : 0 < x < 1\}$. Then $\mathrm{lub}\, S = 1$ and $\mathrm{glb}\, S = 0$. We note that the least upper bound and the greatest lower bound of a set S are not necessarily members of S itself.

Although not every subset of the real numbers has a least upper bound or a greatest lower bound, every pair of elements does. We distinguish posets that behave similarly by calling them lattices.

> **DEFINITION.** Let A be a poset under the relation $\leqslant$. We call A a **lattice** if whenever S is a subset of A containing exactly two elements, then S has both a least upper bound and a greatest lower bound.

Example 16. Consider the poset defined by the Hasse diagram shown in Figure 5.37.

Figure 5.37

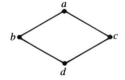

Then we have

$$
\begin{array}{ll}
\mathrm{lub}\{d, b\} = b & \quad \mathrm{glb}\{d, b\} = d \\
\mathrm{lub}\{d, c\} = c & \quad \mathrm{glb}\{d, c\} = d \\
\mathrm{lub}\{d, a\} = a & \quad \mathrm{glb}\{d, a\} = d \\
\mathrm{lub}\{b, c\} = a & \quad \mathrm{glb}\{b, c\} = d \\
\mathrm{lub}\{a, b\} = a & \quad \mathrm{glb}\{a, b\} = b \\
\mathrm{lub}\{a, c\} = a & \quad \mathrm{glb}\{a, c\} = c
\end{array}
$$

Thus A is a lattice because every pair of elements has a least upper bound and a greatest lower bound.

Example 17. The poset defined by the Hasse diagram shown in Figure 5.38 is not a lattice since set $\{s, t\}$ has no upper bound. Also $\{w, q\}$ has no lower bound.

Figure 5.38

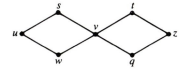

Example 18. The reals, the rationals, the integers, and the natural numbers are all lattices under $\leq$.

Example 19. Consider the natural numbers under the relation $x \, R \, y$ if y is divisible by x. Let m and n be natural numbers. Then lub$\{m, n\}$ is the least common multiple of m and n, and glb$\{m, n\}$ is the greatest common divisor of m and n. Thus **N** is a lattice under this relation.

The operations of finding the least upper bound and the greatest lower bound of a pair of elements of a lattice share many of the characteristics of the "or" and the "and" operations of logic, respectively, as Theorem 1 will show. For this reason we use the notation given in the following definition.

> **DEFINITION.** Let x and y be elements of a lattice A. Then we denote lub$\{x, y\}$ by $x \vee y$ and call it the **join** of x and y. We denote glb$\{x, y\}$ by $x \wedge y$ and call it the **meet** of x and y.

Example 20. Let A be a set. Then $\mathcal{P}(A)$ is a lattice under containment. If S and T are elements of $\mathcal{P}(A)$, then $S \vee T = S \cup T$ and $S \wedge T = S \cap T$. That is to say, the join of S and T is the union of S and T, and the meet of S and T is the intersection of S and T.

Example 21. Let $n > 0$ be an integer and let S be the set of all strings over $\{0, 1\}$ of length n. Then each string in P is of the form $x = x_1 x_2 \cdots x_n$, where each x_j is either 0 or 1. The set S is a lattice under the partial order defined as follows: $x \leq y$ if whenever $x_k = 1$, then y_k is also 1. Thus $1001 \leq 1101$ and $0000 \leq 1000$, but neither $1010 \leq 0101$ nor $0101 \leq 1010$ holds. Let x and y be strings in S. We can find the join of x and y—namely x $\vee$ y—by taking the pairwise disjunction of the symbols in x and y (interpreted as truth values). Thus $1001 \vee 0101 = 1101$. Similarly, we can find the meet of x and y—namely $x \wedge y$—by taking the pairwise conjunction of the symbols in x and y. Thus $1001 \wedge 0101 = 0001$.

In Chapter 3 we saw how to put the elements of the power set of a finite set $A = \{a_1, a_2, \dots, a_n\}$ into one-to-one correspondence with the set of strings of 0's and 1's of length n. If S and T are subsets of A, and x_S and x_T are the strings that represent S and T, then $S \subset T$ if and only if $x_S \lesssim x_T$. The analogy continues: the set $S \cup T$ is represented by the join $x_S \vee x_T$, and the intersection $S \cap T$ is represented by the meet $x_S \wedge x_T$. Forming the join and the meet of strings of 0's and 1's is not difficult to implement on the computer, and it is not difficult to check the relation $x \lesssim y$. Thus the lattice of strings of 0's and 1's provides a convenient data structure for the lattice $\mathscr{P}(A)$.

Theorem 1.   Let A be a lattice and let x, y, and z be elements of A. Then we have

(ai) $x \vee x = x$ (bi) $x \wedge x = x$

(aii) $x \vee y = y \vee x$ (bii) $x \wedge y = y \wedge x$

(aiii) $x \vee (y \vee z) = (x \vee y) \vee z$ (aiii) $x \wedge (y \wedge z) = (x \wedge y) \wedge z$

PROOF: We prove (ai), (aii), and (aiii).

(ai) Since $x \lesssim x$, we know that x is an upper bound of the set $\{x\}$. If y is also an upper bound and if $y \lesssim x$, then we have that $x \lesssim y$ and $y \lesssim x$. By antisymmetry we have $y = x$, so that x is indeed the least upper bound of the set $\{x\}$.

(aii) Let $w = x \vee y$ so that $w = \mathrm{lub}\{x, y\}$. Since the set $\{x, y\}$ is equal to the set $\{y, x\}$, we have that $w = \mathrm{lub}\{y, x\}$—that is, $w = y \vee x$.

(aiii) Let $s = x \vee (y \vee z)$ and let $t = (x \vee y) \vee z$. We have that $x \lesssim s$ and $(y \vee z) \lesssim s$. Since $y \lesssim (y \vee z)$ and $z \lesssim (y \vee z)$, we have that $y \lesssim s$ and $z \lesssim s$ by transitivity. Since $y \lesssim s$ and $z \lesssim s$, we have that s is an upper bound of $\{y, z\}$, and so $y \vee z \lesssim s$. Since $x \lesssim s$ and $y \vee z \lesssim s$, we have that $x \vee (y \vee z) \lesssim s$—that is, $t \lesssim s$. Similarly (starting on the right), we can show that $s \lesssim t$. Thus by antisymmetry we have that $s = t$.

We can ask whether or not other logical properties have analogous statements for the join and meet operations. The reader will indeed find some such analogies in the exercises, but not all the familiar situations extend. For instance, we cannot even state a form of DeMorgan's laws because we do not in general have the analogue of complementation. The distributive laws can be formulated as follows:

(1) $x \wedge (y \vee z) = (x \wedge y) \vee (x \wedge z)$

(2) $x \vee (y \wedge z) = (x \vee y) \wedge (x \vee z)$

but they do not in general hold. For instance, expression (1) fails for the lattice given by the Hasse diagram shown in Figure 5.39. We see that $d \wedge (b \vee c)$ $= d \wedge e = d$, whereas $(d \wedge b) \vee (d \wedge c) = b \vee a = a$.

Figure 5.39

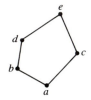

Section 5.5	Exercises

1. Determine which of the following digraphs define a partial order. Draw the Hasse diagram for each that does.

   Figure for Exercise 1

   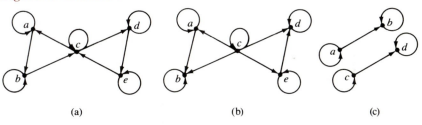

   (a)   (b)   (c)

2. Draw the Hasse diagram for each of the following partially ordered sets.
   (a) The power set of $\{a, b, c, d\}$ under inclusion
   (b) The set $\{1, 2, 3, \ldots, 12\}$ under $x \, R \, y$ whenever $x \mid y$
   (c) The set $\{1, 2, 3, 4, 5\}$ under $x \, R \, y$ whenever $x \geq y$

3. Determine which of the following relations defines a poset, which defines a totally ordered set, and which defines a well-ordered set.
   (a) $\mathbf{N} \times \mathbf{N}$ where $(x, y) \, R \, (s, t)$ whenever $x \leq s$
   (b) $\mathbf{R} \times \mathbf{R}$ where $(x, y) \, R \, (s, t)$ whenever $x^2 + y^2 \leq s^2 + t^2$
   (c) $\mathbf{Z}$ where $m \, R \, n$ whenever $|m| < |n|$ or, if $|m| = |n|$, $m \leq n$

4. Suppose that $E = \{@, \#, \$\}$ and that E is totally ordered as follows: $@ \lesssim \#$ $\lesssim \$$.
   (a) List the following strings in lexicographic order: $\$$, $\#\#$, $@\#\$\$$, $\#@\$\$$, $\$\$@@\#\#$, $@@@@$, $@\#\#\$$.

(b) Let R be the relation defined on E^* as follows: if x and y are words in E^*, then $x\,R\,y$ if the length of x is less than the length of y. If the length of x is equal to that of y, then $x\,R\,y$ if $x = y$ or if x precedes y in lexicographic order. List the words given in part (a) in the order established by R. What is the least element of E^* under R?

5. Consider the poset whose Hasse diagram is as follows:

Figure for Exercise 5

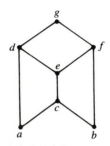

Find the lub and glb of each of the following sets, where possible.

(a) $\{a, b\}$ (b) $\{a, c, f\}$ (c) $\{d, e, f\}$ (d) $\{g, d, f\}$

6. Find the lub and the glb of each of the following subsets.

(a) $S = \{x : -1 < x \leq 2\}$, where S is a subset of R under $\leq$

(b) $S = \{\{a, b\}, \{b, c\}\}$, where S is a subset of $\mathcal{P}(\{a, b, c\})$ under the inclusion relation

(c) $S = \{01101, 11000, 00000, 01011\}$, where S is a subset of the set of all strings of length 5 over $\{0, 1\}$ under the relation described in Example 21 in Section 5.5

7. Decide which of the following Hasse diagrams defines a lattice on the set $\{a, b, c, d, e, f, g\}$.

Figure for Exercise 7

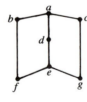 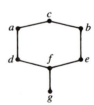

8. A lattice on the set $A = \{a, b, c, d, e, f, g\}$ is given in the accompanying figure. Find:

$f \wedge g$ $d \wedge e$ $d \wedge b$ $a \wedge (b \wedge c)$ $a \wedge (b \vee c)$

$f \vee g$ $d \vee e$ $d \vee b$ $a \vee (b \vee c)$ $a \vee (b \wedge c)$

Figure for Exercise 8

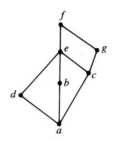

9. Prove that any totally ordered set is a lattice.

10. Let A be a lattice under the relation $\leq$, and let a and b be elements of A. Prove that:

    (a) $a \wedge (a \vee b) = a$ (b) $a \vee (a \wedge b) = a$

Bridge to Computer Science

5.6 Two's-Complement Arithmetic (optional)

In the bridge in Chapter 1 we saw how to design logic circuits for both bit complementation and binary addition. In this bridge we look at "two's-complement" arithmetic, a method used in computer technology to implement the operation of integer subtraction in terms of bit complementation and addition. It is based on the congruence $-x \equiv m - x \pmod{m}$.

With n bits we can represent the numbers from 0 to $2^n - 1$ in base-2 or binary notation. If our strings can be no longer than n bits, we are constrained to do modulo 2^n arithmetic. (In most of the examples that follow, we take $n = 4$ and hence work modulo 2^4. In many computer systems, $n = 16$.)

If x is an n-bit binary number, the **one's complement** of x, denoted x', is the number obtained from x by replacing each of the 1's of x with 0 and each of the 0's with 1. The **two's complement** of x, denoted by x'', is obtained by adding 1 (base 2) to its one's complement x' and ignoring the last (left-most) carry digit where necessary: $x'' = x'' + 1$.

Example 1.

(a) Let $x = 0101$. Then its one's complement $x' = 1010$. Adding 1 to 1010, we obtain $x'' = 1011$.

(b) Let $x = 001100$. Then $x' = 110011$ and $x'' = 110011 + 1 = 110100$.

(c) Let $x = 0000$. Then $x' = 1111$. Adding 1, we obtain $x' + 1 = 10000$. To obtain the two's complement x'', we drop the initial 1 (the carry digit) and we have that $x'' = 0000$. (An extra carry digit is generated only when x is equal to 0.)

Now we show how to subtract binary numbers using two's-complement arithmetic.

Situation I. Suppose that we are working with n bits and that x and y are numbers that satisfy $0 \leq x \leq y < 2^n$. To perform the operation $y - x$ by two's-complement arithmetic, we add $y + x''$ and drop the last (left-most) carry digit.

Example 2.

(a) Let $n = 4$ and suppose that $x = 0101$ and $y = 1010$. Then $x'' = 1011$. To obtain $y - x$, we add: $1010 + 1011 = 10101$. Dropping the initial 1, we claim that $y - x = 0101$. We check our results by converting to the decimal system. They $y = 10$ and $x = 5$, and so $y - x = 5$. The binary representation of 5 is 0101, as we indeed had obtained with two's-complement arithmetic.

(b) Let $y = 1111$ and $x = 0111$. Then $1111 + 1001 = 11000$. Dropping the initial 1, we have $y - x = 1000$. Again checking via the decimal system, we have that $y = 15$, $x = 7$, and $y - x = 8$. The binary representation of 8 is 1000.

Situation II. Now suppose that $x > y$. In this case no extra carry digit is generated. To obtain $y - x$ we set $z = y + x''$, as before. Our final answer is the negative of z''.

Example 3. Let $y = 0101$ and $x = 1010$. We find $y - x$. Setting $z = y + x''$, we have that $z = 0101 + 0110 = 1011$. Then $z'' = 0101$ and we have our final answer, $y - x = -0101$. We check our results by converting to the decimal system. We have $y = 5$, $x = 10$, and $y - x = -5$, confirming our binary results.

We now show that the procedures given in Situations I and II work. First, we note again that $-x \equiv (m - x)$ mod m. If we have n bits to work with—that is, if our string length is n—our modulus is $m = 2^n$. Suppose we are in Situation I and suppose that x and y satisfy the inequality $0 \le x \le y < m$. Let x' be the one's complement of x. Summing, we have that $x + x' = 11111 \ldots 11$, a string of n 1's. Adding 1, we obtain that $x + x'' = 1000 \ldots 00$, a string of n 0's preceded by 1. The latter is the binary representation of $m = 2^n$. Thus $x'' = m - x$ and so $y + x'' = y + (m - x) = m + (y - x)$. To obtain $(y - x)$ we must subtract m. This is done by dropping the initial 1. So we have shown that the procedure given in Situation I works.

Now suppose we are in Situation II and that x and y satisfy the inequality $0 \le y < x < m$. Again letting $z = y + x''$, we have $z = (y - x) + m$. Now z is a positive number and $z + z'' = m$. Thus $-z'' = z - m = y - x$, showing that the procedure given for Situation II also works.

Section 5.6   **Exercises**

---

1. Convert to binary notation and perform each of the indicated subtractions, using two's-complement arithmetic with $n = 4$.

$$7 - 3 \qquad 6 - 1 \qquad 2 - 5 \qquad 1 - 7$$
$$4 - 2 \qquad 5 - 0 \qquad 4 - 6 \qquad 4 - 7$$

2. Use $n = 6$ to perform the following subtractions, using two's-complement arithmetic.

$$19 - 3 \qquad 31 - 17 \qquad 12 - 31$$
$$6 - 4 \qquad 3 - 19 \qquad 14 - 15$$

3. Design a logic circuit for two's-complement arithmetic.

4. Develop the analogue for two's-complement arithmetic for octal (base-8) numbers. This is called eight's-complement arithmetic.

key concepts

5.1 Relations

relation R on a set A: $x\,R\,y$
digraph of a relation
properties of a relation: symmetric, reflexive, transitive, irreflexive, antisymmetric
M_R, the matrix of a relation R

5.2 Composition of Relations

composition of a relation with itself: $R \circ R$
composition of a relation with itself n times: R^n
Boolean addition and multiplication
transitive closure

5.3 Equivalence Relations

equivalence relation: reflexive, symmetric, and transitive
modular relation: $x \equiv y \pmod{m}$; congruence

5.4 Equivalence Classes

equivalence classes
complete set of representatives
orbits
cycle decomposition of a permutation

5.5 Order Relations

partial order $\leq$: reflexive, transitive, and antisymmetric
partially ordered set, or poset
Hasse diagram
totally ordered set
lexicographic order
least element and greatest element of a set
well-ordered set
upper bound, lower bound; least upper bound (lub), greatest lower bound (glb)
lattice: $\{x, y\}$ has a lub and a glb
join $x \vee y$ and meet $x \wedge y$

Chapter 5	Exercises

1. Draw a digraph for each of the following relations on the set $A = \{a, b, c, d, e\}$ and determine which relations are reflexive, which irreflexive, which symmetric, which antisymmetric, and which transitive.

(a) $R_1 = \{(a, a), (a, b), (b, c), (a, c), (b, a), (c, a), (c, b), (d, e)\}$

(b) $R_2 = \{(a, a), (a, b), (b, a), (b, b), (c, c), (d, d), (c, d), (d, c), (e, e)\}$

(c) $R_3 = \{(a, b), (b, c), (b, d), (c, d), (a, e)\}$

2. For each of the following relations, give some related pairs and determine which relations are reflexive, which irreflexive, which symmetric, which antisymmetric, and which transitive.

(a) Let R be the relation on the set of integers $\mathbf{Z}$ given by $x \, R \, y$ if and only if x and y share a common factor other than $+1$ or -1.

(b) Let R be the relation on the set $\mathbf{Z}$ of integers given by $x \, R \, y$ if and only if x divides y.

(c) Let R be the relation on $\mathbf{Z} \times \mathbf{Z}$ given by $(x, y) \, R \, (s, t)$ if and only if $x < s$ and $y < t$.

(d) Let R be the relation on $\mathscr{P}(\mathbf{Z})$ given by $A \, R \, B$ if and only if $A \cap B \neq \emptyset$, where A and B are subsets of $\mathbf{Z}$.

(e) Let C be the relation on the cartesian plane $\mathbf{R} \times \mathbf{R}$ given by $(x, y) \, C \, (s, t)$ if and only if $x^2 + y^2 = s^2 + t^2$. (Describe this relation geometrically.)

3. Prove that the relation given in part (b) of Exercise 2 is transitive.

4. Let A be a set, let f be a function from A to itself, and let R_f be defined as in Section 5.1.

(a) Draw the digraph for R_f for A and f given by the accompanying diagram.

Figure for Exercise 4(a)

(b) Find a set A and a function f (which is not the identity function) such that R_f is symmetric.

5. (a) For each of the relations R given in Exercise 1, draw a digraph for the relations R^2 and R^3.

(b) What properties must a relation R have in order that $R = R \circ R$?

6. Determine which of the following relations are equivalence relations and, for those that are, determine the equivalence classes.

(a) Let $A = \{a, b, c, d, e, f\}$ and let $R = \{(a, a), (a, b), (b, b), (b, a),$
$(e, e), (c, c), (c, d), (d, c), (d, d)\}$.

(b) Let R be the relation on $\mathbf{Z}$ described in part (a) of Exercise 2.

(c) Let $A = \{a, b, c, d\}$, and let R be the relation on $\mathcal{P}(A)$ determined by
$S \mathrel{R} Y$ if and only if S and Y are subsets of A with the same number of
elements.

7. Let X be a set, and let R be the relation on $\mathcal{P}(X)$ defined by $S \mathrel{R} T$ if and only
if there is a bijection from S to T, where S and T are subsets of X. Show that
R is an equivalence relation.

8. Let R be the relation on $\mathbf{Z} \times \mathbf{N}$ defined by $(x, y) \mathrel{R} (s, t)$ if and only if
$xt - sy = 0$.

(a) Give an example of some related pairs.

(b) Prove that this relation is an equivalence relation.

(c) Characterize the equivalence classes of this relation.

(d) Find a complete set of representatives for this relation.

9. Let m be a fixed real number, $m \neq 0$. Let R be the relation on the x-y plane
defined as follows: $(x, y) \mathrel{R} (u, v)$ if $(y - v) = m(x - u)$.

(a) Prove that R is an equivalence relation.

(b) Find the equivalence class of the point $(1, -2)$.

(c) Describe the equivalence classes of R.

(d) Find a complete set of representatives for R.

10. Find the set of all integers x such that:

(a) $x \equiv 3 \pmod 7$ (b) $x \equiv -1 \pmod 2$

(c) $2x \equiv 1 \pmod 3$ (d) $3x \equiv 2 \pmod 5$

11. We say that an integer r is the remainder of x modulo m if we can find an
integer q such that $x = mq + r$ and $0 \le r \le (m - 1)$. Thus 3 is the remainder
of 13 modulo 5 and 1 is the remainder of -13 modulo 7. Prove that, for two
integers x and y, we have $x \equiv y \pmod m$ if and only if x and y have the same
remainder modulo m.

12. Each of the accompanying diagrams defines a permutation f. For each permutation, draw the digraph for the relation $\mathcal{R}_f$ as defined in Example 6 of Section
5.3, find its orbits, and find its cyclic decomposition.

Figure for Exercise 12

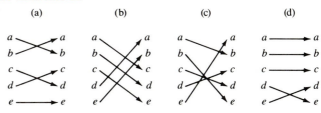

13. Show that the relation $\mathcal{R}_f$ as defined in Example 6 of Section 5.3 is transitive.

14. Determine which of the accompanying digraphs define relations that are partial orders. Draw the Hasse diagrams for those that do.

Figure for Exercise 14

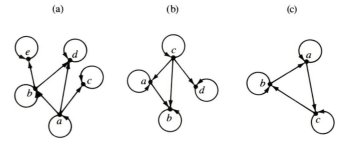

15. Determine which of the following relations establishes a partial order, which establishes a total order, and which establishes a well-ordering.

(a) The relation $\leqslant$ defined on the set **R** of real numbers by $x \leqslant y$ if $|x| \leq |y|$

(b) The relation R on the set $\mathbf{Z} \times \mathbf{Z}$ given by $(m, n)\ R\ (s, t)$ if $m \leq s$ and $n \leq t$

(c) The relation Q on $\mathbf{Z} \times \mathbf{Z}$ given by $(m, n)\ Q\ (s, t)$ if $m < s$ or, whenever $m = s$, if $n \leq t$

(d) The relation Q as described in part (c) restricted to the set $\mathbf{N} \times \mathbf{N}$

(e) Lexicographic order on the set E^* where E is the alphabet $\{a, b\}$ and the symbol a precedes the symbol b

16. Let A be a partially ordered set and let S be a subset of A. Prove that S has at most one greatest element and at most one least element.

17. Prove by induction that every finite totally ordered set is well-ordered.

18. Prove that the relation given in Example 4 of Section 5.5 is transitive, reflexive, and antisymmetric.

19. The set $B = \{0, 1\}$ is well-ordered when 0 precedes 1. Sort the following words over B in lexicographic order: 110, 011, 0001, 0011, 01. Suppose that we let each word in B^* represent the binary expansion of an integer. Determine whether lexicographic order is the same as numerical order.

20. Let E be a finite well-ordered set and define the relation R on E^* as follows: if x and y are words in E^*, then $x\ R\ y$ if the length of x is less than the length of y or, when their lengths are equal, if x precedes y in lexicographic order. (This relation is called the **standard order** on E^*.)

(a) Give an example to show how this order differs from lexicographic order.

(b) Show that E^* is well-ordered under R.

21. For the poset A defined by the accompanying Hasse diagram, find all upper bounds and all lower bounds of each of the following subsets.

(a) $\{c, d\}$ (b) $\{c, d, i\}$ (c) $\{g, d\}$ (d) $\{b, e\}$ (e) $\{f, d, i\}$

Figure for Exercise 21

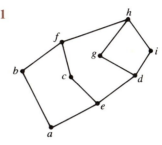

22. Find the least upper bound and the greatest lower bound (where possible) of each of the subsets listed in Example 21.

23. Either draw a Hasse diagram for a lattice that satisfies the following condition or prove that it cannot be done: "Every pair of elements has an upper bound but no lower bound."

24. Find the least upper bound and the greatest lower bound of each of the following subsets of the rational numbers under $\leq$ if possible.

(a) $S = \{x : x = 1/2^n \text{ for } n \geq 1\}$

(b) $Q = \{x : x = n/2^n \text{ for } n \geq 1\}$

(c) $T = \{x : x = m/2^n \text{ for } m \geq 1 \text{ and } n \geq 1\}$

25. Prove that the poset shown in the accompanying diagram is a lattice by directly finding the join and meet of every pair of elements.

Figure for Exercise 25

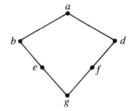

26. Let $D = \{(n, m) : m \text{ and } n \text{ are prime numbers and } m \neq n\}$. Prove that D is a lattice under the following relation:

$$(n, m) \, R \, (s, t) \text{ whenever } nt - sm \leq 0$$

27. Let S denote the set $\{1, 2, 3, 4, 5\}$, and let Q denote the set of functions of the form $f : S \to \mathbf{N}$. Let $\leq$ denote the relation on Q defined as follows: For f and g in Q, let $f \leq g$ if $f(i) \leq g(i)$ for each $i = 1, 2, 3, 4, 5$.

(a) Prove that Q is a poset under $\leq$.

(b) Find the lub and glb for the functions f and g defined as follows: $f(i) = i$ and $g(i) = 6 - i$.

(c) Prove that Q is a lattice.

28. Prove parts (bi), (bii), and (biii) of Theorem 1.

29. Let A be a lattice and suppose that a, b, c, and d are elements of A. Suppose that $a \leq b$ and $c \leq d$. Prove that:

(a) $a \vee c \leq b \vee c$

(b) $a \wedge c \leq a \wedge b$

(c) $a \vee c \leq b \vee d$

(d) $a \wedge c \leq b \wedge d$

GRAPH THEORY

outline

6.1 Basic Concepts
6.2 Paths and Connectivity
6.3 Planar Graphs
6.4 Trees
6.5 Rooted Trees
6.6 Bridge to Computer Science:
      TREESORT (optional)
      Key Concepts
      Exercises

Introduction

EACH of the following four problems can be represented pictorially with a set of dots called nodes and a set of edges that connect various pairs of nodes. Such representations are called **graphs.** The solutions to the given problems can be found in the analysis of their graphs. In some situations we find a solution simply by inspecting the graph, whereas in others a more careful computational analysis is required. In the sections that follow, we shall investigate some of the many aspects and applications of graph theory. Along the way we shall find the answers to these problems.

Problem 1.   The Odd Party Problem. An odd fellow wishes to have an odd party that is attended by an odd number of odd people, each of whom is acquainted with an odd number of other odd people at the party. Can this odd situation occur?

Let us assume that the party is attended by 3 people, A, B, and C, each represented by a node in Figure 6.1. If two people are acquainted we shall connect them with an edge. For us to obtain our Odd Party, each person must be acquainted with exactly 1 other person. So assume that A knows B. But then either C knows no one or either A or B knows 2 people. Thus an Odd Party with 3 people is impossible.

Figure 6.1

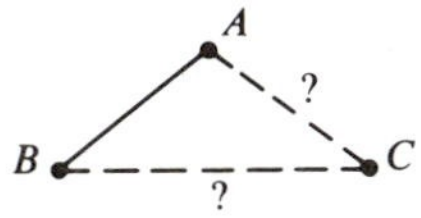

What about a party with 5 people? What about a party with n people (n odd)?

Problem 2.   A Decanting Problem. Suppose we have 3 jugs labeled A, B, and C, with capacities of 8, 5, and 3 gallons, respectively. Suppose jug A is full and the others are empty. We wish to divide the liquid in jug A into 2 equal parts (4 gallons each) by pouring the liquid from one jug to another. We must pour until one jug is empty or the other full. Can this be done? If so, give the most efficient pouring sequence.

We can solve this problem by constructing a graph and inspecting it. The nodes of the graph will be ordered triples, (x, y, z), showing the amount of liquid in each of the jugs A, B, and C, respectively. We shall draw an edge between two nodes if the liquid configuration of one can be obtained from the other in one pouring. Our goal is to obtain a node labeled $(4, 4, 0)$. To ensure efficiency, we shall not repeat a node already appearing in the graph. [If the $(4, 4, 0)$ configuration cannot be obtained, we will come to a dead end.] Thus, starting with $(8, 0, 0)$, we construct the graph shown in Figure 6.2. The necessary sequence of pours is given by the darkened path.

Figure 6.2

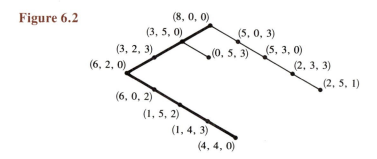

(8, 0, 0)
(3, 5, 0)   (5, 0, 3)
(3, 2, 3)   (0, 5, 3)   (5, 3, 0)
(6, 2, 0)   (2, 3, 3)
(6, 0, 2)   (2, 5, 1)
(1, 5, 2)
(1, 4, 3)
(4, 4, 0)

Problem 3.   The Three Utilities Problem. Suppose we have 3 utility poles (say for telephone, electricity, and cable service) and 3 houses. Suppose each household wishes to be connected to each of the utilities. Can this be done if wires cannot cross each other?

We shall represent the houses with nodes labeled A, B, and C and the utilities with nodes labeled X, Y, and Z. Edges must be drawn from each utility to each house in such a way that they do not intersect. One unsuccessful attempt is shown in Figure 6.3. We urge the reader to try other possibilities.

Figure 6.3

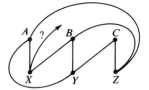

Problem 4.   The Konisberg Bridge Problem. The River Pregel runs through the town of Konigsberg (now Kaliningrad) in the U.S.S.R. There are 2 islands in the river. The larger is connected to each bank by 2 bridges, and the smaller is connected by 1 bridge to each bank. The islands are connected to each other by 1 bridge. (There are 7 bridges in all.) The townspeople wondered whether they could start on one bank, take a walk in which they crossed each bridge exactly once, and return to where they started. We can represent the situation with a graph by letting 4 nodes represent the banks of the river and the 2 islands and by letting 7 edges represent the 7 bridges (see Figure 6.4).

Figure 6.4

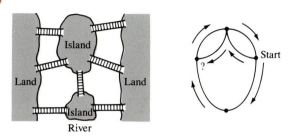

River

The question is whether we can start at a node and find a path in the graph that traverses each edge exactly once and returns to the original node. Again we urge the reader to try. This question was answered by Leonhard Euler in 1736. We shall answer it in Section 6.2.

Section 6.1 | ## Basic Concepts

In this section we give a formal definition of a graph and investigate some simple computational aspects of graphs.

> **DEFINITION.** A **graph** G is a pair of finite sets $\{V, E\}$, where V is called the set of nodes (or vertices) and E (perhaps empty) is called the set of edges. With each edge e in E, we associate an unordered pair of nodes (a, b) that are called the **endnodes** of e.

We say that the pair of endnodes (a, b) of an edge e is unordered because in a graph (unlike in the digraphs of Chapter 5), it does not matter whether we list the endnodes as (a, b) or as (b, a).

Example 1. Suppose that the graph G has the set $V = \{a, b, c, d\}$ as its set of nodes and the set $E = \{e_1, e_2, e_3, e_4\}$ as its set of edges, where $e_1 = (a, b)$, $e_2 = (b, c)$, $e_3 = (b, d)$, and $e_4 = (a, c)$. A diagram for G is shown in Figure 6.5.

Figure 6.5

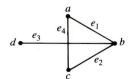

Example 2. Suppose that a graph G has the set $V = \{x, y, z\}$ as its set of nodes and the set $E = \{e_1, e_2, e_3, e_4, e_5\}$ as its set of edges, where $e_1 = (x, y)$, $e_2 = (y, z)$, $e_3 = (z, x)$, $e_4 = (x, x)$, and $e_5 = (x, y)$. A diagram for G is shown in Figure 6.6.

Figure 6.6

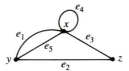

The edges e_1 and e_5 in Figure 6.6 have the same set of endnodes. They are said to be **parallel.** The edge e_4 is called a **self-loop** at x. Its two endnodes are identical. *Note:* Some authors reserve the word "graph" for those configurations with no parallel edges and no self-loops and use the term "multigraph" for the more general situation. We shall denote a graph with no self-loops and no parallel edges by the term **simple graph.**

Example 3. The graph used to model the Konigsberg Bridge problem is not a simple graph. It has two sets of parallel edges.

We now look for some aspects of a graph that admit quantification and hence computation. The most obvious quantities to count are the number of nodes, $|V|$, and the number of edges, $|E|$. For each node x in V we can count the number of edges **incident** to x; that is, we can count the number of edges of which x is an endnode. This number is called the **degree of x** and is denoted by $\deg(x)$. When an edge e is a self-loop at x, we count e twice in $\deg(x)$, once for each end incident to x.

Example 4. Consider the graph with the diagram shown in Figure 6.7. We have that $\deg(x) = 2$ and that $\deg(y) = 1$. The degree of z is 5. The self-loop at z is counted twice, once for each end incident to z.

Figure 6.7

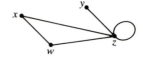

The quantities $|V|$ and $|E|$ and the degrees of the nodes are related in the following theorem.

Theorem 1. Let $G = \{V, E\}$ be a graph. The sum of the degrees of all the nodes in V is twice the number of edges in E.

PROOF: If the edge e is a self-loop at node x, it is counted twice in $\deg(x)$. If e is an edge with distinct endnodes x and y, it is also counted twice: once in $\deg(x)$ and again in $\deg(y)$.

We can verify Theorem 1 on Figure 6.7 by noting that the number of edges is 5 and that the sum of the degrees over all the nodes is $1 + 2 + 2 + 5 = 10$.

Corollary to Theorem 1. In any graph, the number of nodes of odd degree is even.

PROOF: Let $G = \{V, E\}$ be a graph. Let A be the sum of the degrees of the nodes of even degree and let B be the sum of the degrees of the nodes of odd degree. By Theorem 1 we have that $2|E| = A + B$. Since A is the sum of even numbers, we know that A is an even number. Solving for B, we find that it too is an even number. Thus B must be the sum of an even number of odd numbers. We conclude that there must be an even number of nodes of odd degree.

Solution to the Odd Party Problem. It is impossible to have an Odd Party as described in Problem 1 of the introduction to this chapter. If each person at the party is represented by a node and each pair of acquaintances is connected by an edge, an Odd Party would be represented by a graph with an odd number of nodes of odd degree. But this is impossible by the corollary to Theorem 1.

Example 5. We wish to determine how many nodes are necessary to construct a graph with exactly 6 edges in which each node is of degree 2. Let n represent the number of nodes. By Theorem 1 we have $2 \cdot 6 = 2n$. Solving for n, we find that we must have 6 nodes.

Example 6. We wish to determine whether it is possible to construct a graph with 12 edges such that 2 of the edges have degree 3 and the remaining edges have degree 4. We let x represent the number of nodes of degree 4. We must satisfy the equation $24 = (2 \cdot 3) + (4x)$. Solving, we find that $x = 9/2$, which is impossible. Thus no such graph can be constructed.

Example 7. The **complete graph on n nodes** is the simple graph with n nodes in which every pair of distinct nodes is connected by an edge. It is denoted by K_n. The diagrams for K_5 and K_3 are as shown in Figure 6.8. In K_n the degree of each node is $(n - 1)$. Letting x be the number of edges in K_n, we have by Theorem 1 that $2x = n(n - 1)$. Thus there are $n(n - 1)/2$ edges in K_n. (Note that we can count $(5 \cdot 4)/2 = 10$ edges in the diagram for K_5.) We can see this by reasoning combinatorially as well. First we note that $n(n - 1)/2 = C(n, 2)$. Now an edge

Figure 6.8

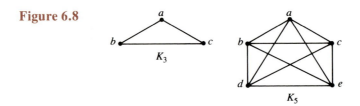

in a simple graph is determined by choosing two distinct nodes. There are exactly $C(n, 2)$ ways of choosing 2 nodes from n nodes. Thus there are $C(n, 2)$ edges in K_n because there is exactly one edge for each pair of nodes.

Example 8. We wish to determine the number of edges in a graph with 5 nodes, 2 of degree 3 and 3 of degree 2. Let x be the number of edges. Then we have $2x = (2 \cdot 3) + (3 \cdot 2)$. Solving for x, we find that we must have 6 edges in such a graph. The graphs shown in Figures 6.9 and 6.10 both have 5 nodes of the degrees described. The configurations shown in these figures are not merely different sketches of the same graph with its nodes renamed. They are intrinsically different graphs. In Figure 6.9 there is an edge between the two nodes of degree 3, whereas in Figure 6.10 there is not. When specifying a graph, it is not enough simply to indicate the number of edges, the number of nodes, and the degrees of the nodes. In most cases, the edge incidence must also be specified.

Figure 6.9   **Figure 6.10**

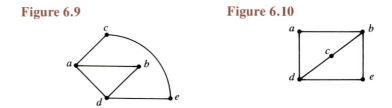

Example 9. We wish to determine the number of edges needed to construct a simple graph on 4 nodes: 2 of degree 2, and 2 of degree 3. To do this, we let x represent the number of edges. Then we have that $2x = (2 \cdot 2) + (2 \cdot 3)$. Solving for x, we find that we must have 5 edges. In this special case there is only one simple graph satisfying these numbers. (It may vary in its drawing and labeling but not in its structure.) To see this we note that, since there are only 4 nodes, each of the nodes of degree 3 must be connected by an edge to each of the 3 other nodes in the graph. But this set of connections gives us 5 edges, thus completing the graph. Figure 6.11 shows two different sketches of the *same graph*.

Figure 6.11

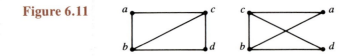

Examples 8 and 9 lead us to ask how we can determine when two graphs are the same except for the names of their nodes and edges. We call two such graphs **isomorphic.** Two isomorphic graphs must have the same number of nodes and the same number of edges. Thus we must be able to establish a one-to-one correspondence between their sets of nodes and another one-to-one correspondence between their sets of edges. But this is not enough. The graphs G_1 and G_2 shown in Figure 6.12 have the same number of nodes and edges, but they differ in many ways. For example, G_1 has a node of degree 1 whereas G_2 does not.

Figure 6.12

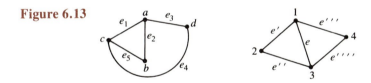

We must further be able to make the nodes and edges correspond in such a way that edges between corresponding nodes also correspond. To see this, let us consider the isomorphic graphs shown in Figure 6.13. If we let nodes a, b, c, and

Figure 6.13

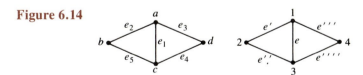

d correspond to nodes 1, 2, 3, and 4, respectively, then edge e_2 must correspond to edge e' because the edge between nodes a and b must correspond to the edge between nodes 1 and 2. Similarly, edges e_1, e_3, e_4, and e_5 must correspond to edges e, e''', e'''', and e'', respectively. When we redraw the graphs of Figure 6.13 as shown in Figure 6.14, their isomorphism becomes apparent.

Figure 6.14

We are led to the following formal definition of isomorphism.

> **DEFINITION.** Let $G = \{V, E\}$ and $G' = \{V', E'\}$. The graphs G and G' are **isomorphic** if we can find bijections $f_V : V \to V'$ and $f_E : E \to E'$ such that if e is an edge in E with endnodes a and b, then the endnodes of $f_E(e)$ in G' are $f_V(a)$ and $f_V(b)$.

The graphs G and G' shown in Figure 6.15 are isomorphic. The bijection f_V between the node sets and f_E between the edge sets given in Figure 6.16 satisfies our definition of isomorphism.

Figure 6.15

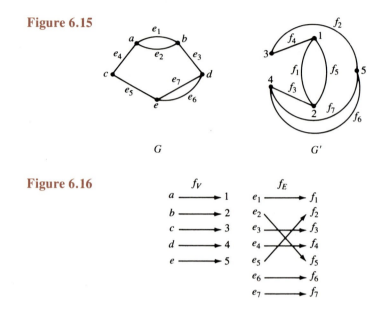

G G'

Figure 6.16

We note that for simple graphs G and G' we need only establish the correspondence f_V between their node sets and check that, if there is an edge between nodes a and b in G, then there is an edge between nodes $f_V(a)$ and $f_V(b)$ in G'. This is because if e is the unique edge between nodes a and b, then $f_E(e)$ must be the unique edge between $f_V(a)$ and $f_V(b)$ in G'. (For simple graphs we shall delete the subscript and refer only to the bijection f between the node sets.)

Example 11. The simple graphs shown in Figure 6.17 are isomorphic. The bijection shown in Figure 6.18 establishes the isomorphism.

Figure 6.17   **Figure 6.18**

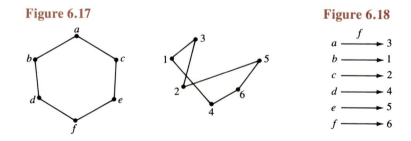

Example 12. The graphs shown in Figure 6.19 are not isomorphic. Any bijection between the edges of G_3 and the edges of G_4 must send edges e_1, e_2, and e_3 to three edges in G_4 that share a common endnode. But no such triple exists in G_4.

Figure 6.19

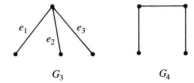

G_3 G_4

In general, it is quite a time-consuming process to determine whether two graphs are isomorphic, especially if the number of nodes and edges is large. If our graphs have n nodes and m edges, there are potentially $(n!)(m!)$ different pairs of bijections to investigate. (In simple graphs we can restrict our attention to the $n!$ bijections between the node sets—still no small task.) So we are led to look for aspects of a graph that must be reflected in any graph isomorphic to it. There are many such invariants. The following proposition gives us some examples.

Proposition 1. Suppose that $G = \{V, E\}$ and $G' = \{V', E'\}$ are isomorphic under the bijections $f_V : V \rightarrow V'$ and $f_E : E \rightarrow E'$.

(a) Let x be a node in V. Then the degree of x is equal to the degree of $f_V(x)$.

(b) The graphs G and G' must possess the same number of nodes of any given degree.

PROOF: We shall prove part (a) and leave the proof of part (b) as an exercise. Since f_E is a bijection that makes an edge e of G that is incident to x correspond to an edge e' in G' if and only if e' is incident to $f_V(x)$, the nodes x and $f_V(x)$ must have the same degree.

Example 13. The graphs shown in Figure 6.20 cannot be isomorphic because G_5 has a node of degree 4 whereas G_6 does not.

Figure 6.20

G_5 G_6

However, it is not enough to check that two graphs have the same number of nodes of any given degree. The graphs shown in Figures 6.9 and 6.10 are not isomorphic; there is an edge between the two nodes of degree 3 in Figure 6.9, whereas there is no such edge in Figure 6.10. We are led to the following continuation of Proposition 1:

Proposition 1 (continued).

(c) If e is an edge in G with endnodes a and b, then the endnodes of $f_E(e)$ in G' have the same degrees as a and b.

The proof of part (c) is also left as an exercise.

Section 6.1   **Exercises**

1. Draw each of the following graphs. Indicate which are simple.
   (a) $G = \{V, E\}$, where $V = \{a, b, c, d, e\}$ and $E = \{e_1, e_2, \ldots, e_7\}$ with $e_1 = (c, d)$, $e_2 = (a, b)$, $e_3 = (d, c)$, $e_4 = (a, a)$, $e_5 = (b, c)$, $e_6 = (a, e)$, and $e_7 = (e, e)$
   (b) The graph T with node set $\{1, 2, 3, 4, 5, 6, 7\}$ and exactly one edge between the members of each of the following pairs of nodes: (1, 2), (1, 3), (2, 4), (2, 5), (3, 6), (3, 7)
   (c) $G = \{V, E\}$, where $V = \{aa, ab, bc, ac\}$ and the edges are given as follows: $e_1 = (aa, ab)$, $e_2 = (ab, ac)$, $e_3 = (ac, bc)$, $e_4 = (bc, bc)$, and $e_5 = (bc, aa)$

2. Verify Theorem 1 for each of the graphs given in Exercise 1.

3. Draw at least one graph that satisfies each of the following sets of specifications, or indicate why it is impossible to do so.
   (a) A simple graph with 4 nodes and 3 edges
   (b) A graph with 4 nodes and 7 edges
   (c) A simple graph with 4 nodes and 7 edges
   (d) A graph with 5 nodes, 11 edges, and no parallel edges
   (e) A graph with 3 nodes, 10 edges, and no parallel edges
   (f) A simple graph with 6 nodes all of degree 2 or greater and with at least 2 nodes of degree 3

4. (a) Determine the number of edges in a graph with 6 nodes, 2 of degree 3 and 4 of degree 2.
   (b) Draw two distinct (nonisomorphic) such graphs.

5. Determine the number of edges in a graph with 6 nodes, 2 of degree 4 and 4 of degree 2. Draw 2 such graphs, one simple, one not.

6. Are there 2 nonisomorphic simple graphs with 6 nodes, 2 of degree 5, 2 of degree 3, and 2 of degree 2?

7. A graph has 12 edges and 6 nodes, each of which has degree 2 or 5. How many nodes are there of each degree?

8. Find the maximal number of edges in a simple graph with 7 nodes.

9. Determine which of the following pairs contain isomorphic graphs.

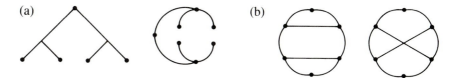

10. Draw all simple graphs with 4 nodes.

11. Either find a graph that models the following or show that none exists: Each of 102 students will be assigned the use of 1 of 35 computers, and each of the 35 computers will be used by exactly 1 or 3 students.

Section 6.2   Paths and Connectivity

The Konigsberg Bridge problem posed in the introduction to this chapter requires us to find a particular kind of path in the graph associated with the problem. In this section, we give a formal definition of a path and discuss some related concepts.

> **DEFINITION.** Let $G = \{V, E\}$ be a graph, and let v_0 and v_n be nodes in V. A **path** P of length n from v_0 to v_n is a sequence of nodes and edges of the form $(v_0, e_1, v_1, e_2, \ldots, e_n, v_n)$, where each e_j is an edge between v_{j-1} and v_j. The nodes v_0 and v_n are called the **end points** of the path, and the other nodes $v_1, \ldots, v_{n-1}$ are called its **interior nodes.** If the nodes $v_0, v_1, \ldots, v_n$ are all distinct, we say that P is a **simple path.**

Example 1. Let $G = \{V, E\}$ be a graph with $V = \{a, b, c, d, e\}$ and $E = \{e_1, e_2, e_3, e_4, e_5, e_6\}$, where $e_1 = (a, b)$, $e_2 = (b, c)$, $e_3 = (c, d)$, $e_4 = (b, d)$, $e_5 = (b, e)$, and $e_6 = (b, a)$. A diagram for G is shown in Figure 6.21. A path P from a to e of length 5 is given by the sequence $(a, e_6, b, e_2, c, e_3, d, e_4, b, e_5, e)$.

Figure 6.21

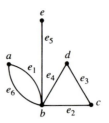

A shorter path P' from a to e is found by avoiding the loop at b so that $P' = (a, e_6, b, e_5, e)$, which is simple. The path $(b, e_2, c, e_3, d, e_4, b)$ begins and ends at the node b. Such a path is called a circuit.

DEFINITION. Suppose that $C = (v_0, e_1, v_1, \ldots, v_n)$ is a path in a graph G. If $v_0 = v_n$—that is, if C begins and ends at the same node—then C is called a **circuit**. If no interior nodes of C are repeated and if no interior node is equal to v_0, then C is said to be a **simple circuit**. A **proper circuit** is a simple circuit of length 1 or greater in which no edge is repeated.

Example 1 (continued). In the graph shown in Figure 6.21, the sequence $(a, e_1, b, e_4, d, e_3, c, e_2, b, e_6, a)$ defines a circuit that begins and ends at the node a, but it is not a simple circuit because the interior node b is visited twice. The sequence $(b, e_4, d, e_3, c, e_2, b)$ is a proper circuit. The sequence (a, e_1, b, e_6, a) is a proper circuit because the edges e_1 and e_6 are distinct. The circuit (a, e_1, b, e_1, a) is simple but not proper.

Example 2. The graph shown in Figure 6.22 has 4 distinct proper circuits. They are $C_1 = \{a, e_1, b, e_2, a\}$, $C_2 = \{a, e_1, b, e_5, c, e_4, a\}$, $C_3 = \{a, e_2, b, e_5, c, e_4, a\}$, and $C_4 = \{a, e_3, a\}$.

Figure 6.22

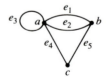

Proposition 1. Let $G = \{V, E\}$ be a graph, and let x and y be nodes of G. If there is a path in G from x to y, then there is a simple path from x to y.

PROOF: Suppose that $P = (x, e_1, v_1, \ldots, v_j, \ldots, v_k, \ldots, e_n, y)$ is a path from x to y, and suppose also that only one interior node b is repeated.

Suppose that $b = v_j$ is the first occurrence of b in the sequence defining P and that $b = v_k$ is its last occurrence. Then the sequence $(x, e_1, \ldots, v_{j-1}, e_j, v_k, \ldots, y)$ is a simple path from x to y. We leave as an exercise the inductive proof for the case in which the number of distinct repeated interior nodes is more than one.

Proposition 2. Let x and y be nodes in a graph G. If there are two different paths in G from x to y, then G contains a proper circuit.

PROOF: Suppose that P and Q are two different paths from x to y. Suppose that $P = \{x, e_1, v_1, \ldots, y\}$ and $Q = \{x, e_1', v_1', \ldots, y\}$. Let k be the smallest integer such that $e_k \neq e_k'$. Let $a = v_{k-1}$. (Hence a is the node where the paths P and Q first part. See Figure 6.23.) Let m be the smallest integer greater than k such that $v_m = v_q'$ for some $q > k$. Let $b = v_q'$. (Node b is where the paths P and Q first rejoin.) We can now find two simple paths P' and Q' from a to b that share no common edges or interior nodes. Suppose that $P' = (a, e_k, \ldots, e_{k+j}, b)$ and $Q' = (a, e_k', \ldots, e_{k+i}', b)$. Then the sequence $(a, e_k, \ldots, e_{k+j}, b, e_{k+i}', \ldots, e_k', a)$, whereby we have gone from a to b along P' and back again to a along Q', defines a proper circuit.

Figure 6.23

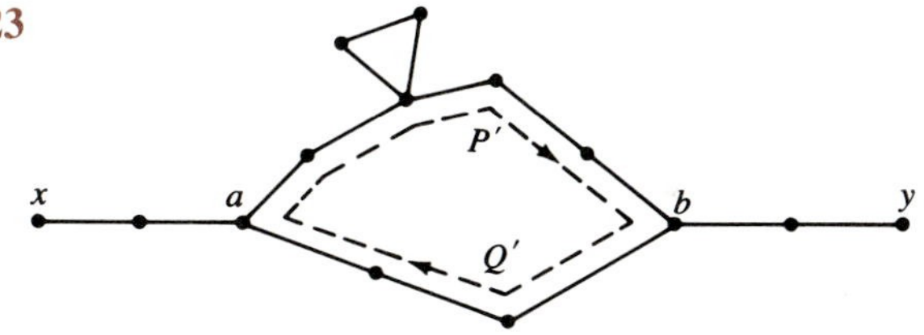

In a simple graph with no self-loops or parallel edges, a path may be described by giving only the sequence of nodes traversed in the path. For example, the sequence (a, b, c, d, b, e, a) is a circuit in Figure 6.24, and (b, c, d, e, a, b) is a simple circuit.

Consider the graph given in Figure 6.25 with the node set $\{a, b, c, d, e, f\}$. There is no path from node a to node f. This graph has two parts or **components**, and it is not **connected**. We say that two nodes are in the **same component** of a graph if there is a path between them. A graph is **connected** if it has only one component—that is, if there is a path between any pair of distinct nodes.

Figure 6.24

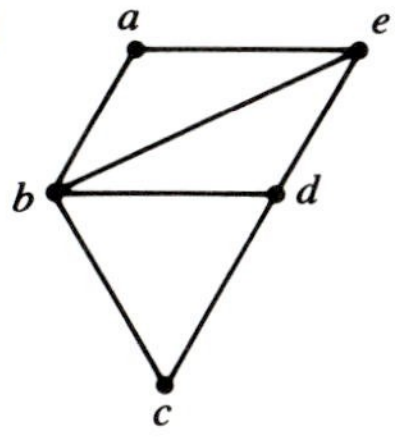

Figure 6.25

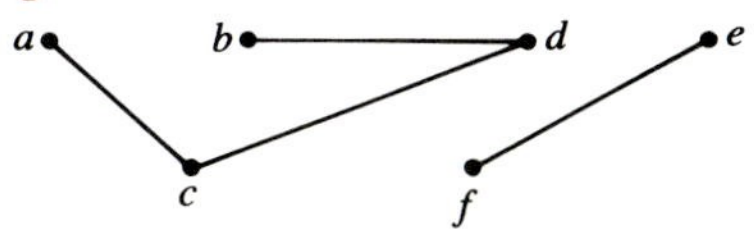

Example 3. The graph G_1 shown in Figure 6.26 has two components. There is no path from node a to node d. The graph G_2 has one component and is connected.

Figure 6.26

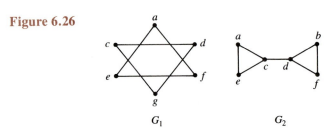

G_1 G_2

Example 4. The graph shown in Figure 6.27 is connected but contains no proper circuit. The removal of any edge disconnects the graph. Such graphs are called **trees.** Many applied situations can be modeled with trees. We shall study these further in Sections 6.4 and 6.5.

Figure 6.27

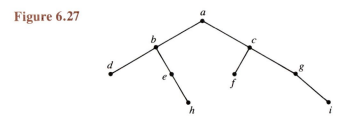

The **edge connectivity** of a connected graph is the cardinality of the smallest set of edges the removal of which results in a disconnected graph. The edge connectivity of the graph shown in Figure 6.28 is 1. The removal of the edge between nodes c and d results in a disconnected graph. The edge connectivity of the graph shown in Figure 6.29 is 2, because it requires the removal of at least 2 edges to disconnect it.

Figure 6.28   **Figure 6.29**

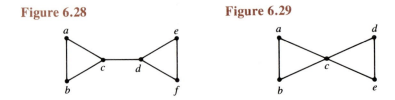

Let $G = \{V, E\}$ be a connected graph, and suppose that $|V| = n$ and that $|E| = e$. Let x be the node in V of smallest degree and let $d = \deg(x)$. How these numbers are related to the edge connectivity of G is established in the following propositions.

Proposition 3. The edge connectivity of G is less than or equal to d.

> PROOF: Let x be a node of G of degree d. Removing all the edges incident to x disconnects G. (The node x is left isolated. See Figure 6.30.)

Figure 6.30

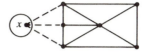

Proposition 4. The edge connectivity is less than or equal to $\lfloor 2e/n \rfloor$.

> PROOF: The degree of each of the n nodes of G is greater than or equal to d. Thus $nd \leq 2e$; or, solving for d, $d \leq 2e/n$. Since d must be an integer, we have $d \leq \lfloor 2e/n \rfloor$. By Proposition 3, it follows that the edge connectivity is less than or equal to $\lfloor 2e/n \rfloor$.

Example 5. Suppose that 8 spies whose code names are $a1$, $a2$, $a3$, . . . , $a8$ can communicate with each other on radio links as shown in Figure 6.31. Suppose that their effectiveness depends on each spy's being able to communicate with all the others (perhaps through intermediaries). The enemy wants to determine the number of the smallest set of radio links that must be jammed in order to disrupt operations.

Figure 6.31

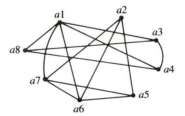

What must be determined is the edge connectivity of the graph in Figure 6.31. There are 13 edges and 8 nodes. By Proposition 4, the edge connectivity is less than or equal to $\lfloor 2 \cdot 14/8 \rfloor = 3$. Because the removal of no one edge disconnects the graph, the edge connectivity must be 2 or 3. It is 2. Jamming the radio link between spies $a1$ and $a7$ and the link between spies $a1$ and $a6$ will disrupt operations.

Example 6. Suppose that we have 6 houses and enough wire to establish telephone links between exactly 10 pairs of houses. How should this be done so as to make it possible for any 2 houses to communicate (perhaps through intermediaries) and so as to minimize the danger of disrupted communications due to severed lines? We shall represent the houses by nodes and the wires by edges. Thus what we wish to do is find a connected graph with 6 nodes, 10 edges, and maximal edge connectivity. By Proposition 3, the edge connectivity is less than or equal to $\lfloor 2 \cdot 10/6 \rfloor = 3$. Thus we should try to draw a graph in which each node has degree at least 3. The graph shown in Figure 6.32 accomplishes our task and has edge connectivity 3.

Figure 6.32

We now return to the Konigsberg Bridge problem. In this problem we are asked to find a particular kind of circuit—namely a circuit in which every edge in the graph appears exactly once. Such a circuit is called an **Euler circuit**. A path in a graph in which every edge appears exactly once is called an **Euler path**.

Example 7. Consider the graphs G_1, G_2, and G_3 with diagrams as shown in Figure 6.33. There is an Euler circuit in G_1 given by (b, a, c, b, d, e, b). There is no Euler circuit in G_2, but there is an Euler path given by the sequence (s, u, v, w, u, t). There is neither an Euler circuit nor an Euler path in G_3.

Figure 6.33

 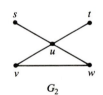 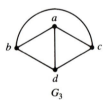

G_1 G_2 G_3

Solution to the Konigsberg Bridge Problem. The solution is actually quite simple when translated into nodes and edges. In order to find an Euler circuit in a connected graph, one must be able to find a path that enters or leaves every node along each of the edges incident to it. Because the path must leave the node along a different edge from that on which it entered, each node must be of even degree. All the nodes in the graph that models the Konigsberg Bridge problem are of odd degree. Thus there is no Euler circuit in this graph, and no walk crosses each bridge exactly once and returns the walker to his or her starting point. The following theorem shows that we also have the converse: a connected graph in which every node is of even degree must have an Euler circuit.

Theorem 1. Let $G = \{V, E\}$ be a connected graph. We can find an Euler circuit on G if and only if all the nodes of G are of even degree.

PROOF: We have already seen that if an Euler circuit exists, then all the nodes must be of even degree. To prove the converse, assume that $G = \{V, E\}$ is a connected graph in which all the nodes are of even degree. We will show how to construct an Euler circuit. First choose any node a_0 in V. Then alternately choose a sequence of edges and nodes so as to obtain a path that starts at a_0 and repeats no edges. Continue until it is no longer possible to choose an edge not already appearing in the sequence. The last node chosen must be a_0 because all nodes are of even degree, and so whenever the path under construction enters a node different from a_0, it can leave by a different edge. The resulting path is thus a circuit. If it is not an Euler circuit, some excluded edge e must be incident to one of the nodes already on the path, say node a_j. We can then use the edge e to build a subcircuit at a_j. We continue this process until all the edges of the graph appear in our circuit.

Example 8. Consider the graph G_4 shown in Figure 6.34. Each node is of even degree, and thus Theorem 1 guarantees the existence of an Euler circuit. Following the procedure detailed in the theorem, we might obtain it as follows: We start at node a and obtain the circuit (a, b, e, f, a). We note that the edge from b to c is missing from this circuit. We thus adjoin at b the subcircuit (b, c, d, b) to obtain the circuit (a, b, c, d, b, e, f, a). However, the edge between e and g is still missing, so we adjoin the subcircuit (e, g, h, e) to obtain the Euler circuit $(a, b, c, d, b, e, g, h, e, f, a)$.

Figure 6.34

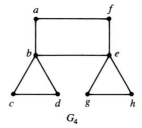

G_4

The graph G_5 shown in Figure 6.35 has an Euler path but no Euler circuit. It contains exactly 2 nodes of odd degree. The Euler path must begin and end at these nodes.

Figure 6.35

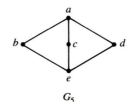

G_5

Theorem 2.   A connected graph with exactly 2 nodes of odd degree has an Euler path.

We leave the proof as an exercise.

Example 9. Suppose a traveler starting at city *A* wishes to visit each of 8 other cities (designated by the nodes in the graph shown in Figure 6.36) exactly once, travel only on certain good roads (as designated by the edges), and return home to city *A*. Can he do this? Here we ask for a circuit in which every node in the graph appears exactly once. Such a circuit is called a **Hamiltonian circuit.** (A **Hamiltonian path** is a path in which every node appears exactly once.) The answer to the traveler's problem is no. Every route to city *C* must enter and leave through city *B*. Note that the notions of Euler and Hamiltonian circuits are distinctly different. Figure 6.36 *does* have an Euler circuit because each node is of even degree.

Figure 6.36

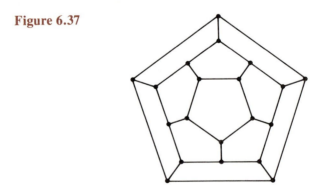

Example 10.   Around the World. Suppose that each of the 20 nodes in the graph shown in Figure 6.37 represents a famous international city and that each edge represents a major transportation route. Can a person travel around the world along these routes, visiting each city exactly once, and end her trip in the city from which she started? The answer is yes, and we urge the reader to plan a trip.

Figure 6.37

Sir William Hamilton presented the Around the World problem as a newspaper puzzle in 1859. He posed a more general question as well: What are the necessary and sufficient conditions under which a graph possesses a Hamiltonian circuit? To date, no satisfactory characterization of such graphs has been formulated.

Example 11. The graph G_6 shown in Figure 6.38 does not possess a Hamiltonian circuit. Any circuit must enter and leave node C and proceed to either D or E—let us say D. But such a circuit could not then include E without passing through either A or B twice. However, the sequence (C, B, D, A, E) gives us a Hamiltonian path in G_6. The graph G_7 has no Hamiltonian path (and hence no Hamiltonian circuit).

Figure 6.38

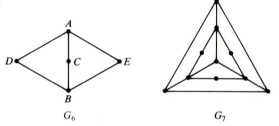

G_6 G_7

Example 12. The complete graph on n nodes K_n always has a Hamiltonian circuit. In fact, if we ignore the point of origin, K_n has $(n - 1)!$ distinct Hamiltonian circuits. If we also ignore the direction in which the circuit is traversed, K_n has $(n - 1)!/2$ Hamiltonian circuits. The distinct Hamiltonian circuits of K_4 as shown in Figure 6.39 are (A, B, C, D, A), (A, C, B, D, A), and (A, B, D, C, A), where we have ignored both the node of origin and the direction traversed.

Figure 6.39

Example 13.   Traveling Sales Representative. Suppose that a sales representative based at city A wishes to visit cities B, C, D, and E exactly once and return home. The representative also wants to minimize the total distance traveled. The distances between the cities are given by the following chart. Which Hamiltonian circuit should this person make?

	A	B	C	D	E
A	0	8	6	5	10
B	8	0	7	4	3
C	6	7	0	9	5
D	5	4	9	0	6
E	10	3	5	6	0

We verify by checking all possible Hamiltonian circuits ($4!/2 = 12$ in all) that the circuit (A, D, B, E, C, A), with a total distance of 23 miles covered, is the most efficient. If we increase the number of cities to be visited, the number of calculations required to determine the shortest route grows rapidly. For instance, 20 cities would require us to check $19!/2$ paths. Currently, no algorithm exists that guarantees the problem could be solved in fewer steps. This important problem continues to be actively investigated.

| Section 6.2 | Exercises |

The first three exercises refer to the graph G_1 shown in Figure 6.40.

Figure 6.40

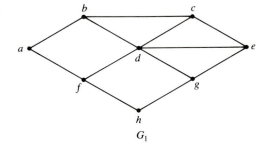

G_1

1. List all the simple paths in G_1 from a to e. What is the length of the shortest such path? What is the length of the longest such path?

2. Find a simple path with the same endnodes as the following path: $\{a, b, d, f, h, g, d, c, b, d, e\}$.

3. Find all proper circuits of length 3 in G_1. Find all proper circuits of length 4 in G_1. Are there any other proper circuits in G_1?

4. What is the minimal number of edges in a proper circuit? What is the minimal number of edges in a graph that has at least 2 distinct proper circuits? Depict your results.

5. Draw all simple, connected graphs with 4 nodes and no proper circuits, and count the number of edges in each. Repeat the problem for 5 nodes. Generalize your results as to the number of nodes in a simple, connected graph with n nodes and no proper circuits. Prove your results by induction.

6. State the contrapositive of Proposition 2.

7. Find the number of components in each of the following graphs.

(a) 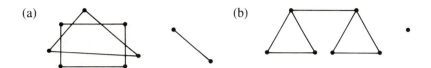 (b)

(c) The simple graph G_2 with nodes {2, 3, 4, 6, 7, 8, 9, 10, 11, 12}, where there is an edge between nodes x and y if $x \neq y$ and if either x divides y or y divides x

(d) The graph G_3 for which the node set is the set of words of length 1, 2, or 3 over the alphabet {a, b} and where there is an edge between two words x and y if either x is a prefix of y or y is a prefix of x. (Is this a simple graph?)

8. Find the edge connectivity of each of the following graphs.

(a) 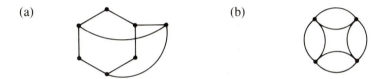 (b)

9. Use Proposition 4 to determine the maximal edge connectivity of a connected graph with 5 nodes and 8 edges. Draw a graph in which the connectivity is met and one in which it is not.

10. Find an Euler circuit in each of the following graphs or prove that none exists.

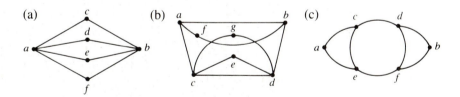

11. For each of the graphs given in Exercise 10, find an Euler path from a to b or show that none exists.

12. Draw a graph that has exactly one Euler circuit. Characterize all such graphs.

13. Find a Hamiltonian circuit in the following graph.

Figure for Exercise 13

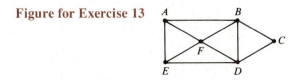

14. Is there a Hamiltonian circuit in the following graph? What about a Hamiltonian path?

Figure for Exercise 14

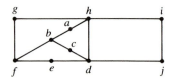

15. For each of the following complete bipartite graphs, either find a Hamiltonian circuit or show that none is possible. Generalize your results and characterize all complete bipartite graphs $K_{m,n}$ that admit Hamiltonian circuits. (*Note*: The node set of the **complete bipartite graph** $K_{n,m}$ is the union of two disjoint subsets A and B, where $|A| = n$ and $|B| = m$. There is exactly one edge from each node in A to each node in B, and there are no other edges.)

(a)

(b)

Section 6.3   **Planar Graphs**

In the Three Utilities problem given in the introduction to this chapter, we asked whether the graph with which we modeled the problem could be drawn with no intersecting edges. We now address this problem.

> **DEFINITION.** A graph is said to be **planar** if it can be drawn in the plane with no intersecting edges.

Example 1. The graph shown in Figure 6.41 appears not to be planar. However, it is redrawn in Figure 6.42 and we find that it is indeed planar.

Figure 6.41

Figure 6.42

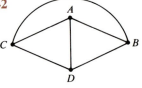

Solution to the Three Utilities Problem. Recall that the graph for this problem has 6 nodes—{A, B, C, X, Y, Z}—and 9 edges that connect each of A, B, and C to each of X, Y, and Z. This graph is $K_{3,3}$ and it is not planar. Every depiction must contain a pair of intersecting edges. To see this geometrically, draw in all the edges of the graph except one—say, the edge from B to X—in such a way that the resulting graph is planar. (See Figure 6.43.) The node B is then on two circuits, (Z, C, Y, B, Z) and (A, Z, B, Y, A). Any new edge incident to B must enter one of the regions enclosed by these two circuits. But to reach the node X, any such edge would have to intersect an edge on one of these circuits. Thus $K_{3,3}$ is not planar.

Figure 6.43

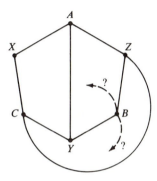

Example 2. The complete graph with 5 nodes K_5 is not planar. We leave the geometric argument (similar to that given for $K_{3,3}$) as an exercise. A computational argument will be presented later in this section.

It is not particularly difficult to ascertain that the graphs $K_{3,3}$ and K_5 are not planar. What is remarkable (and more difficult to prove) is that these are essentially the only two nonplanar graphs in the following sense: if G is a nonplanar connected graph, then we can find a copy of either $K_{3,3}$ or K_5 in G if we agree to ignore nodes of degree 2 where necessary. (We replace a node of degree 2 and its two incident edges by one long edge. So $\overset{a}{\bullet}\underset{}{\bullet}\overset{b}{\bullet}$ is replaced by $\overset{a}{\bullet}\rule{3cm}{0.4pt}\overset{b}{\bullet}$.) The proof of this fact is beyond the scope of this text. The reader will find a careful statement of the theorem and a complete proof in F. Harary's *Graph Theory* (Reading, Mass.: Addison-Wesley, 1969, p. 108ff).

Example 3. Each of the graphs shown in Figure 6.44 is nonplanar. In each, we

Figure 6.44

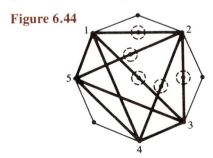

trace the copy of K_5 or $K_{3,3}$ in bolder edges. (Note that we ignore the nodes of degree 2 that we have circled.)

The graphs shown in Figures 6.45 and 6.46 are two different depictions of the same planar graph on 5 nodes. Although they are shaped differently, they both define the same number of regions—namely 4. (We always include the unbounded region outside the graph.)

Figure 6.45   **Figure 6.46**

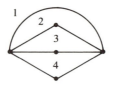 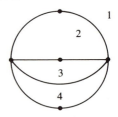

Although a planar graph may be drawn in many different ways, every planar depiction defines the same number of regions. The following theorem relates the number of regions R defined by a planar graph to the number of nodes and edges in the graph. It is due to Leonhard Euler.

Theorem 1.   Euler's Formula. Let $G = \{V, E\}$ be a connected planar graph, and let R be the number of regions defined by any planar depiction of G. Then we have $R = |E| - |V| + 2$.

PROOF: We shall construct a planar depiction of G by adding one edge (and its endnodes) at a time. At each step the resulting graph will be planar and connected, and it will satisfy the formula given by the theorem. First we pick any edge e. If e has two endnodes, it defines only the unbounded region and satisfies the equation $1 = 1 - 2 + 2$, as required. If e is a self-loop at one node, then it defines two regions and satisfies $2 = 1 - 1 + 2$, again as required. Now we proceed inductively. Assume we have chosen n edges in such a way that the resulting graph is planar and connected, and satisfies the equation $r = n - m + 2$, with r equal to the number of regions it defines and n and m the numbers of its edges and nodes, respectively. Now we choose another edge e_1 incident to one of the nodes already on the graph. If both the endnodes of e_1 are already on the graph, then we complete a circuit by adding e_1 and thus add another region. No new nodes have been added, so the equation $(r + 1) = (n + 1) - m + 2$ is satisfied as required. However, if only one of the endnodes of e_1 is on the graph, then no new regions are defined by adding e_1 but the number of nodes is increased by 1. Thus the equation $r = (n + 1) - (m + 1) + 2$ is satisfied. Since G is connected, we can proceed this way until all the edges of G have been added.

Example 4. Each of the graphs shown in Figures 6.47 and 6.48 is planar and defines 4 regions with 7 edges and 5 nodes. They are different graphs, however. The graph in Figure 6.47 has a node of degree 4, whereas that in Figure 6.48 does not.

Figure 6.47   **Figure 6.48**

   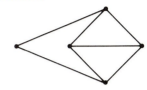

Example 5.

(a) How many edges must a planar graph have if it defines 5 regions and has 6 nodes? To answer this, we solve for $|E|$ in the equation $5 = |E| - 6 + 2$ and find that there must be 9 edges.

(b) A planar graph with 11 edges and 7 nodes must define 6 regions.

Corollary to Theorem 1. If $G = \{V, E\}$ is a simple, connected planar graph with more than one edge, it must satisfy the following inequalities:

(a) $2|E| \geq 3R$

(b) $|E| \leq 3|V| - 6$

PROOF:

(a) We assume that $|E| > 1$. If G defines only one region (the unbounded region), then $R = 1$ and part (a) certainly holds. Now if $R > 1$, then each region is bounded by at least 3 edges. But each edge in a planar graph touches at most 2 regions. Thus we have

$$2|E| \geq 3R$$

To prove part (b), we use Theorem 1 and substitute for R in part (a) of this corollary to obtain

$$2|E| \geq 3(|E| - |V| + 2)$$

Solving this inequality for $|E|$, we obtain

$$|E| \leq 3|V| - 6$$

Example 6. The graph K_5 is not planar because it fails to satisfy the second inequality given in the corollary to Theorem 1. It has 5 nodes and 10 edges, but $10 \leq 15 - 6$ is false.

Example 7. The converse to part (b) of the corollary is not true. That is, a graph may satisfy the inequality $|E| \leq 3|V| - 6$ and yet fail to be planar. For instance, the 6 nodes and 9 edges of the nonplanar graph $K_{3,3}$ satisfy $9 \leq 18 - 6$.

Section 6.3 **Exercises**

1. Redraw each of the following graphs as a planar graph.

(a) 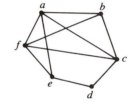   (b)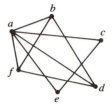

2. Verify Euler's formula for each of the graphs obtained in Exercise 1.

3. Find a copy of K_5 in the following nonplanar graph.

     Figure for Exercise 3

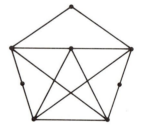

4. Find a copy of $K_{3,3}$ in the following nonplanar graph.

     Figure for Exercise 4

5. Determine the number of regions defined by a connected planar graph with 4 nodes and 8 edges. Draw such a graph.

6. Determine the number of regions defined by a connected planar graph with 6 nodes and 10 edges. Draw a simple and a nonsimple example.

7. How many edges must be drawn in order to obtain a planar graph with 5 nodes that defines 7 regions? Draw such a graph.

8. Draw 2 nonisomorphic, simple planar graphs with 6 nodes and 9 edges.

9. Draw a simple planar graph and a simple nonplanar graph, each with 6 nodes and 11 edges.

10. Can we add a self-loop or a parallel edge to a graph and change its planarity? Is there a maximal number of edges in a planar graph with n nodes? Explain. Is there a maximal number of regions that can be defined by a graph with n nodes? Explain.

11. On a certain circuit board, each of the 7 resistors must be connected to at least 5 others. Can this be done in such a way that the wires do not cross? Explain. (A full explanation will be somewhat difficult. Try it.)

Section 6.4 Trees

A simple, connected graph is called a **tree** if it contains no proper circuits. The graphs shown in Figure 6.49 are trees with 1, 2, 3, and 4 nodes, respectively.

Figure 6.49

Trees are extremely useful as modeling tools. In the nineteenth century Sir Arthur Cayley used trees to study the structure of hydrocarbons. Today trees are used in applications that range from the analysis of algorithms to game theory. In the next two sections, we shall study some of the properties of trees and look at some of their uses.

Example 1. Suppose that in a certain northern county in Maine there are five towns, A, B, C, D, and E, and that each pair of towns is connected by a road. If we represent the towns by nodes and the roads by edges, then the county road map is isomorphic to K_5 (see Figure 6.50). Now suppose the county wishes to plow as

Figure 6.50

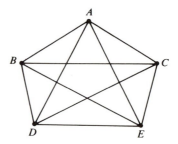

few roads as possible in winter and yet still maintain communication between any two towns. Which roads should they plow?

There are many solutions to this problem, some of which are given in Figure 6.51. Any solution can be represented by deleting those edges in the graph given in Figure 6.50 that represent roads that will not be plowed. But in order that communication between any two towns be maintained, the resulting graph must be connected. Also, the resulting graph should contain no simple circuits since we can delete any edge on such a circuit, reduce the number of roads to be plowed, and still maintain communication between each pair of towns. The graph representing any solution must be a tree. We can see that although there are many solutions to the plowing problem, each requires that exactly 4 roads be plowed.

Figure 6.51

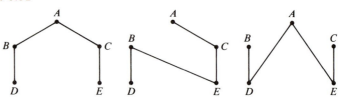

Theorem 1.   Let G be a connected graph with n nodes.

(a) G is a tree if and only if G has exactly 1 simple path between any 2 nodes.

(b) G is a tree if and only if G has exactly $n - 1$ edges.

PROOF:

(a) First, we note that any connected graph contains at least one path between any pair of nodes. But recall that if a graph G contains two distinct simple paths between a pair of nodes, then it contains a proper circuit and cannot be a tree.

(b) We shall use the Second Principle of Induction to prove that if G is a tree and G has n nodes, then G has $n - 1$ edges. (We leave the proof of the converse as an exercise.) Let $n = 1$. If G has only 1 node, then any edge must be a self-loop and hence a circuit. Thus a tree with 1 node has no edges. Now suppose that the theorem holds for trees with n or fewer nodes, and let G be a graph with $n + 1$ nodes. We will show that G has n edges. By deleting one edge e from G, we obtain a graph with exactly two connected components, one component containing those nodes of G with paths between them that contain the edge e, and the other containing those nodes that do not. Each component is a tree with n or fewer nodes, and so we may use our induction hypothesis. Suppose that the first component has r nodes and hence $r - 1$ edges and that the second component has q nodes and hence $q - 1$ edges. Since $r + q = n + 1$, the components contain a total of $n - 1$ edges. Restoring the one deleted edge e, we find that G has n edges.

Example 2. In the 1850s Arthur Cayley found that he could represent the saturated hydrocarbon molecules C_kH_{2k+2} by graphs. Each carbon atom and each hydrogen atom in C_kH_{2k+2} is represented by a node. An edge is placed between bonded atoms. Reflecting their respective valences, carbon atoms must have degree 4 and hydrogen atoms must have degree 1. The total number of nodes in the graph representing C_kH_{2k+2} is $3k + 2$. Let E be the number of edges in the graph. Summing the degrees of the nodes, we obtain

$$2E = 4k + 2k + 2 = 6k + 2$$

Thus solving for E, we find that $E = 3k + 1$, which is 1 less than the number of nodes. The graphs that represent the saturated hydrocarbons are thus all trees. The graphs for CH_4 and C_2H_6 are shown in Figure 6.52.

Figure 6.52

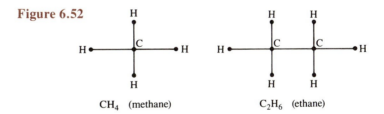

CH₄   (methane)   C₂H₆   (ethane)

Suppose that a and b are nodes in a graph G and that we wish to determine whether there is a path in G from a to b. When G has only a small number of nodes, this problem is readily solved by a look at a depiction of G. However, a more systematic approach is generally required when G has a large number of nodes. In the paragraphs that follow, we describe two different procedures, both of which involve the construction of trees. We begin by describing how to construct a **depth-first** search tree. Starting with node a, we build a tree T inductively from the nodes and edges of G. At each step we adjoin a new edge e and its endnodes to T. Suppose that the endnodes of e are the nodes a_j and a_k. To ensure that T remains connected, one of the endnodes of e must already be a node of T. To ensure that we form no circuits, the other endnode must be a new node not already appearing in T. At each step in the procedure, we shall denote the node at which we seek to adjoin the new edge by the letter x. We begin by letting T contain only the single node a and no edges; that is, $T = \{\{a\}, \emptyset\}$. The procedure for building T is as follows:

(1) Let $T = \{\{a\}, \emptyset\}$ and let $x = a$.

(2) Look for a node y such that y is not already a node of T and such that there is an edge e in G with endnodes x and y.

    (i) If y is found and $y = b$, adjoin e to T and stop. The tree is done and there is a path from a to b.

    (ii) If y is found and $y \neq b$, adjoin e to T and let $x = y$. Return to the beginning of step 1. (We resume our search for b from the last node adjoined.)

    (iii) If y cannot be found and if z is the node adjoined before x, backtrack to z. Let $x = z$ and return to the beginning of step 1. (We resume our search for b from the node adjoined just before the node at which we failed to be able to continue.) If z cannot be found, stop. There is no path from a to b.

Consider the graph given in Figure 6.53. Suppose that we wish to see whether there is a path from 1 to 9. (We can see that the answer is yes.) We construct two different depth-first search trees to answer our question. These are given in Figures 6.54 and 6.55, respectively. In the first, we adjoin the node with the smallest possible label at each step; in the second, we adjoin the largest at each step.

Figure 6.53

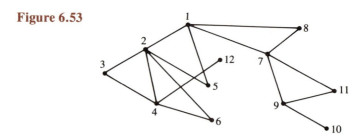

Figure 6.54   **Figure 6.55**

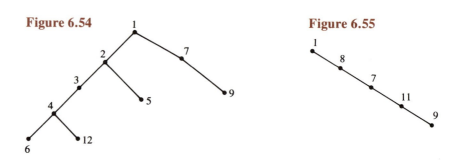

Example 3. Suppose that an office system has 16 computer terminals numbered 1 through 16. Suppose that some—but not all—pairs of terminals are directly linked. We wish to determine whether terminal 1 can communicate with terminal 15, perhaps through intermediaries. We have placed a 1 in the following 16×16 matrix in the (i, j)th position if terminals i and j are directly linked and have placed a 0 there otherwise.

	1	2	3	4	5	6	7	8	9	10	11	12	13	14	15	16
1	0	1	0	0	0	1	0	0	0	0	0	0	0	0	0	0
2	1	0	0	0	1	0	1	0	0	0	0	0	0	0	0	0
3	0	0	0	1	0	0	0	1	0	1	0	0	0	0	0	0
4	0	0	1	0	0	0	0	1	0	0	0	0	0	0	0	0
5	0	1	0	0	0	1	0	0	1	1	0	0	0	0	0	0
6	1	0	0	0	1	0	0	0	0	0	0	0	0	0	0	0
7	0	1	0	0	0	0	0	0	0	0	0	0	0	0	1	0
8	0	0	1	1	0	0	0	0	0	0	0	0	0	0	0	0
9	0	0	0	0	1	0	0	0	0	0	0	0	0	0	0	0
10	0	0	1	0	1	0	0	0	0	0	0	0	1	0	0	0
11	0	0	0	0	0	0	0	0	0	0	0	1	0	1	0	0
12	0	0	0	0	0	0	0	0	0	0	1	0	0	0	0	1
13	0	0	0	0	0	0	0	0	0	1	0	0	0	0	1	0
14	0	0	0	0	0	0	0	0	0	0	1	0	0	0	0	1
15	0	0	0	0	0	0	1	0	0	0	0	0	1	0	0	0
16	0	0	0	0	0	0	0	0	0	0	0	1	0	1	0	0

We construct a depth-first search tree by investigating the rows of the preceding matrix according to the following rule: If we are searching from node i, we adjoin the edge (i, j) if the column labeled j is the first column we encounter such that the node j is not already a node on the tree and such that there is a 1 in the (i, j)th position of the matrix. The tree generated is shown in Figure 6.56.

Figure 6.56

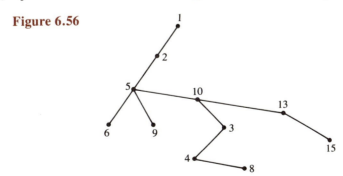

We added the nodes in the following order: 1, 2, 5, 6, 9, 10, 3, 4, 8, 13, 15. We can link computer 1 to computer 15 by going through computers 2, 5, 10, and 13.

A depth-first search tree generally does not contain the shortest path from a to b. To find the shortest path, we construct a **breadth-first** tree. The procedure is again inductive, and again we begin by letting our tree T be the single node a.

(1) Let $T = \{\{a\}, \emptyset\}$.

(2) To each of the nodes x in T, adjoin all the edges incident to x that do not form circuits. Stop if b is found. If b is not found, return to the beginning of step 1 unless there are no edges that can be adjoined that do not form circuits. In this case, also stop and conclude that there is no path from a to b.

Returning to the graph shown in Figure 6.53, we use a breadth-first tree to search for node 9 from node 1. We obtain the tree shown in Figure 6.57.

Figure 6.57

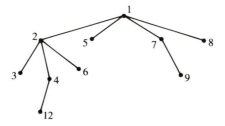

Example 3 (continued). Suppose we wish to link terminal 1 to terminal 15, using the fewest possible intermediaries. We construct the breadth-first tree shown in Figure 6.58 and find that this can be done using only two intermediary terminals, numbers 2 and 7.

Figure 6.58

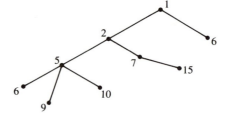

Either a depth-first or a breadth-first tree will find node b from node a successfully if and only if a and b are in the same component. If a and b are not in the same component, the resulting trees will contain exactly those nodes that are in the same component as a. Thus if G is connected and we ignore the command to stop the construction of our depth-first (respectively, breadth-first) tree after a certain node is encountered, then our procedures will halt only after all the nodes of G have been adjoined. The resulting tree is called a **depth-first** (respectively, **breadth-first**) **spanning tree** of G.

Example 4. In Figure 6.60, a depth-first spanning tree and a breadth-first spanning tree are given for the graph G appearing in Figure 6.59. In each case we started our construction at the node labeled z.

Figure 6.59

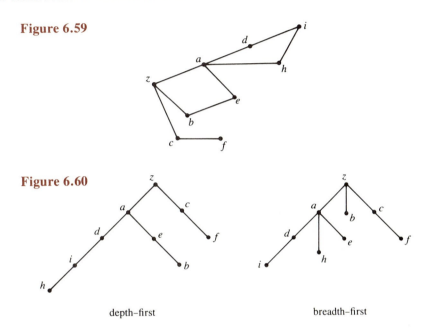

Figure 6.60

depth–first   breadth–first

If G is a connected graph with n nodes and T is any tree obtained from G by deleting all but $n - 1$ edges of G, then T is called a **spanning tree** of G. In general, a connected graph has many different spanning trees. The graph K_4 has 16 different spanning trees, some of which are shown in Figure 6.61.

Figure 6.61

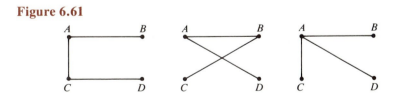

In Example 1 the road map of the 5 towns in Maine was modeled by K_5 and K_5 has 125 different spanning trees, each of which provides a solution as to which roads should be plowed. Of course, the county officials realize that, for the sake of efficiency, the county should choose that solution which involves plowing the fewest miles. Suppose that the distance between two towns is indicated on the edge between them, as given in Figure 6.62. By inspecting the possibilities, we find that an optimal solution involves plowing 24 miles (see Figure 6.63).

Figure 6.62   **Figure 6.63**

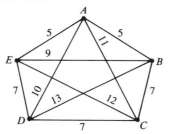

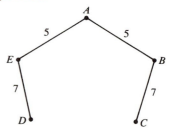

A **weighted graph** is a graph in which each of the edges has been assigned a non-negative real number. The graph shown in Figure 6.62 is a weighted graph. The weight of an edge between nodes X and Y is the distance between the towns X and Y. If G is a connected weighted graph and T is a spanning tree of G, then T is a **minimal spanning tree** if the sum of the weights of the edges of T is less than or equal to the sum of the weights of the edges of any other spanning tree. The tree given in Figure 6.63 is a minimal spanning tree. The algorithm given in the following paragraph finds a minimal spanning tree for a given connected weighted graph G. It is known as Prim's algorithm.

Prim's Algorithm.   Let G be a connected weighted graph and let a_0 be a node of G. Construct a minimal spanning tree T for G inductively as follows:

(1) Let $T = \{\{a_0\}, \varnothing\}$.

(2) Suppose that the nodes of T form the set $S = \{a_0, a_1, \ldots, a_k\}$. Find the edge e of G with endnodes x and y such that x is in S and y is not in S and such that the weight of e is as small as possible. Adjoin e to T. If T now contains all the nodes of G, stop, because T is a minimal spanning tree. If not, return to the beginning of step 2.

At each iteration of step 2 in Prim's algorithm, we scan all the nodes currently in T and adjoin an edge with the smallest weight possible that does not form a circuit. Let us apply Prim's algorithm to the graph given in Figure 6.62, starting with node A. We can adjoin either the edge to node B or the edge to node E. Choosing to adjoin the edge to node B, we obtain Figure 6.64. Now, among the edges incident to either A or B, the edge from A to E is the shortest. Thus we adjoin it and obtain Figure 6.65. Now we may adjoin either the edge from B to C or the edge from E to D. Choosing the edge from E to D, we obtain Figure 6.66. Finally, adjoining the edge from D to C, we obtain Figure 6.67. The weight of this minimal

Figure 6.64   **Figure 6.65**

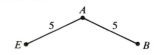

Figure 6.66   **Figure 6.67**

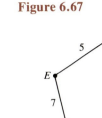

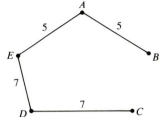

spanning tree is 24, but we note that it is not the same solution as presented in Figure 6.63. A graph may have several minimal spanning trees, but they all have the same weight.

Example 5. Suppose that the nodes of G are $\{a, b, c, d, e, f, g, h\}$, and suppose that the weights of the edges in G are given in the following matrix. (We have entered the symbol $*$ when there is no edge between a given pair of nodes.)

	a	b	c	d	e	f	g	h
a	0	4	5	*	1	2	1	5
b	4	0	1	3	*	1	8	15
c	5	1	0	9	2	*	4	7
d	*	3	9	0	8	2	4	6
e	1	*	2	8	0	*	*	10
f	2	1	*	2	*	0	5	8
g	1	8	4	4	*	5	0	*
h	5	15	7	6	10	8	*	0

We apply Prim's algorithm, starting at node a, to construct a minimal spanning tree for G as shown in Figure 6.68. The weight of the minimal spanning tree is 13.

We now present a proof that Prim's algorithm yields a minimal spanning tree.

PROOF: Suppose that T is the tree obtained by Prim's algorithm and that the edges were adjoined in the following order: $e_1, e_2, e_3, \ldots, e_n$. Let T_m be a

Figure 6.68

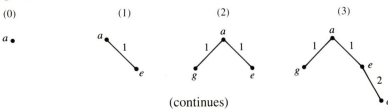

(0)   (1)   (2)   (3)

(continues)

Figure 6.68 (*continued*)

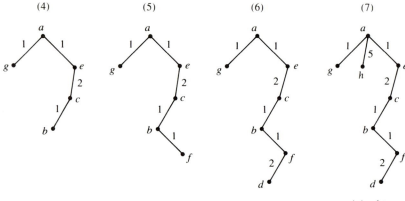

(4)   (5)   (6)   (7)

minimal tree

minimal spanning tree with the maximal number of edges in common with
T. (We shall prove that $T = T_m$ by contradiction.) Suppose that $T \neq T_m$ and
that e_i is the first edge in T that is not also in T_m. Suppose that the endnodes
of e_i are a and b and that b is the endnode of e_i that is not already one of the
endnodes of the edges in the set $S = \{e_1, e_2, \ldots, e_{i-1}\}$. Because T_m is a
tree, there is exactly one path P in T_m from a to b. Let e be the first edge on
the path P that is not in the set S. Then e shares an endnode with one of the
edges in S (see Figure 6.69). Thus the weight of e must be greater than or
equal to that of e_i (otherwise, e would have been adjoined by Prim's algorithm
instead of e_i). Now the path P, together with the edge e_i, forms a circuit.
Thus we can replace e in T_m with e_i to obtain a new spanning tree with weight
equal to (or less than) that of T_m but with one more edge in common with T.
This contradicts our choice of T_m. Thus $T = T_m$, and T itself must be a
minimal spanning tree.

Figure 6.69

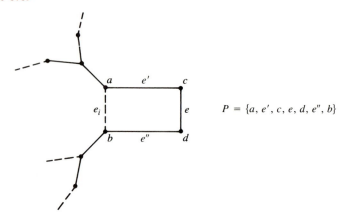

$P = \{a, e', c, e, d, e'', b\}$

Section 6.4	Exercises

1. Draw all nonisomorphic trees with 5 nodes.

2. Draw a tree with 6 nodes, exactly 3 of which have degree 1.

3. Can you draw two nonisomorphic trees with 7 nodes, each of which has either degree 1 or degree 2?

4. Draw a tree with 10 nodes each of which has either degree 1 or degree 3. Show that it is impossible to draw such a tree with 11 nodes. Repeat the problem by showing that you can draw such a tree with 16 nodes but that it is impossible to do so for 15 nodes. Characterize the size of the node set for which such a tree exists.

5. Prove that if a tree has at least 2 nodes, then it has at least 2 nodes of degree 1. If a tree has n nodes, what is the maximal number of nodes of degree 1?

6. Draw a tree with 12 nodes at least half of which have degree 1.

7. Model the following situation with a tree. A tennis tournament is to be played by 9 players designated by $P1, P2, \ldots, P9$. The tournament must be designed so that, in order to win, player Pi must play i games for $i = 1, \ldots, 8$. (So player $P1$ plays 1 game, player $P2$ plays 2 games, etc.) How many games must $P9$ play in this tournament in order to win?

8. Find all the spanning trees for the following graph.

Figure for Exercise 8

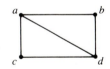

A graph G has node set $\{1, 2, 3, 4, 5, 6, 7, 8, 9, 10\}$. We have placed a 1 in the ith row, jth column of the following matrix if there is an edge between the nodes i and j. Exercises 9, 10, and 11 refer to this graph.

	1	2	3	4	5	6	7	8	9	10
1	0	1	0	1	1	0	0	0	0	0
2	1	0	1	0	0	0	0	0	0	0
3	0	1	0	0	1	0	0	0	0	1
4	1	0	0	0	1	0	1	0	0	0
5	1	0	1	1	0	1	0	1	0	0
6	0	0	0	0	1	0	0	0	1	0
7	0	0	0	1	0	0	0	1	0	0
8	0	0	0	0	1	0	1	0	1	1
9	0	0	0	0	0	1	0	1	0	1
10	0	0	1	0	0	0	0	1	1	0

9. Construct a depth-first search tree to determine whether there is a path in G from node 1 to node 10.

10. Construct a depth-first spanning tree for G, starting from node 1. Construct another starting from node 5.

11. Construct a breadth-first spanning tree for G, starting from node 1, and find the shortest path from node 1 to node 10.

12. Use Prim's algorithm to construct a minimal spanning tree for the weighted graph that follows, starting from node a. Repeat the process starting from node b. Verify that both trees have the same weight.

Figure for Exercise 12

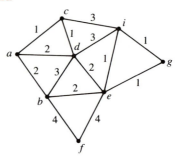

Section 6.5 | Rooted Trees

Many of the trees that appeared in the examples in the previous section had one node distinct from the rest. For instance, the node at which a depth-first search tree is started determines the shape and size of the rest of the tree. Such trees are called **rooted trees.** We shall look at their structure and some of their applications in this section.

> **DEFINITION.** A **rooted tree** is a simple connected graph with no proper circuits in which exactly one node has been designated as its **root**.

A rooted tree is simply a tree in which we have singled out a node to call a root. Each of the graphs in Figure 6.70 is a rooted tree. The root is circled.

Figure 6.70

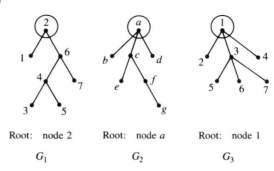

Root: node 2   Root: node a Root: node 1

G_1 G_2 G_3

If there is a path of length n from the root to node x, then x is said to be at **level n**. In G_2 in Figure 6.70, b is at level 1 and g is at level 3. We shall always draw a rooted tree according to the following convention: the root will be drawn at the top; the nodes at level 1 will be drawn below the root; the nodes at level 2 will be drawn below those at level 1; and so forth. The graphs in Figure 6.70 have been drawn this way. Our drawing convention leads to the following terminology for nodes x and y in the rooted tree T.

    (i) If the level of y is greater than the level of x, then we say y is **below** x.

    (ii) If y is below x and there is an edge from x to y, then we say y is the **son** of x.

    (iii) If $P = (v_0, v_2, \ldots, v_n)$ is a path from $x = v_0$ to $y = v_n$ and, for each i from 0 to $n - 1$, v_{i+1} is the son of v_i, then y is called a **descendant** of x.

    (iv) The node x, together with all its descendants, is called the **subtree of T rooted at x**.

    (v) A node with no sons is called a **leaf**.

    (vi) All nodes (including the root) that are not leaves are called **interior nodes**.

    (vii) A tree in which every interior node has exactly m sons is called an **m-ary tree**.

Graph G_1 in Figure 6.70 is a 2-ary, or binary, tree. It has 3 interior nodes and 4 leaves. The descendants of node 6 are nodes 3, 4, 5, and 7. The sons of node 6 are nodes 4 and 7. Graph G_3 is a 3-ary, or ternary, tree whereas G_2 is not an m-ary tree for any m.

The graphs shown in Figure 6.71 are isomorphic as trees (nodes a and a' correspond), but their structures as rooted trees are quite different. For instance, T_1 is a binary tree whereas the root in T_2 has 3 sons. In order that 2 trees T_1 and T_2 be isomorphic as rooted trees, their roots must correspond. That is, the bijection f between their nodes sets that establishes isomorphism must map the root of T_1 to the root of T_2.

Figure 6.71

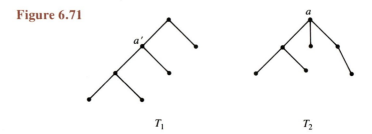

T_1 T_2

Example 1. The graphs shown in Figure 6.72 are all isomorphic as rooted trees.

Figure 6.72

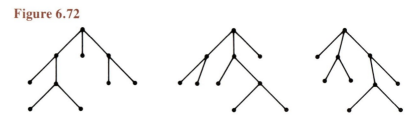

Example 2. The binary trees shown in Figure 6.73 have the same number of nodes and leaves, but they are not isomorphic as rooted trees. (In fact, they are not isomorphic as graphs.)

Figure 6.73

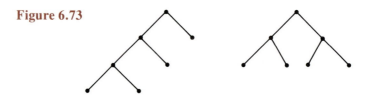

Example 3. We wish to determine all routes through the maze shown in Figure 6.74 that begin at A, end at $*$, and have no circuits. To do this, we construct a

Figure 6.74

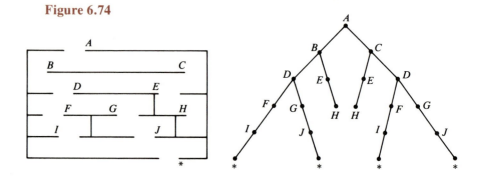

rooted tree in which each of the nodes is labeled by a gate in the maze. We proceed inductively. The root is gate A. Suppose gate x is on the tree. Then gate y is the son of gate x if y is one of the next gates that can be crossed after crossing x and if there is no gate labeled y already on the path from A to x. (This way we avoid looping.) The routes are the paths from A to $*$.

Example 4. Arithmetic expressions can be diagrammed with binary rooted trees called **arithmetic trees**. If $\#$ is a binary operation like addition or exponentiation, we shall represent $x \# y$ by the configuration

Thus the expression $((2 + 4) * 7) - ((3 * (-4)) ** 4)$ is represented by Figure 6.75. The interior nodes are the operations, and the leaves are the numbers. (Note

Figure 6.75

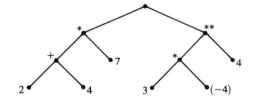

that a^x is denoted by $a ** x$ and $x \cdot y$ is denoted by $x * y$.)

The following propositions describe certain computational aspects of rooted trees.

Proposition 1. A rooted tree with n nodes has $n - 1$ edges.

PROOF: Each of the nodes except for the root is the lower endnode of exactly one edge.

Proposition 2. An m-ary tree with i interior nodes has $n = mi + 1$ nodes in all.

PROOF: Each of the i interior nodes has m sons, and there are thus mi sons on the graph. Only the root is not a son. Thus we have a total of $mi + 1$ nodes in all.

Example 5. A binary tree with 15 nodes has 14 edges by Proposition 1. By Proposition 2, we have $15 = 2i + 1$. Thus we have 7 interior nodes and 8 leaves.

These numbers may be verified on the graphs G_4 and G_5 shown in Figure 6.76, two rooted trees that are not isomorphic.

Figure 6.76

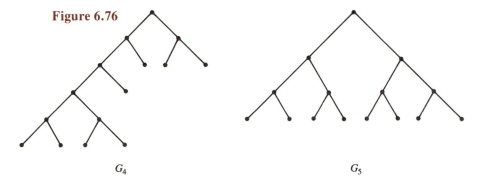

G_4 G_5

Example 6.

    (a) What is the total number of nodes in a binary tree with 20 leaves? To answer this, we first let n represent the total number of nodes, q the number of leaves, and i the number of interior nodes. Thus $q = n - i$. Using Proposition 2 with $m = 2$ and substituting for n, we find that $20 = (2i + 1) - i$. Solving for i, we find that there are 19 interior nodes and thus 39 nodes in all.

    (b) Does there exist a ternary tree with exactly 21 nodes? The answer is no. To see this, we use Proposition 2 with $m = 3$. The solution to the equation $21 = 3i + 1$ is not an integer, and thus no such tree exists.

Example 7. How many matches are played in a tennis tournament with 20 players in which any player is eliminated after his first loss? Such a tournament can be modeled with a binary tree in which the players are represented by leaves and each of the matches is represented by an interior node. Two such tournaments are modeled in Figure 6.77. The winner of each match moves up the tree until the tournament winner is found at the root. As we saw in Example 3, a binary tree with 20 leaves must have 19 interior nodes. Thus 19 matches must be played in order to determine a winner. (This figure can be arrived at more simply by noting that each match eliminates exactly 1 player and that 19 players must be eliminated.)

Example 8. A telephone network is established among 100 people. Information received by the first person is passed along to the 99 others as follows: the first person calls exactly 3 people, and each of these people calls 3 others, and so on until there are no others to call. If each call takes 5 minutes, how long does it take for a message to be relayed from the first person to receive the message to everyone else? How many people make no calls? We can model this situation with a ternary tree with a total of 100 nodes, each representing 1 person in the telephone network.

Figure 6.77

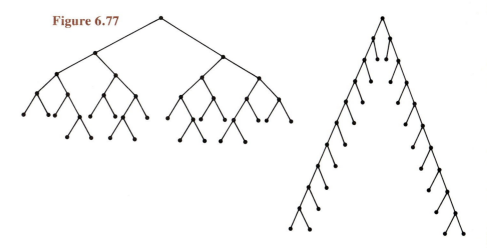

Solving for i in $100 = 3i + 1$, we find there are 33 interior nodes and thus 67 leaves. These 67 persons make no calls. Now after 5 minutes, $1 + 3 = 4$ people have received the message. After 10 minutes, the total is $1 + 3 + 9 = 13$. After 15 minutes, we have $1 + 3 + 9 + 27 = 40$. After 20 minutes, $1 + 3 + 9 + 27 + 81 = 121$ could be reached. (We need to reach only 100.) Thus it takes 20 minutes for the message to be relayed to everyone.

> **DEFINITION.** The **height** of a rooted tree is the length of the longest simple path from the root to a leaf.

Example 9. In Figure 6.78, the height of G_6 is 4 and the height of G_7 is 3.

Figure 6.78

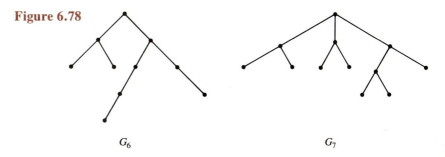

G_6 $\qquad\qquad\qquad\qquad\qquad\qquad$ G_7

> **DEFINITION.** A rooted tree of height h is said to be **balanced** if each leaf is at level h or $h - 1$.

The graph G_6 in Figure 6.78 is not balanced because it has a leaf at level 2 and a leaf at level 4. The graph G_7 is balanced.

We return to the tennis tournament with 20 players. We wish to determine how many matches will be played by the winner. We assume that the tournament is modeled by a balanced binary tree of height h so that any winner will play either h or $h - 1$ games. What we ask, then, is the height of a balanced binary tree with 20 leaves. A binary tree with all its leaves at height h will have 2^h leaves. Thus if a balanced binary tree has q leaves, we must have that $2^{h-1} < q \leq 2^h$. (To see this, simply prune off the leaves at level h to obtain a tree with 2^{h-1} leaves. Then graft additional leaves onto the tree so that all the leaves are at level h for a total of 2^h. The number of leaves on the original tree is somewhere in between.) We test various values of h and find that, for $h = 5$, we have that $16 < q \leq 32$. Since $q = 20$ satisfies this inequality, we conclude that a balanced binary tree with 20 leaves has height 5. Thus the winner of the tournament will play either 4 or 5 games. Similar reasoning will prove the following proposition.

Proposition 3.

(a) An m-ary tree of height h has at most m^h leaves.

(b) If an m-ary tree is balanced and has q leaves, then we have $m^{h-1} < q \leq m^h$.

Example 10. In Example 8 we wished to determine how long it would take for a message to be relayed to all of 100 people. The network model is a balanced ternary tree with 67 leaves. Thus $3^{h-1} < 67 \leq 3^h$. This inequality is satisfied by taking $h = 4$. Since the height of the modeling tree is 4, it takes $4 \cdot 5 = 20$ minutes for a message to be relayed to everyone.

We may take $\log_m$ of each of the terms appearing in the inequality given in part (b) of Proposition 3 to obtain the equivalent inequality:

$$h - 1 < \log_m(q) \leq h$$

Thus the height of a balanced m-ary tree with q leaves is the least integer that is greater than or equal to $\log_m(q)$:

$$h = \lceil \log_m(q) \rceil$$

(See Exercise 6 in Section 3.1 for notation.) On most calculators, only log (that is, $\log_{10}$) and the natural logarithm function ln are available. To compute $\log_m(x)$, remember that

$$\log_m(x) = \log(x)/\log(m) \quad \text{or} \quad \log_m(x) = \ln(x)/\ln(m)$$

Example 11. We find the height of a balanced ternary tree with 3001 nodes. First we find that the number of leaves is 2001 and so $q = 2001$. Thus $\log_3(2001)$

= log(2001)/log(3) = 3.301/0.477 = 6.919. Thus the height of the tree is ⌈6.919⌉ = 7.

Example 12.   A Counterfeit Coin Problem. Suppose we have a set of n pennies, one of which is known to be a lighter counterfeit coin. Using a balance, we can determine whether two stacks of coins have the same weight or one stack is lighter. We wish to identify the counterfeit coin in the fewest possible number of weighings. To do this, we divide the original set of n coins into three equal or nearly equal stacks. (If n is not divisible by 3, we can obtain two stacks with the same number of coins and a third stack with either one more or one less coin.) Now we place the two equal stacks on the balance pans. If one stack is lighter, it must contain the counterfeit coin. If the weight of the stacks is the same, the counterfeit coin is in the omitted stack. We repeat our divide-and-weigh procedure on the stack known to contain the counterfeit coin and continue to repeat until the counterfeit coin is isolated. We can model this procedure with a balanced ternary tree with n leaves. We do this for $n = 8$ in Figure 6.79. We see that we need at most two weighings (the height of the tree) to accomplish our task. In general, because the number of coins n is equal to the number of leaves on our balanced ternary tree, the number of weighings is h, where h satisfies $3^{h-1} < n \le 3^h$.

Figure 6.79

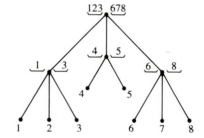

One of the primary tasks of data processing is to sort or arrange data (names, numbers, inventory, and the like) into ordered lists. The models for most sorting procedures are binary trees. We shall demonstrate with a procedure known as Bubblesort for rearranging a list of numbers into ascending order. [For instance, the list (7, 4, 6, 2) should appear as (2, 4, 6, 7).] Given a list of numbers $(x_1, x_2, x_3, \ldots, x_n)$, we can compare any two elements x_j and x_k and determine whether $x_j > x_k$ is true. The efficiency of our procedure will be measured by the number of times we use this comparison. We now describe **Bubblesort**. Suppose we are given a list of n numbers $(x_1, x_2, x_3, \ldots, x_n)$.

(1) Starting with x_1, we compare each adjacent pair (x_j, x_{j+1}) and interchange their values if $x_j > x_{j+1}$. When the members of the last pair have been compared and their values interchanged if necessary, the largest number in the list will be in the final position.

(2) Repeat the procedure given in step 1 on the smaller sublist $(x_1, x_2, \ldots, x_{n-1})$ and continue to repeat on progressively shorter sublists, ending with the sublist (x_1, x_2).

We sort the list (10, 4, 2, 6) using Bubblesort and indicate which pair is being compared and what the result is after any necessary interchange.

$$(10, 4, 2, 6) \rightarrow (4, 10, 2, 6)$$
$$(4, 10, 2, 6) \rightarrow (4, 2, 10, 6)$$
$$(4, 2, 10, 6) \rightarrow (4, 2, 6, 10)$$

Now that 10 is in the proper position, we repeat the process with the sublist (4, 2, 6). (We shall retain the 10 in its position.)

$(4, 2, 6, 10) \rightarrow (2, 4, 6, 10)$ Although this list is now sorted, our procedure does not recognize it and so we continue.

$$(2, 4, 6, 10) \rightarrow (2, 4, 6, 10)$$
$(2, 4, 6, 10) \rightarrow (2, 4, 6, 10)$ End.

For our list of 4 numbers, we used 6 comparisons. For a list of n numbers, $(n - 1) + (n - 2) + \cdots + 1 = (n - 1)n/2$ comparisons are used. (A pass through all adjacent pairs of a list of length n requires $n - 1$ comparisons. Each subsequent pass through the list requires one less comparison.) The tree diagram shown in Figure 6.80 is a model for Bubblesort. In it we use $x_1, x_2, \ldots, x_n$ to

Figure 6.80

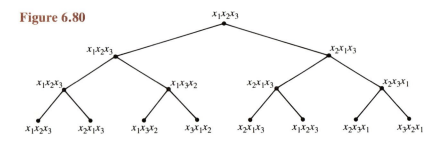

represent the numbers to be sorted. We branch left if no interchange is made in a comparison; we branch right otherwise. Our tree diagram is for lists of length 3 and thus has height $(2 \cdot 3)/2 = 3$. All possible rearrangements of the set $\{x_1, x_2, x_3\}$ appear as leaves of this tree. Any sorting procedure for a list of n numbers that can be modeled by a binary tree must have at least $n!$ leaves, one for each possible rearrangement of the list. The theoretical lower limit to the number of comparisons needed to sort a list of length n is the height h of a balanced binary tree with $n!$ leaves. That is, the height h satisfies $2^{h-1} < n! \le 2^h$ or $h = \lceil \log_2(n!) \rceil$. The following list compares h and $n(n - 1)/2$ for various values of n.

n	$n(n-1)/2$	h
2	1	1
3	3	3
4	6	5
8	28	16
100	4950	525

For large values of n, the value of h is significantly smaller than $n(n-1)/2$. There are sorting procedures that come closer to using the theoretical lower limit of comparisons and are thus faster. ["Mergesort," described in Tucker's *Introduction to Combinatorics* (New York: John Wiley & Sons, 1980), is one such procedure.]

We now describe three systematic methods of enumerating the nodes in a binary tree. Each is an inductive procedure: at each stage the node in question is considered the root of the subtree below it.

   (i) **Preorder.** Visit and label the root, then the left branch, then the right branch.

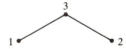

   (ii) **Inorder.** Visit and label the left branch, then the root, then the right branch. (This is also called **symmetric** order.)

   (iii) **Postorder.** Visit and label the left branch, then the right branch, then the root.

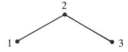

Below we label the nodes of the same tree under each of the orders.

   Preorder:

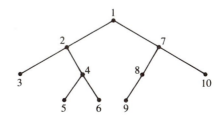

Inorder:

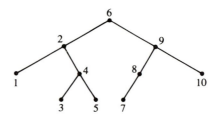

Postorder:

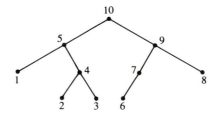

Example 13. In Figure 6.81, we enumerate the nodes of the arithmetic tree for the expression $((8 + 2) * 5) - (3 ** 2)$ in both preorder and inorder.

Figure 6.81

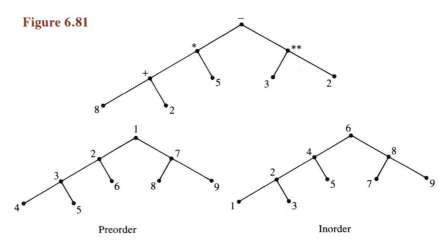

Preorder   Inorder

We now list the symbols of the expression in the order of the nodes on the enumerated tree.

Preorder: $-, *, +, 8, 2, 5, **, 3, 2$

Inorder: $8, +, 2, *, 5, -, 3, **, 2$

The inorder enumeration gives the symbols in the order in which they appear in the original expression (given in the standard algebraic in-fix notation). But without parentheses, this sequence does not define a unique arithmetic expression. For instance, the symbols in the expression $8 + (2 * 5 - 3) * * 2$ occur in the same order. However, because a preorder enumeration must start at the root and because only numbers appear as leaves, the sequence of symbols given in preorder defines only one arithmetic expression. It is a **parenthesis-free** notation also called Polish notation (after the Polish logician Lukasiewicz). For example, the list of symbols given in preorder by $(+, *, 5, -, 3, 2, -, * *, 6, 7, 1)$ defines only the tree shown in Figure 6.82. In in-fix notation, this expression is $(5 * (3 - 2)) + ((6 * * 7) - 1)$.

Figure 6.82

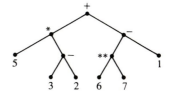

Section 6.5   **Exercises**

---

1. Draw all rooted trees with 5 nodes.

2. Draw all binary trees with 7 nodes.

3. Draw all binary trees with 6 leaves.

4. Draw two ternary trees with 11 leaves.

5. Draw an arithmetic tree for each of the following expressions.
   (a) $(2 + 3)^2 * [(3 - 9)/(7 + 2)]$
   (b) $\{(x + y) * [(z - w) + (x - 5)]\} - [(3 - z) - 8]$

6. (a) Determine the total number of nodes on a binary tree with 5 interior nodes.
   (b) Determine the number of interior nodes on a ternary tree with 1000 nodes.

7. (a) Determine the number of leaves on a binary tree with 21 nodes.
   (b) Determine the number of nodes on a ternary tree with 21 leaves.
   (c) Determine the number of leaves on a 4-ary tree with 20 interior nodes.

8. Is there a 4-ary tree with 100 nodes?

9. Prove that the number of leaves on a ternary tree is always odd.

10. Prove that the number of leaves on a binary tree can be any number greater than 1.

11. Interpret: If $m > 1$, then most of the nodes on an m-ary tree are leaves.

12. (a) What is the height of a balanced binary tree with 51 nodes?

(b) What is the height of a balanced ternary tree with 73 nodes?

(c) What is the height of a balanced binary tree with 50 leaves?

(d) What is the height of a balanced 4-ary tree with 100 leaves?

13. What are the maximal and the minimal numbers of nodes on a balanced binary tree of height 8?

14. Label each of the following trees according to preorder, inorder, and postorder enumeration.

(a)   (b)

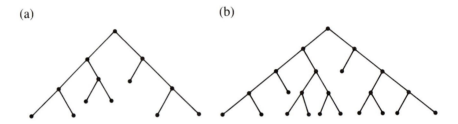

15. Label each of the arithmetic trees obtained for Exercise 5 in preorder. Rewrite each of the expressions in the parenthesis-free notation obtained from your preordered tree.

16. Find the standard "in-fix" notation for each of the following expressions, which are given in the parenthesis-free notation determined by preorder enumeration of their arithmetic trees.

(a) $+, *, +, 3, 5, -, 7, 1, -, -, 8, 2, 5$

(b) $-, /, +, x, y, +, 7, z, /, +, 5, z, +, 3, y$

Bridge to Computer Science

6.6 TREESORT (optional)

In Section 6.5 we analyzed the sorting algorithm Bubblesort. Though the algorithm itself did not explicitly involve a tree structure, we modeled its use of the comparison "$<$" with a binary tree. Then we analyzed the tree to get a measure of the relative efficiency of Bubblesort. Trees are an important tool in the analysis of algorithms. In this bridge section, we take another look at the sorting problem. This time we develop an algorithm that sorts a list by explicitly constructing a tree from the numbers on the list. These numbers will be the nodes of the tree. An inorder (symmetric) traversal of the tree produces the sorted list.

The type of tree we will need is an **almost binary** tree. Each interior node of an almost binary tree may have 1 or 2 sons: a left son and/or a right son. The tree shown in Figure 6.83 is almost binary. The left son of node 7 is node 9, and the right son of node 7 is node 11. Node 45 is an interior node but it has no right son.

Figure 6.83

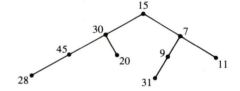

Suppose that T is an almost binary tree and that its nodes are s_1, s_2, . . . , s_n. We can characterize T completely if we know the right and left sons of each node. To do this, we construct two sequences l_1, l_2, . . . , l_n and r_1, r_2, . . . , r_n. We set $l_i = j$ if s_j is the left son of s_i. Similarly, we set $r_i = k$ if s_k is the right son of s_i. (Note that we have entered the positions of the nodes, not the nodes themselves, in the sequences $\{l_i\}$ and $\{r_i\}$.) If a node s_i has no left son (respectively, right son) we set $l_i = 0$ (respectively, we set $r_i = 0$). For example, suppose we list the nodes of the tree shown in Figure 6.84 in the following order:

$$23, 35, 11, 15, 14, 13, 19, 27, 12, 41$$

Then the sequence of left sons $\{l_i\}$ must be

$$2, 3, 0, 5, 6, 0, 8, 0, 0, 0$$

The sequence of right sons $\{r_i\}$ must be

$$7, 4, 0, 0, 0, 0, 9, 0, 10, 0$$

Figure 6.84

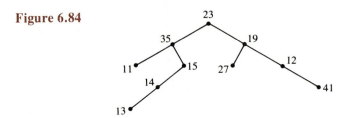

Now suppose that $s_1, s_2, \ldots, s_n$ is a list of distinct numbers to be sorted into ascending order. To do this, we construct a **sorting tree** T with the following properties:

(1) The nodes of T will be the numbers to be sorted.

(2) The right subtree of any node x (that is, the right son of x and all its descendants) will contain all the nodes greater than x. Similarly, the left subtree of x (the left son of x and all its descendants) will contain all the nodes less than x.

(3) An inorder (symmetric) traversal of the tree will list the nodes in ascending numerical order.

We construct the sorting tree T inductively. We let T_1 be the sorting tree with only one node, s_1. Now suppose that we have constructed a sorting tree T_i from the nodes $s_1, s_2, \ldots, s_i$ and that we want to place s_{i+1} on the tree in order to obtain T_{i+1}. First we compare s_{i+1} with the root of T_i. If s_{i+1} is greater, we branch to the right and compare s_{i+1} to the root of the right subtree. If it is smaller, we branch to the left and compare s_{i+1} to the root of the left subtree. We continue comparing and branching until we are at a node x that does not have the appropriate right or left son. If s_{i+1} is greater than x, then s_{i+1} becomes the right son of x. If not, it becomes its left son. We continue until we have placed all n nodes. Then $T = T_n$.

For example, suppose that we wish to construct a sorting tree from the sequence of eight numbers 7, 8, 2, 3, 5, 10, 1, 9. Suppose also that we have constructed T_7 as shown in Figure 6.85 and want to place the last entry 9 on this tree. We proceed as follows. Since $9 > 7$, branch right. Since $9 > 8$, branch right again. Since $9 < 10$, and 10 has no left son, 9 becomes the left son of 10.

Figure 6.85

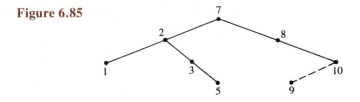

The inorder labeling of the sorting tree shown in Figure 6.85 appears in Figure 6.86. When we list the nodes of T according to this inorder traversal, we

Figure 6.86

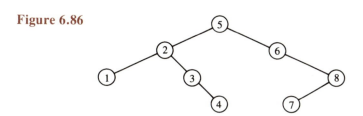

obtain 1, 2, 3, 5, 7, 8, 9, 10. This is our original list, correctly sorted into ascending numerical order.

In Algorithm 1, which follows, we begin with a sequence of numbers s_1, s_2, . . . , s_n to be sorted. The algorithm produces the sorting tree T by constructing the two sequences $l_1, l_2, . . . , l_n$ and $r_1, r_2, . . . , r_n$ in which we store the positions of left and right sons of each node.

Algorithm 1.   Sort Tree.

(1) (Initialize.) Let $i = 1$. For $j = 1$ to n, let $l_j = 0$ and let $r_j = 0$.

(2) Increment i by 1. (Replace i by $i + 1$.)

(3) (Done?) If $i + 1 > n$, stop.

(4) Let $k = 1$. (Start by comparing at the root.)

(5) If $s_i < s_k$, go to step 7. (Go left.)

(6) (Going right.)

    (i) If $r_k = 0$, set $r_k = i$ and go to step 2. (Node s_i is placed; get the next node to be placed.)

    (ii) If $r_k \neq 0$, set $k = r_k$ and go to step 5. (Resume comparison at the root of the right subtree of s_k.)

(7) (Going left.)

    (i) If $l_k = 0$, set $l_k = i$ and go to step 2. (Node s_i has been placed; get the next node to be placed.)

    (ii) If $l_k \neq 0$, set $k = l_k$ and go to step 5. (Resume comparison at the root of the left subtree of s_k.)

Next we develop an algorithm that produces an inorder traversal of the nodes of an almost binary tree, given the sequence of nodes $\{s_i\}$ and the sequences of left and right sons $\{l_i\}$ and $\{r_i\}$.

When we label the nodes of a tree inorder, we must keep a record of the nodes we have visited but did *not* label because they had left subtrees. We return to label such a node immediately after all the nodes in its left subtree have been labeled. We shall keep track of nodes we have visited, but have not labeled, in a sequence called the **stack**. Initially, the stack is empty. Each time we visit but do not label a node x, we shall enter x at the end (or top) of the stack. When we backtrack during the inorder traversal, we backtrack to the last (top) entry in the stack sequence, at which time we label it and then delete it from the stack.

For example, the nodes of the tree shown in Figure 6.87 are labeled inorder in Figure 6.88. In Table 6.1 we give the list of entries in the stack during the iterations of Algorithm 1. The top of the stack is on the right.

Figure 6.87   **Figure 6.88**

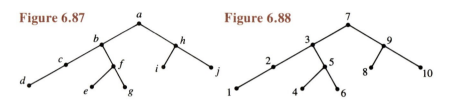

Table 6.1

		Stack Contents
1.	Initialize.	Ø
2.	Visit a.	a
3.	Visit b.	a, b
4.	Visit c.	a, b, c
5.	Label d.	a, b, c
6.	Backtrack to and label c; delete c from stack.	a, b
7.	Backtrack to and label b; delete b from stack.	a
8.	Visit f.	a, f
9.	Label e.	a, f
10.	Backtrack to and label f; delete f from stack.	a
11.	Label g.	a
12.	Backtrack to and label a; delete a from stack.	Ø
13.	Visit h.	h
14.	Label i.	h
15.	Backtrack to and label h; delete h from stack.	Ø
16.	Label j and stop.	Ø

Algorithm 2, which follows, produces an inorder listing of the nodes of an almost binary tree T by setting $x_k = s_i$ if s_k is the kth node encountered during an inorder traversal. The nodes of T are given by the sequence $\{s_i\}$, the positions of the left sons by $\{l_i\}$, and the positions of the right sons by $\{r_i\}$. We denote the ith entry of the stack sequence by stack (i), and we denote the position of the last entry in the stack by the number called "top." If the stack is empty, then top = 0.

Algorithm 2.   Inorder Tree Traversal.

(1) (Initialize.) Let top $= 0$, $i = 1$, and $k = 1$.

(2) (Going left.) If $l_i = 0$, go to step 3.
Otherwise:

    (i) Increment top by 1 (top $:=$ top $+ 1$).

    (ii) Stack (top) $= i$

    (iii) Set $i = l_i$.

    (iv) Go to the beginning of step 2.

(3) (Left subtree done; label root.)

    (i) Set $x_k = s_i$.

    (ii) Increment k by 1. ($k := k + 1$).

    (iii) If $k > n$, stop.

(4) (Going right.)
If $r_i \neq 0$, then set $i = r_i$ and go to step 2.
Otherwise:

    (i) (Backtrack.) Set $i =$ stack (top).

    (ii) (Delete.) Decrease top by 1 (top $:=$ top $- 1$).

    (iii) Go to step 3.

If we store the sequences $\{s_i\}$, $\{l_i\}$, and $\{r_i\}$, it is a simple matter to determine whether a number x is already on the tree and, if not, to add it to the tree as $x = s_{n+1}$. (We need only adjust the appropriate l_i or r_i. We leave the details as an exercise.) Thus, if no additional demands are made on it, the sorting tree is a data structure that is easy to construct and update.

In Section 6.5 we analyzed Bubblesort and determined that, for each list of numbers of length n, Bubblesort uses the comparison ">" exactly $n(n - 1)/2$ times. But the number of times "<" is used when we construct a sorting tree with Algorithm 1 depends on how far from being sorted the initial list is. Here's a rough analysis. Suppose that at each position i, approximately half the numbers following s_i are greater than s_i and half are smaller, so that the original list is far from being sorted. Then, at each iteration, the algorithm produces a sorting tree T_i that is balanced. To place s_{i+1} on the tree, we make approximately $\log_2(i)$ comparisons. (Recall that $\lceil \log_2(i) \rceil$ is the height of the tree.) Thus to place all n elements on the tree, we make approximately $\log_2(1) + \log_2(2) + \log_2(3) + \cdots + \log_2(n) = \log_2(n!)$ comparisons. For large values of n, this number is significantly smaller than $n(n - 1)/2$. However, if the original list is already sorted, then the resulting tree is linear (all sons being right sons or all sons being left sons). In this case, the number of comparisons used is again $n(n - 1)/2$.

We end this section with a Pascal routine that sorts a list s_1, s_2, $\ldots$, s_n by constructing a sorting tree and then printing its contents according to an inorder traversal. We use two procedures. The first constructs the sequences of left and right sons. The second produces the inorder traversal.

Program Treesort(Input,Output);
```
{Sorts a sequence of numbers with a sorting tree}
const
  maxvals=50;
type
  seqtype=array[1..maxvals] of integer;
var
  seqindex,nodeindex,nrofvalues:integer;
  seq,leftsonptr,rightsonptr:seqtype;
{------------------------------------------------------------------}
procedure buildtree(nrofvalues:integer;seq:seqtype;
                    var leftsonptr,rightsonptr:seqtype);
var
  seqindex,nodeindex:integer;
begin {buildtree}
  seqindex:=1; {Point to first value in sequence}
  nodeindex:=1; {Point to root node}
  {A value from the sequence is not physically put in a tree structure.}
  {Instead, the INDEX or position of that value is placed in the array }
  {of leftson pointers or the array of rightson pointers.Together,      }
  {these "pointer" arrays hold all the information about the position  }
  {of the value in the (abstract) tree structure.    The following    }
  {WHILE statement searches the "tree" for each value of the sequence, }
  {looking for an "empty" node in the proper branch (left or right) of }
  {the tree. When such a node is located, the index of the seq value is}
  {stored in rightson or leftson pointer arrays to fix its position in }
  {the abstract tree relative to the other values already in the tree. }
  while seqindex < nrofvalues do
    if seq[seqindex+1] > seq[nodeindex] then
      {ASSERT: Curr val of seq  > val at curr node,so it goes to right.}
      if rightsonptr[nodeindex] <> 0 then
        {ASSERT: There is already a right son of the current node.}
        nodeindex:=rightsonptr[nodeindex] {Make it the current node.}
      else begin {Curr node has no right son.}
        {Make the curr value of seq the right son of the curr node:}
        rightsonptr[nodeindex]:=seqindex+1; {Put cur seq val in tree.}
        seqindex:=seqindex+1; {Prepare to do next value in sequence.}
        nodeindex:=1 {Begin each search at root node.}
        end {Curr node has no right son.}
    else
      {ASSERT: Curr val of seq  <= val at curr node, so it goes to left.}
      if leftsonptr[nodeindex] <> 0 then
        {ASSERT: There is already a left son of the current node.}
        nodeindex:=leftsonptr[nodeindex] {Make it the current node.}
      else begin {Curr node has no left son.}
        {Make the curr value of seq the left son of the curr node:}
        leftsonptr[nodeindex]:=seqindex+1; {Put cur seq val in tree.}
        seqindex:=seqindex+1; {Prepare to do next value in sequence.}
        nodeindex:=1 {Begin each search at root node.}
        end {Curr node has no left son.}
end; {buildtree}
```

```
{------------------------------------------------------------------}
procedure inordertraversal(nrofvalues:integer;
                           seq,leftsonptr,rightsonptr:seqtype);
var
  nodeindex,seqindex,counter,top:integer;
  stack:array[1..maxvals] of integer;
begin {inordertraversal}
  writeln;
  writeln('The sorted sequence is');
  top:=0; {Make stack pointer indicate empty stack}
  nodeindex:=1; {Make root node the current node}
  for counter:=1 to nrofvalues do begin
    {While there is a left son of curr node}
    while leftsonptr[nodeindex] <> 0 do begin {Left son exists}
      top:=top+1;                 {"Push" leftson onto stack.}
      stack[top]:=nodeindex;
      nodeindex:=leftsonptr[nodeindex] {Make the left son curr node.}
      end; {Left son exists}
    {ASSERT: nodeindex is pointing to a node that has no left son.}
    write(seq[nodeindex]:1,' ');            {Show the current node, and}
    if rightsonptr[nodeindex] <> 0 then {if it has a right son, make   }
      nodeindex:=rightsonptr[nodeindex] {that the current node instead.}
    else begin {Reached a leaf node}
      {ASSERT: nodeindex is pointing to a node that has neither a right}
      {son nor a left son }
      nodeindex:=stack[top]; {"Pop" last nodepointer pushed.}
      top:=top-1;
      {ASSERT: Left son of node just popped has already been shown.}
      leftsonptr[nodeindex]:=0 {"Remove" it from the tree.}
      end {Reached a leaf node}
    end
end; {inordertraversal}
{==================================================================}
begin {treesort}
  writeln('Number of values in sequence? (Maximum ',maxvals:1,'):');
  readln(nrofvalues);
  writeln('Enter  ',nrofvalues:1,' values:');
  for seqindex:=1 to nrofvalues do begin
    read(seq[seqindex]);
    rightsonptr[seqindex]:=0;
    leftsonptr[seqindex]:=0
    end;

  buildtree(nrofvalues,seq,leftsonptr,rightsonptr);
  writeln;
  writeln('inordertraversal will receive:');
  for seqindex:=1 to nrofvalues do
    write(seq[seqindex]:1,' ');
  writeln;
  inordertraversal(nrofvalues,seq,leftsonptr,rightsonptr);
  writeln
end. {treesort}
```

Section 6.6   Exercises

1. Construct a sorting tree for each of the following lists. Show that an inorder traversal then produces the list sorted into ascending order.

   (a) 1, 4, 3, 6, 2, 8

   (b) 7, 5, 4, 3, 2, 1

   (c) 12, 34, 5, 4, 7, 3, 8, 2, 9, 1, 13, 6, 2

2. Algorithm 1 given in this section assumed that the lists to be sorted contained no repeated elements. Modify the algorithm in two different ways in order to accommodate repeated entries. First, ignore repeats so that the sorting tree has only distinctly labeled nodes. Second, modify Algorithm 1 so that repeated numbers are also placed on the sorting tree.

3. Modify Algorithm 1 so that an inorder traversal sorts the nodes into descending order.

4. Write an algorithm to test whether a number x is on a given sorting tree and, if not, places it in the correct position.

5. Show the contents of the stack sequence at each iteration of Algorithm 2 when it is applied to each of the sorting trees constructed in Exercise 1. (See Table 6.1.)

6. Show that Algorithm 1 uses "$<$" $n(n-1)/2$ times if the original list of n numbers is already sorted into ascending order.

7. Write an algorithm that produces the preorder traversal of an almost binary tree and one that produces its postorder traversal.

key concepts

6.1 Basic Concepts	graph $G = \{V, E\}$; V is its set of nodes and E its set of edges
	edge $e = (a, b)$; nodes a and b are its endnodes
	edge e incident to node a
	parallel edges
	self-loops
	simple graph
	degree of node a, deg (a)
	isomorphism
	complete graph K_n
6.2 Paths and Connectivity	path of length n from a to b
	simple path
	circuit
	proper circuit
	components
	connected graph
	edge connectivity
	Euler circuit
	Euler path
	Hamiltonian circuit
	Hamiltonian path
6.3 Planar Graphs	planar graph
	nonplanar graphs: $K_{3,3}$ and K_5
	regions defined by a planar graph
	Euler's formula
6.4 Trees	tree
	depth-first search tree
	breadth-first search tree
	spanning tree: depth-first, breadth-first
	weighted graph
	minimal spanning tree
	Prim's algorithm
6.5 Rooted Trees	rooted tree: root, level of a node, descendants, interior node, leaf
	m-ary tree (binary and ternary trees)
	isomorphic rooted trees
	arithmetic tree
	height of a tree
	enumerated tree: preorder, inorder, and postorder
	parenthesis-free notation (Polish notation)

| Chapter 6 | Exercises |

1. Model each of the following situations with a graph.

   (a) A news-flash telephone network is to be established so that news received by individual A can be passed on to 49 other persons. Each person receiving the news will call 3 other people until all have been informed. How many calls will be made? How many people will make no calls?

   (b) There are 7 teams within each of 2 divisions. Find a schedule such that each team plays 4 games with teams in its own division and 2 games with teams in the other division.

   (c) There are telephone lines between every 2 of 5 look-out towers. How many lines must be severed in order that 2 of the towers can communicate only with each other?

   (d) Model the process of putting any permutation of the letters A, B, and C back into alphabetical order.

2. Draw each of the following graphs and indicate which are simple.

   (a) $G_1 = \{V, E\}$, where $V = \{a, b, c, d, e, f\}$ and $E = \{e_1, e_2, e_3, e_4\}$ with $e_1 = (a, b)$, $e_2 = (b, c)$, $e_3 = (c, a)$, and $e_4 = (d, e)$

   (b) $G_2 = \{V, E\}$, where $V = \{a, b, c\}$ and $E = \{e_1, e_2, e_3, e_4, e_5\}$ with $e_1 = (a, b)$, $e_2 = (b, c)$, $e_3 = (c, a)$, $e_4 = (a, a)$, and $e_5 = (c, b)$

   (c) $G_3 = \{V, E\}$, where $V = \{a, b, c, d, e\}$ and $E = \{e_1, e_2, e_3, e_4, e_5, e_6, e_7\}$ with $e_1 = (a, b)$, $e_2 = (b, c)$, $e_3 = (c, d)$, $e_4 = (d, a)$, $e_5 = (a, c)$, $e_6 = (a, e)$, and $e_7 = (e, b)$

3. Verify Therorem 1 in Section 6.1 for each of the graphs given in Exercise 2.

4. Let G be the graph given in the accompanying figure.

   (a) Find two different paths from x to y.

   (b) Find all the proper circuits in G.

   (c) Find the length of the longest simple path in G.

   Figure for Exercise 4

   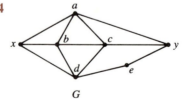

   G

5. Find all possible simple paths from point A to point B in each of the following graphs.

Figure for Exercise 5

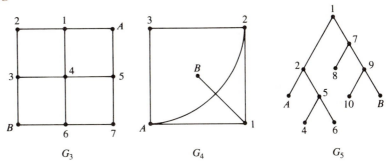

G_3 G_4 G_5

6. Let G be a graph. Let x and y be nodes in G, and let P be a path from x to y. Prove Proposition 1 in Section 6.2— namely, prove that there is a simple path from x to y. (Prove this by induction on the number of distinct repeated nodes in P.)

7. Let x and y be distinct nodes in a graph G that has n nodes. Let P be a simple path from x to y. Prove that the length of P is at most $n - 1$.

8. Find the maximal and the minimal number of edges in a simple connected graph on n nodes.

9. Draw all the simple, connected graphs that have 5 nodes and 6 edges.

10. Draw all the simple, connected graphs with 6 nodes and 15 edges.

11. How many edges does a graph with 4 nodes have if 2 nodes have degree 2 and 2 nodes have degree 3? Draw two different graphs with these specifications.

12. Draw all possible simple, connected graphs with 7 edges and 5 nodes exactly 2 of which have degree 3.

13. Suppose that there are 8 teams in Division A and 7 teams in Division B. Can a schedule be devised such that each team plays exactly 5 games with teams in its own division and 2 games with teams in the other division?

14. Is it possible to have a graph with 17 edges such that 3 of its nodes have degree 3 and the rest have degree 5? What about 4 nodes of degree 3 and the rest of degree 5?

15. A graph is to have 6 nodes and 9 edges, and each of the nodes is to be of degree 2, 3, or 4. What are the possibilities for the number of nodes of each degree?

16. Prove that a graph is connected if and only if there is a path that contains each node. Is there always a simple path?

17. We may represent an equivalence relation R on a set A by a graph if we take the set of nodes to be A itself and draw an edge between any two elements *a* and *b* whenever *a R b*. Let $V = \{0, 1, 2, \ldots, 20\}$ be the set of nodes for a graph *G*, and draw an edge between two nodes *x* and *y* whenever $(x - y)$ is divisible by 5. What do the components of *G* represent?

18. (a) Find the maximal number of edges in a simple, disconnected graph with 5 nodes. Repeat the question for graphs with 4 and 7 nodes.

   (b) Find a formula for the maximal number of edges in a simple, disconnected graph with *n* nodes and prove your result.

19. Find the maximal number of edges in a graph with *n* nodes and no parallel edges. Prove your result.

20. Prove that any two graphs that result from deleting exactly 1 edge from K_n are isomorphic.

21. Suppose that the graphs *G* and *G'* are isomorphic. Prove:

   (a) *G* has no proper circuits if and only if *G'* has no proper circuits.

   (b) *G* and *G'* both have the same number of components.

22. (a) For each of the following graphs, find an Euler circuit or demonstrate that none exists.

   (b) Do the same for Euler paths.

Figure for Exercise 22

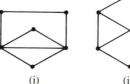 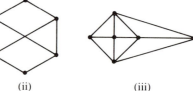

(i)   (ii)   (iii)

23. Prove Theorem 2 in Section 6.2.

24. For what values of *n* does K_n, the complete graph on *n* nodes, have an Euler circuit? For which does it have an Euler path?

25. Both Euler Park and Non-Euler Park have 6 lakes. Each park must have 9 paths connecting various pairs of lakes, with at least 2 paths going to each lake and with at most 1 path between any given pair of lakes. Design a path system for Euler Park for which there exists an Euler circuit and a path system for Non-Euler Park for which no Euler circuit exists.

26. There are 2 bus routes between cities *A* and *B*, 1 between *A* and *D*, 2 between *C* and *E*, 1 between *B* and *D*, 1 between *A* and *C*, and 1 between *A* and *E*. Can a traveler take a trip that goes over each route exactly once? Can he make a round trip? Can he make a round trip that visits each city exactly once?

27. Find a Hamiltonian circuit for each of the following graphs.

Figure for Exercise 27

(a)   (b)

28. Find Hamiltonian paths for each of the following graphs and show that no Hamiltonian circuits exist.

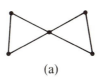   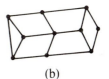

(a)   (b)

29. Show that no Hamiltonian path exists in the following graph.

Figure for Exercise 29

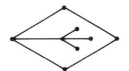

30. Let A be a nonempty set containing m elements, and let B be a nonempty set containing n elements. The **complete bipartite graph** $K_{m,n}$ is the graph with $A \cup B$ as its set of nodes and an edge between each pair (a, b) where a is in A and b is in B. (The Three Utilities problem was modeled by $K_{3,3}$.)

(a) Find Hamiltonian circuits in $K_{3,3}$ and $K_{4,4}$.

(b) Find a Hamiltonian path in $K_{3,4}$.

(c) Characterize the relation between m and n for which Hamiltonian paths and circuits exist, and prove your result.

31. Redraw each of the following as a planar graph or trace a copy of K_5 or $K_{3,3}$.

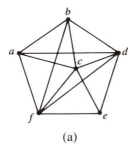   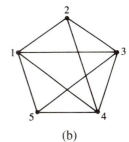

(a)   (b)

32. Verify Theorem 1 in Section 6.3 (Euler's formula) for each of the planar graphs in Exercise 22.

33. Give a geometric argument to show that K_5 is nonplanar. Why is every graph with 4 nodes planar? Can a nonplanar graph be drawn in which every node has degree at most 2?

34. Determine the number of edges and regions in a simple, connected planar graph with 6 nodes each of degree 3. Draw such a graph and show how it differs from $K_{3,3}$.

35. (a) Prove that if G is a connected planar graph in which each proper circuit contains at least 4 edges, then $2|E| \geq 4R$.

   (b) Use the result of part (a) to prove that $K_{3,3}$ is nonplanar.

36. How many edges does a tree with 4 nodes have? Draw 4 different such trees.

37. Prove that for a graph with n nodes to be connected, it must have at least $(n - 1)$ edges.

38. Prove the converse of part (b) of Theorem 1 in Section 6.4.

39. Suppose that the set of nodes of a graph G is the set $\{1, 2, 3, 4, 5, 6, 7, 8, 9, 10\}$. In the following matrix M, we have placed a 1 in the (j, k)th position if there is an edge from the node j to the node k in G, and we have placed a 0 there otherwise.

   (a) Construct a depth-first spanning tree of G to determine whether there is a path in G from node 5 to node 7.

   (b) Construct a breadth-first search tree and find the shortest path from node 1 to node 7.

M:

	1	2	3	4	5	6	7	8	9	10
1	0	1	0	1	0	0	0	0	1	0
2	1	0	0	0	1	1	0	0	0	0
3	0	0	0	0	1	1	0	1	0	0
4	1	0	0	0	0	0	0	1	0	0
5	0	1	1	0	0	0	0	0	1	0
6	0	1	1	0	0	0	1	0	0	1
7	0	0	0	0	0	1	0	0	0	0
8	0	0	1	1	0	0	0	0	0	1
9	1	1	0	0	0	0	0	0	0	0
10	0	0	0	0	0	1	0	1	0	0

   (c) Construct a depth-first spanning tree for G to determine whether G is connected.

40. Construct depth-first and breadth-first spanning trees for each of the following graphs, starting at the node labeled x. Repeat the problem starting at the node labeled y.

Figure for Exercise 40

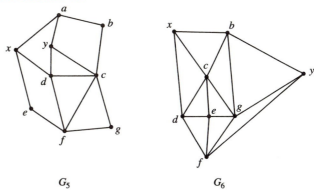

G_5 G_6

41. Use Prim's algorithm to find a minimal spanning tree for the accompanying weighted graph G_7.

Figure for Exercise 41

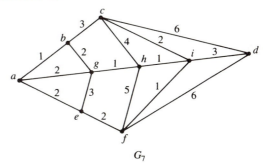

G_7

42. Suppose that G is a simple weighted graph with 10 nodes $\{1, 2, 3, 4, 5, 6, 7, 8, 9, 10\}$. Suppose that the weight of the edge from node j to node k is entered in the (j, k)th position of the following matrix. (If there is no edge from j to k, we have entered a *.) Use Prim's algorithm to find a minimal spanning tree for G.

	1	2	3	4	5	6	7	8	9	10
1	*	4	*	5	*	*	*	*	7	*
2	4	*	*	*	2	3	*	*	*	*
3	*	*	*	*	*	2	*	6	*	*
4	5	*	*	*	*	*	*	7	*	*
5	*	2	*	*	*	*	*	*	9	*
6	*	3	2	*	*	*	9	*	*	5
7	*	*	*	*	*	9	*	*	*	*
8	*	*	6	7	*	*	*	*	*	8
9	7	*	*	*	9	*	*	*	*	*
10	*	*	*	*	*	5	*	8	*	*

43. (Kruskal's algorithm) Let G be a weighted, connected graph with n nodes. Kruskal's algorithm enables us to construct a minimal spanning tree for G as follows: List the edges of G in order of increasing weight and then adjoin, in order, the first $n - 1$ edges that do not form a circuit with the previous edges.

(a) Apply Kruskal's algorithm to find minimal spanning trees for the graphs given in Exercises 41 and 42.

(b) Prove that Kruskal's algorithm works.

44. Model each of the following problems with a rooted tree.

(a) Find all paths through the following maze that start at point A, end at point B, and go through no gate twice.

Figure for Exercise 44(a)

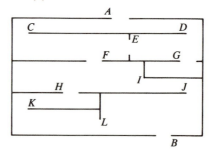

(b) Suppose a telephone network among n people is such that each person makes either no calls or 2 calls. Suppose that each call takes 5 minutes. Find the maximal and minimal time for a message received by one person to be conveyed to the rest. (Draw a model for $n = 20$ people and generalize.)

45. (a) How many edges does a tree with 19 nodes have? Draw a binary and a ternary tree, each with 19 nodes, and verify your result.

(b) How many interior nodes does a binary tree with a total of 19 nodes have?

(c) How many leaves does a ternary tree with 12 edges have?

46. (a) Draw all binary trees with 13 nodes.

(b) Show that the number of nodes on a binary tree is always odd.

(c) Show that the number of leaves on a binary tree is always 1 more than the number of interior nodes.

47. Show that the number of leaves on an m-ary tree with i interior nodes is $(m - 1) i + 1$. How many leaves are there on a 5-ary tree with 26 interior nodes?

48. (a) How many nodes are there on a ternary tree with 23 leaves?

(b) Show that the number of leaves on a ternary tree is always odd and, further, that if n is any odd number, then there is a ternary tree with n leaves.

49. Show that an m-ary tree with q leaves has a total of $(mq - 1)/(m - 1)$ nodes. What is the total number of nodes on a 4-ary tree with 64 leaves?

50. (a) Determine the height of a balanced binary tree with 50 leaves.

(b) Determine the height of a balanced ternary tree that has a total of 43 nodes.

(c) Find the maximal and minimal height for a binary tree with 101 nodes.

(d) Find the upper and lower limit for the total number of nodes on a ternary tree of height 5.

51. Find a formula for the maximal and minimal numbers of nodes on an m-ary tree of height h.

52. Prove Proposition 3 in Section 6.5.

53. Label each of the following binary trees in preorder, inorder, and postorder.

Figure for Exercise 53

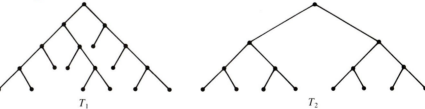

T_1 T_2

54. (a) Draw an arithmetic tree for each of the following expressions. Each is given in standard in-fix notation.

(i) $((2 - 3) * 5) - (6 - (7 + 3))$

(ii) $(5 * (7 + (3 - 4))) ** 3$

(iii) $(a + (b * c)) - (d ** (e - f))$

(b) Label each of the trees you drew in part (a) in preorder, and find a parenthesis-free expression for each of the expressions given in part (a).

(c) Find the standard in-fix notation for each of the following expressions given in parenthesis-free (preorder) notation.

(i) $*, -, +, 2, 3, -, 5, 2, 7$

(ii) $+, *, **, f, g, -, d, e, +, *, b, c, a$

55. "Reverse Polish notation" is obtained by listing the symbols of an arithmetic tree in postorder.

(a) Find the reverse Polish notation for the expressions listed in part (a) of Exercise 54.

(b) Reverse Polish notation is another parenthesis-free notation. The following expressions are given in reverse Polish notation. Find their standard in-fix expressions by reconstructing the arithmetic tree. Note that the root in postorder is always the last node enumerated and that all operations are interior nodes. (*Hint*: It is easier to go from right to left.)

(i) $a, b, +, c, *, d, e, a, b, -, *, +, -$

(ii) $a, b, c, d, +, *, -, a, e, +, c, e, -, *, +$

Chapter 7

DIGRAPHS

outline

7.1 Digraphs
7.2 Paths and Connectivity
7.3 Weighted Digraphs
7.4 Acyclic Digraphs
7.5 Finite State Machines
7.6 Bridge to Computer Science:
     Kleene's Theorem (optional)
     Key Concepts
     Exercises

Introduction

A PROGRAMMER finds that he must read a long string of 0's and 1's, and so he designs a register that will tell him the last two symbols read. What the register must change to, upon reading a symbol, depends both on what it currently registers and on the new symbol being read. For instance, if the register reads "0, 1" where 1 was read last and 0 next to last, and if the next symbol read is 1, then the register will change to "1, 1." If the next symbol is 0, however, it will change to "1, 0."

The transitions of the register can be modeled with the digraph shown in Figure 7.1. In this model, the nodes represent the possible displays or states of the register. To find how the register will change, we simply find the node with the current display of the register and follow the arrow labeled with the next symbol read. (We always start with the register set at "0, 0.") The arrow leads to the next display of the register.

Figure 7.1

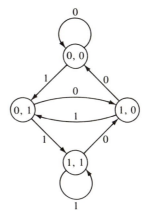

Figure 7.1 models a device that must change to one of several possible states according to what is input and according to a certain well-defined rule. In this context, the digraph is called a "finite state machine." Such models have proved to be valuable tools in computer and industrial design. (We shall investigate them further in Section 7.5.)

We have already studied digraphs as models of relations (see Chapter 5). The preceding example hints at the broad application of digraphs beyond this already rich scope. In this chapter we shall investigate digraphs in the context of more general graph theory, and we shall see some of their varied applications in such areas as operations research and automata theory.

Section 7.1
Digraphs

In Chapter 5 we represented a relation R on a finite set A by a digraph. We represented the elements of A by a set of nodes, and we drew a directed edge from node a to node b whenever $a R b$. If we ignore the orientation of the directed edges in such a digraph—that is, if we erase the arrowheads and consider a directed edge from a to b as simply an edge between a and b—then what results is a graph with node set A. Thus the underlying structure of every digraph is a graph, and all the graph-theoretic results that we have obtained so far apply to digraphs as well.

Example 1. Let A be the set $\{2, 3, 4, 9, 36\}$, and define the relation R on A as follows: $x R y$ if $x \neq y$ and y is divisible by x. The digraph for the relation R and its underlying graph are shown in Figures 7.2(a) and (b). We see, for instance, that the graph in Figure 7.2 is a connected planar graph. The degree of each node is even, and so we know that we can find an Euler circuit. The sequence $(2, 4, 36, 9, 3, 36, 2)$ defines such a circuit, but if we trace this circuit on Figure 7.1, we find that the path sometimes fails to follow the direction of the arrows. In fact, no Euler circuit on this digraph can follow the arrows. We need to reformulate some of the definitions and refine some of the results obtained for undirected graphs to reflect the additional structure imposed on digraphs by the oriented edges. We begin with some definitions.

Figure 7.2

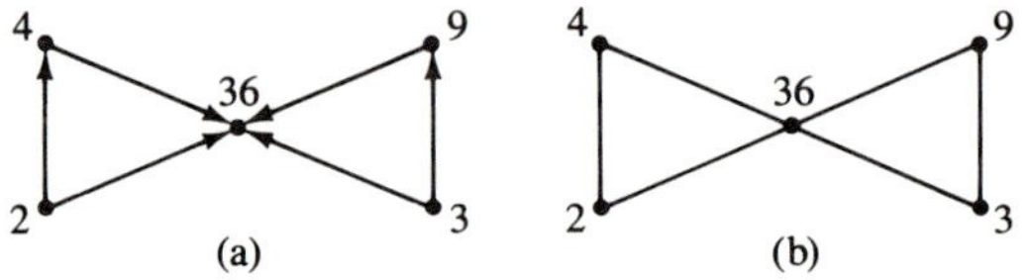

(a)   (b)

DEFINITION. A **digraph** D is a finite set A (called the set of nodes or vertices of D) together with a subset E of $A \times A$. Each ordered pair (a, b) in E is called a **directed edge** from a to b.

If $e = (a, b)$ is a directed edge in D, we say that e starts at a and ends at b. Note that our definition of a digraph allows self-loops, but for any given pair of nodes a and b, we allow at most one directed edge from a to b. (Some authors do not allow self-loops; others allow self-loops and more than one directed edge from a to b.)

Example 2. Let the node set of a digraph D be the set $\{a, b, c, d\}$, and let the set of directed edges of D be $\{(a, b), (b, a), (c, a), (c, c), (b, c), (b, d)\}$. Digraph D is depicted in Figure 7.3. Note that the directed edge (a, b) from a to b is distinct from the directed edge (b, a) from b to a. There is a self-loop at c.

Figure 7.3

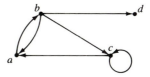

Suppose that D is a digraph with n nodes $a_1, a_2, \ldots, a_n$. We can represent D by an $n \times n$ **adjacency matrix** M as follows: we place 1 in the (j, k)th position of M if there is a directed edge from a_j to a_k in D, and we place 0 there otherwise. The adjacency matrix for the digraph shown in Figure 7.3 is given in Table 7.1.

Table 7.1

	a	b	c	d
a	0	1	0	0
b	1	0	1	1
c	1	0	1	0
d	0	0	0	0

> **DEFINITION.** Let D be a digraph and let x be a node in D. The **out-degree** of x is the number of directed edges in D of the form (x, y). The **in-degree** of x is the number of directed edges in D of the form (y, x). The **total degree** of x is the sum of its in-degree and its out-degree.

The out-degree of x is the number of arrows leaving x, and the in-degree of x is the number of arrows entering x.

Example 3. Consider the digraph given in Figure 7.4 and its adjacency matrix. The in-degree of node a is 2 and the out-degree of node a is also 2. We count the directed edge (a, a) as both entering and leaving node a. The total degree of a is 4. The in-degree of node e is 1 and its out-degree is 0. The out-degree of any node

Figure 7.4

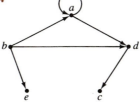

	a	b	c	d	e
a	1	0	0	1	0
b	1	0	0	1	1
c	0	0	0	0	0
d	0	0	1	0	0
e	0	0	0	0	0

x is the number of 1's in the row of the adjacency matrix labeled x. Similarly, the in-degree of x is the number of 1's appearing in the column labeled x. We can verify this in Figure 7.4 by noting that indeed there are no 1's appearing in the row labeled e but that there is exactly one 1 in the column labeled e.

As with undirected graphs, we can find a relation between the number of edges in a digraph and the degrees of its nodes. We do this in Theorem 1.

Theorem 1. The sum of the out-degrees of all the nodes in a digraph D is equal to the number of edges of D. Similarly, the sum of the in-degrees of all the nodes is equal to the number of edges of D.

PROOF: When we sum the out-degrees of the nodes, we count each edge exactly once at its initial node. Similarly, when we sum the in-degrees of the nodes, we count each edge exactly once at the node at which it terminates.

We can verify the results of the theorem on the digraph given in Example 3. The number of edges is 6 and the sum of the out-degrees of nodes in the order a, b, c, d, e is $2 + 3 + 0 + 1 + 0 = 6$. The sum of in-degrees in the same order is $2 + 0 + 1 + 2 + 1 = 6$.

Corollary to Theorem 1. The sum of the total degrees of the nodes in a digraph D is equal to twice the number of edges in D.

Example 4. The number of edges in a digraph with 4 nodes in which each node has in-degree exactly 2 is 8. Three such configurations are shown in Figure 7.5.

Figure 7.5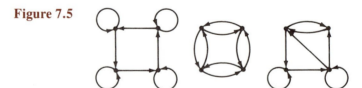

Example 5. We wish to determine whether it is possible to construct a digraph that has 10 edges and 5 nodes, each of which has out-degree equal to either 1 or 3. To do this, we must see whether we can solve the equation $x_1 + x_2 + x_3 + x_4 + x_5 = 10$ with each x_j equal to either 1 or 3. The answer is no, because the sum of 5 odd numbers must be odd and thus cannot equal 10.

We have studied digraphs as models of relations given on finite sets, but digraphs also arise quite naturally in many other applications. Here are two examples.

Example 6. Suppose a certain code is given as a string of a's and b's, and suppose that a scanner wishes to detect the presence of the substring *aabb* in a given string. Suppose the scanner reads the letters of the string from left to right. The scanner must keep track of the state of its progress toward detecting the substring. The possibilities may be summarized as follows:

1. no a detected
2. a detected; looking for a second a to follow
3. aa detected; looking for b to follow
4. aab detected; looking for a second b
5. $aabb$ found

The scanner will change states according to the digraph shown in Figure 7.6. The scanner will start in state 1 or state 2, depending on whether the word it is scanning begins with b or a. Then, if it is in state i and the next letter read is a, we follow the arrow labeled a to its next state. If the next letter is b, we follow the arrow labeled b to its next state. A word contains the pattern *aabb* if and only if it defines a path that ends in state 5. For example, the word *babaabba*, which contains the pattern *aabb*, defines the path with node sequence (1, 2, 1, 2, 3, 4, 5, 5). But the word *ababab* defines the path (2, 1, 2, 1, 2, 1), which does not end at state 5. The scanner has thus been unsuccessful in its search for *aabb*.

Figure 7.6

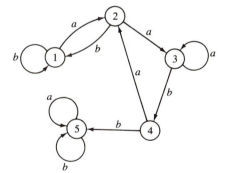

Example 7. A digraph is called a **complete tournament** if it has no self-loops and has exactly 1 directed edge between any 2 distinct nodes. For example, 2 distinct tournaments on 4 nodes are shown in Figure 7.7. There are $n(n - 1)/2$ directed edges in a complete tournament on n nodes. (The total degree of each of the n nodes is $n - 1$. Thus $n(n - 1)$ is twice the number of edges.) Because an edge between any 2 distinct nodes can be oriented in 2 different ways, there are

Figure 7.7

2^{n(n-1)/2} different complete tournaments on *n* nodes. In a round-robin tennis tournament, every player plays every other player exactly once. Suppose we represent each player by a node and place a directed edge from player *A* to player *B* if *A* defeats *B*. The digraph we obtain is a complete tournament. If there are 4 players, there are $2^6 = 64$ possible ways for the tournament to turn out. But if there are 5 players, there are $2^{10} = 1{,}024$ possibilities!

Section 7.1 | **Exercises**

---

1. (a) Find the in-degree, the out-degree, and the total degree of each node in the digraph shown in the accompanying figure.

    (b) Verify Theorem 1 in Section 7.1 for the digraph shown in this figure, using the in-degrees of each node. Do the same for the out-degrees.

Figure for Exercise 1

2. Repeat Exercise 1 for the digraph given in the accompanying figure. Also verify the corollary to Theorem 1 for this digraph.

Figure for Exercise 2

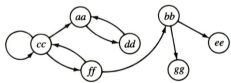

3. Suppose that the nodes of a digraph *D* are {*a*, *b*, *c*, *d*, *e*, *f*} and that its directed edges are {(*a*, *b*), (*b*, *a*), (*c*, *c*), (*a*, *c*), (*e*, *d*), (*e*, *a*), (*f*, *c*), (*c*, *f*), (*f*, *a*)}.

    (a) Draw *D*.

    (b) Find the degrees of each of the nodes in *D*.

(c) Find the adjacency matrix of D.

4. What is the maximal number of edges in a digraph with n nodes? Draw such a digraph when $n = 3$.

5. Let $A = \{2, 3, 4, 6, 9, 18, 36\}$ be the set of nodes of a digraph D, and let (x, y) be a directed edge in D if and only if x divides y. Draw D. Characterize the in-degree of a node x in D, and characterize the out-degree of x.

6. Let $A = \{2, 3, 4, 5, \ldots, 200\}$ be the set of nodes of a digraph D, and let (x, y) be an edge in D if and only if x is prime and x divides y. Don't draw D. Find the in-degree of the nodes 12, 36, and 105. Find the out-degree of the nodes 5, 6, and 19. If x is prime what is its in-degree? Characterize the in-degree and the out-degree of a typical node x.

7. Consider the digraph D with the following adjacency matrix:

	a	b	c	d	e	f
a	1	0	0	1	1	0
b	1	0	1	0	0	1
c	0	1	0	1	1	0
d	0	0	0	0	0	1
e	1	1	0	0	0	0
f	1	1	0	0	0	1

Without drawing D, find each of the following. Afterwards, verify your results on a depiction of D.

(a) The out-degree of nodes a, d, and f

(b) The in-degree of nodes a, e, and d

(c) The number of edges in D (use Theorem 1)

8. Draw a digraph with 7 nodes in which each node has out-degree 2.

9. How many edges are there in a digraph with 5 nodes, each of which has out-degree 2? Draw such a digraph.

10. A digraph D has 6 nodes and 7 edges in which each node has out-degree equal to either 1 or 2. Determine the number of nodes in D of each out-degree, and draw such a digraph.

11. Construct a digraph that models the process of scanning a word over $\{a, b\}$ for the purpose of counting the number of occurrences of the pattern *bab* in the case that:

(a) We allow the final b of *bab* to start the next appearance of the pattern (so we count the pattern as appearing twice in the word *babab*).

(b) We don't allow the final b to start the next appearance of the pattern (so we count *bab* as appearing only once in *babab*).

Section 7.2   Paths and Connectivity

Suppose 6 ham radio operators are assigned broadcasting frequencies and receiving frequencies, but suppose not all the assignments are compatible. Suppose we label 6 nodes with the names of the operators—say A, B, C, D, E, and F—and suppose we place a directed edge from operator x to operator y if y can receive the signal broadcast by x. Their situation is represented by the digraph shown in Figure 7.8. We can see that the underlying graph is connected and that there is an undirected path between any two nodes in our graph. However, there is no way for operator D to receive a message from operator F because there is no path from F to D that follows the arrows. In fact, there is no way in which C, D, or E can receive messages from A, B, or F. (A message from C to B, however, can be relayed through D, E, F, and A.)

Figure 7.8

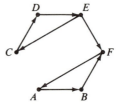

The preceding example points out the need to define a directed path in a digraph and the need to refine our notion of connectivity for digraphs. We address these problems in this section.

> **DEFINITION.** Let D be a digraph and let A be the set of nodes of D. A **directed path** in D from node a_0 to node a_n of length n is a sequence of nodes of the form $(a_0, a_1, a_2, \ldots, a_n)$, where each pair (a_k, a_{k+1}) in the sequence is a directed edge in D. The nodes $a_1, a_2, \ldots, a_{n-1}$ are called the interior nodes of P, and the nodes a_0 and a_n are called its end points. We call P **simple** if all the interior nodes of P are distinct and none is equal to an end point.

A path is "directed" if it follows the orientation of the edges. In Figure 7.8 there is no directed path from any node in the set $\{A, B, F\}$ to any node in the set $\{C, D, E\}$. But the sequence (C, D, E, F, A, B) defines a directed path from C to B. Operator F can receive his own messages as relayed through A and B. The sequence (F, A, B, F) defines a **directed circuit** from F to itself.

> **DEFINITION.** A directed path $(a_0, a_1, a_2, \ldots, a_n)$ is called a **directed circuit** if $a_0 = a_n$. A directed circuit is **proper** if it is simple, has length at least equal to 1, and repeats no edges.

Example 1. Consider the digraph D shown in Figure 7.9. The sequence (b, c, d, f, g, d, a, b) defines a directed circuit in D. Because the path visits the interior node d twice, it is not a simple circuit. However, the sequence (b, c, d, a, b) defines a simple circuit at b. We have taken the short cut at node d and skipped the subcircuit (d, f, g, d).

Figure 7.9

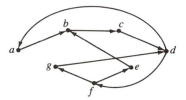

A directed path (or circuit) that is not simple contains a subcircuit at the repeated interior node. By deleting all such subcircuits, we obtain a simple path (or circuit). Suppose that D is a digraph with n nodes and suppose that a and b are distinct nodes in D. If P is a simple directed path from a to b, then the sequence defining P contains at most n nodes (no member of the sequence can be repeated) and hence at most $n - 1$ edges. If P is a simple circuit, then P contains at most n edges. We summarize these facts in the following proposition.

Proposition 1. Let D be a digraph. If there is a directed path from a to b, then there is a simple directed path from a to b. The maximal length of a simple path P in a digraph containing n nodes is $n - 1$ if P is not a circuit and it is n if P is a circuit.

Example 2. Digraph D_1 in Figure 7.10 contains 7 nodes. The sequence (b, c, a, e, f, g, d, b) is a simple directed circuit of length 7. Although the circuit visits all the nodes (it is a directed Hamiltonian circuit), it is not an Euler circuit. No simple, directed Euler circuit exists because there are 10 edges in this digraph and Proposition 1 tells us that the length of the longest simple circuit in D_1 is at most 7. There is no directed Euler circuit at all in D_1. Digraph D_2 in Figure 7.10 has 5 nodes. Although it has a directed circuit of length 6 that visits each of its nodes, it has no such simple directed circuit. It does, however, contain a directed Euler circuit. Digraph D_3 in Figure 7.10 does not contain a directed circuit that visits each of its nodes. In fact, there is no directed path from node a to node b.

Figure 7.10

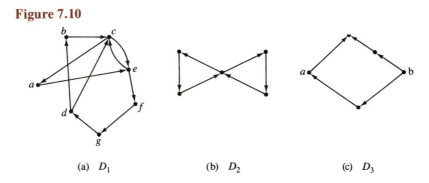

(a) D_1 (b) D_2 (c) D_3

We say that a digraph D is **connected** if its underlying undirected graph is connected. We say that it is **strongly connected** if, for any pair of distinct nodes a and b in D, there is a directed path from a to b. Digraph D_3 in Figure 7.10 is connected, but it is not strongly connected. Even though there is a directed path from b to a, there is no directed path from a to b. Digraph D_1 in Figure 7.10 is strongly connected.

Example 3. Suppose the ham operators we discussed at the beginning of this section wish to acquire additional receiving frequencies so that each can receive messages from every other operator (perhaps through relays). This can be done if operator E can receive messages from operator F. The resulting digraph (Figure 7.11) is then strongly connected.

Figure 7.11

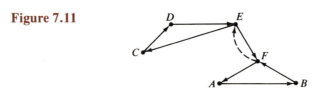

A connected directed digraph in which we can find a directed Euler circuit must be strongly connected. Every node is an endnode of some edge. Hence, to find a directed path from node x to node y, we need only find x on the Euler circuit and then follow the circuit until we encounter y. But, as digraph D_1 in Figure 7.10 shows, a strongly connected digraph does not necessarily have a directed Euler circuit. (Recall that each edge in an Euler circuit appears exactly once.) However, as with undirected graphs, the determination of just which digraphs have directed Euler circuits is simple and easily verified.

Theorem 1.   Euler's Theorem for Digraphs. Let D be a strongly connected digraph. Then D has a directed Euler circuit if and only if the in-degree of each node x in D is equal to its out-degree.

The proof of the undirected version of Euler's theorem requires only a few modifications to serve as a proof of the directed case as well. We thus leave it as an exercise.

We note that digraph D_1 in Figure 7.10 fails the criterion of Theorem 1 at nodes c and d. The in-degree of c is 3; the out-degree is 2. The in-degree of d is 1; the out-degree is 2.

A directed Euler circuit can be used to solve the following old problem, which is known as the "teleprinter's problem."

The Teleprinter's Problem. Given a positive integer n, find the longest possible cyclic sequence of 0's and 1's such that no subsequence of length n is repeated.

For $n = 3$, the sequence $(0, 1, 1, 0, 1)$ fails to solve the teleprinter's problem because the subsequence $(1, 0, 1)$ that starts at the third position repeats beginning at the fifth position and continuing around to the first and second positions. (Hence the word "cyclic.") The sequence $0\ 1\ 1\ 0\ 0\ 1$ has no repeated subsequence of length 3, but it is not of maximal length, as we shall see. Let us note that for a given value of n, there are 2^n different possible subsequences of length n. Thus any sequence in which no subsequence of length n repeats is at most 2^n symbols long, because we must be able to start a new subsequence at each of its 2^n (or fewer) positions.

Now we construct a digraph D through which we can solve the teleprinter's problem for subsequences of length n. We let the nodes of D be the set of strings of 0's and 1's of length $n - 1$. Thus if x is a node of D, then $x = x_1 x_2 \ldots x_{n-1}$, where each x_k, $k = 1, 2, \ldots, n - 1$, is either 0 or 1. Suppose that $y = y_1 y_2 \ldots y_{n-1}$ is also a node of D. We shall place a directed edge e from x to y if $x_2 x_3 \ldots x_{n-1} = y_1 y_2 \ldots y_{n-2}$. We label e with $x_1 x_2 \ldots x_{n-1} 0$ if y ends with 0. If y ends with 1, then we label e with $x_1 x_2 \ldots x_{n-1} 1$. (The digraph for the case $n = 3$ is given in Figure 7.12.) We note the following facts about D:

(i) The out-degree of each node is 2.

(ii) The in-degree of each node is also 2. If $x = x_1 x_2 \ldots x_n$, then there is a directed edge from y to x if and only if $y = y_1 x_1 x_2 \ldots x_{n-2}$, where $y_1 = 0$ or 1.

(iii) If e_j and e_{j+1} are two adjacent edges in any directed path in D, then the first $n - 1$ symbols appearing in the label of e_{j+1} agree with the last $n - 1$ symbols appearing in the label of e_j.

Figure 7.12

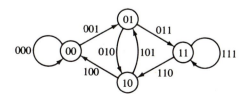

(iv) Each edge in D is labeled with a different string of 0's and 1's of length n. There are 2^n edges in D.

Because the in-degree of each node equals its out-degree, we can find a directed Euler circuit that contains each of the 2^n edges of D. Suppose that we traverse the edges of that Euler circuit in the following order: $e_1, e_2, \ldots, e_{2^n}$. Consider the subsequence of length n that starts at position j—namely $e_j, e_{j+1}, \ldots, e_{j+n-1}$. The first symbol of the label of e_{j+1} is the second symbol of the label of e_j. Similarly, the first symbol of e_{j+2} is the second symbol of e_{j+1}, but (more importantly) it is the third symbol of e_j. In fact, when we take the first symbols of each of the edges e_j through e_{j+n-1}, we spell out the label of the initial edge e_j. This is the key to solving the teleprinter's problem. Let S be the sequence of 0's and 1's formed by taking the first symbol of each of the labels of each of the edges of D in the order defined by our Euler circuit. The length of S is 2^n. Suppose $S = s_1 s_2 \ldots s_{2^n}$. If we take the subsequence of length n starting at position j, we spell out the label of the edge e_j. Because each of the edges of D appears somewhere in our Euler circuit, we can spell out each possible label by starting at the appropriate position in S. Since each label is different, no subsequence of length n repeats in S. We have thus solved the teleprinter's problem and found that the maximal length is indeed 2^n.

We can form a directed Euler circuit in the graph shown in Figure 7.12 by traversing the edges labeled in the following order: 000, 001, 011, 111, 110, 101, 010, 100. Taking the first symbol from each label, we have $S = 00011101$. The sequence S is of the maximal length 8 for $n = 3$. We see that no subsequence of length 3 is repeated. We have solved the teleprinter's problem for $n = 3$.

For purposes of illustration, we have considered only digraphs with a small number of nodes and edges. To answer questions about the existence of a directed path between two given nodes, for instance, we have simply looked at a depiction of the digraph in question. This approach is not feasible when the numbers of nodes and edges are large. For the rest of this section, we investigate the use of the adjacency matrix of a digraph in answering questions about paths and connectivity. The matrix operations needed are readily implemented on the computer, making this approach particularly valuable when we are working with large numbers of nodes and edges.

As in Chapter 5, we shall use Boolean addition and multiplication for our matrix operations. (We urge the reader to review this material.)

Proposition 2. Let D be a digraph and suppose that the nodes of D are $\{a_1, a_2, \ldots, a_n\}$. Let M be the adjacency matrix of D and let a_j and a_k be nodes in D. Then there is a directed path of length q, $q \geq 1$, in D from a_j to a_k if and only if the (j, k)th entry of M^q is equal to 1.

PROOF: We prove this by induction on q. The theorem holds when $q = 1$ by definition of the adjacency matrix M. Let us assume that the theorem holds for q. First we prove that if there is a path P in D of length $q + 1$ from a_j to a_k, then the (j, k)th entry of M^{q+1} is equal to 1. Let P be such a path. We can find m such that $P = \{a_j, \ldots, a_m, a_k\}$. Therefore, stopping at the node before a_k, we have a path of length q from a_j to a_m. By the induction hypothesis, we may assume that the (j, m)th entry of M^q is 1. Because there is an edge from a_m to a_k, the (k, m)th entry in M is also equal to 1. Since $M^{q+1} = M^q \times M$, we see that there is a 1 in the (j, k)th entry of M^{q+1}, as was to be shown.

Now we must show the converse for $n > 1$: if the (j, k)th entry of M^{q+1} is equal to 1, then there is a path of length $q + 1$ in D from a_j to a_k. Again we note that $M^{q+1} = M^q \times M$. Because the (j, k)th entry of M^{q+1} is 1, we know that we can find m such that the (j, m)th entry of M^q is 1 and such that the (m, k)th entry of M is 1. Thus, by the induction hypothesis, we can find a path P of length q from a_j to a_m and an edge e from a_m to a_k. Adjoining the edge e to P, we have found a path of length $q + 1$ from a_j to a_k, concluding our proof.

Example 4. Suppose that the adjacency matrix M for a digraph D with 5 nodes $\{1, 2, 3, 4, 5\}$ is as follows:

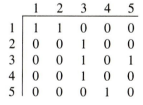

	1	2	3	4	5
1	1	1	0	0	0
2	0	0	1	0	0
3	0	0	1	0	1
4	0	0	1	0	0
5	0	0	0	1	0

Figure 7.13

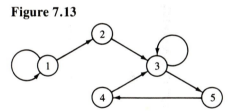

We wish to determine which pairs of nodes are connected by directed paths of length exactly 3. To do this, we compute M^3:

	1	2	3	4	5
1	1	1	1	0	1
2	0	0	1	1	1
3	0	0	1	1	1
4	0	0	1	1	1
5	0	0	1	0	1

Thus we see that, for instance, there is a directed path of length 3 from node 1 to node 2 because the entry in the first row, second column of M^3 is 1. We can verify this on the depiction of digraph D shown in Figure 7.13.

> **DEFINITION.** Let a and b be two nodes in a digraph D. We say that b is **reachable** from a if there is a directed path from a to b of length at least 1.

Suppose that the set of nodes of a digraph D is $\{a_1, a_2, \ldots, a_n\}$ and that m is its adjacency matrix. The **reachability matrix** of D is the $n \times n$ matrix M^R, where the entry in the (j, k)th position is 1 if a_k is reachable from a_j and 0 otherwise. Proposition 2 tells us that if a_k is reachable from a_j, then it can be reached by a directed path of length n or less. Thus the (j, k)th position of M^R is 1 if and only if the (j, k)th position is 1 in one of matrices M^q for $1 \leq q \leq n$. Thus we have

$$M^R = M + M^2 + M^3 + \cdots + M^n$$

where the addition and multiplication operations are Boolean.

Example 5. Suppose that the nodes of a digraph D are $\{a_1, a_2, a_3, a_4, a_5\}$ and that its adjacency matrix M is

	a_1	a_2	a_3	a_4	a_5
a_1	0	1	0	0	0
a_2	0	0	1	0	0
a_3	0	0	0	1	1
a_4	1	0	0	0	0
a_5	0	0	0	1	0

We compute the reachability matrix M^R of D by summing $M + M^2 + M^3 + M^4 + M^5$.

$$
\begin{pmatrix}
0 & 1 & 0 & 0 & 0 \\
0 & 0 & 1 & 0 & 0 \\
0 & 0 & 0 & 1 & 1 \\
1 & 0 & 0 & 0 & 0 \\
0 & 0 & 0 & 1 & 0
\end{pmatrix}
+
\begin{pmatrix}
0 & 0 & 1 & 0 & 0 \\
0 & 0 & 0 & 1 & 1 \\
1 & 0 & 0 & 1 & 0 \\
0 & 1 & 0 & 0 & 0 \\
1 & 0 & 0 & 0 & 0
\end{pmatrix}
+
\begin{pmatrix}
0 & 0 & 0 & 1 & 1 \\
1 & 0 & 0 & 1 & 0 \\
1 & 1 & 0 & 0 & 0 \\
0 & 0 & 1 & 0 & 0 \\
0 & 1 & 0 & 0 & 0
\end{pmatrix}
$$

$$
+
\begin{pmatrix}
1 & 0 & 0 & 1 & 0 \\
1 & 1 & 0 & 1 & 0 \\
0 & 1 & 1 & 0 & 0 \\
0 & 0 & 0 & 1 & 1 \\
0 & 0 & 1 & 0 & 0
\end{pmatrix}
+
\begin{pmatrix}
1 & 1 & 0 & 0 & 0 \\
0 & 1 & 1 & 0 & 0 \\
0 & 0 & 1 & 1 & 0 \\
1 & 0 & 0 & 1 & 0 \\
0 & 0 & 0 & 1 & 1
\end{pmatrix}
$$

$$
=
\begin{pmatrix}
1 & 1 & 1 & 1 & 1 \\
1 & 1 & 1 & 1 & 1 \\
1 & 1 & 1 & 1 & 1 \\
1 & 1 & 1 & 1 & 1 \\
1 & 1 & 1 & 1 & 1
\end{pmatrix}
$$

Each entry in the reachability matrix of D is equal to 1. That is, for each pair of nodes a_j and a_k we have a directed path from a_j to a_k. The digraph D is thus strongly connected.

Proposition 3. A digraph D with two or more nodes is strongly connected if and only if each of the entries in its reachability matrix is equal to 1.

In the paragraph that follows, we present **Warshall's algorithm,** an efficient method for finding the reachability matrix of a digraph and a method that by-passes finding the powers of the adjacency matrix. Much research has been done both in finding efficient algorithms to solve the problems modeled by digraphs and in improving the way in which a digraph can be represented in a computer. These are some of the topics pursued further in the study of the analysis of algorithms and in the study of data structures.

Warshall's Algorithm. Let D be a digraph. Suppose that $\{a_1, a_2, \ldots, a_n\}$ is the set of nodes of D and that M is the adjacency matrix of D. We construct a sequence of matrices $W_0, W_1, \ldots, W_n$ inductively as follows:

(a) Let $W_0 = M$.

(b) Suppose that W_k has been constructed and that $0 \le k < n$. Let $_kW_{h,j}$ be the entry in the (h, j)th position of W_k. To find the matrix W_{k+1}, set
$_{k+1}W_{h,j} = {}_kW_{h,j} \vee ({}_kW_{h,k+1} \wedge {}_kW_{k+1,j})$.

(c) The reachability matrix of D is W_n.

The (h, k)th entry in W_{k+1} is equal to 1 if either the (h, k)th entry in the previously obtained matrix W_k is 1 or both the $(h, k + 1)$th entry and the $(k + 1, j)$th entry in W_k are 1. Thus we need check a total of only three positions in W_k to determine an entry in W_{k+1}.

To see that W_n is the reachability matrix of D, we note the following fact: For $k \ge 1$, the (h, j)th entry in W_k is equal to 1 if and only if there is a path from a_h to a_j whose interior nodes are a subset of the set $\{a_1, a_2, \ldots, a_k\}$. Because there are only n nodes in D, the matrix W_n is the reachability matrix of D. We leave the proof (by induction) of this fact as an exercise.

Example 6. We compute W_1 through W_4 for the digraph D (Figure 7.14) whose adjacency matrix is as follows:

Figure 7.14

$$M = \begin{pmatrix} 1 & 1 & 0 & 0 \\ 0 & 0 & 1 & 1 \\ 1 & 0 & 0 & 0 \\ 0 & 0 & 1 & 0 \end{pmatrix}$$

$$W_1 = \begin{pmatrix} 1 & 1 & 0 & 0 \\ 0 & 0 & 1 & 1 \\ 1 & 1 & 0 & 0 \\ 0 & 0 & 1 & 0 \end{pmatrix} \qquad W_2 = \begin{pmatrix} 1 & 1 & 1 & 1 \\ 0 & 0 & 1 & 1 \\ 1 & 1 & 1 & 1 \\ 0 & 0 & 1 & 0 \end{pmatrix}$$

$$W_3 = \begin{pmatrix} 1 & 1 & 1 & 1 \\ 1 & 1 & 1 & 1 \\ 1 & 1 & 1 & 1 \\ 1 & 1 & 1 & 1 \end{pmatrix} \qquad W_4 = \begin{pmatrix} 1 & 1 & 1 & 1 \\ 1 & 1 & 1 & 1 \\ 1 & 1 & 1 & 1 \\ 1 & 1 & 1 & 1 \end{pmatrix}$$

Because all the entries in W_4 are equal to 1, we know immediately that D is strongly connected.

As a measure of the efficiency of using Warshall's algorithm versus obtaining the reachability by computing the powers of the adjacency matrix, we can count the number of times each process uses the conjunction operation for a digraph with n nodes. To multiply M by M^k, we use the conjunction operation n times for each of the n^2 positions of M^{k+1} for a total of n^3 times for each matrix multiplication. To obtain $M^2, \ldots, M^n$, we must perform $n - 1$ matrix multiplications. Thus to find the reachability matrix by finding the powers of M, we use conjunction $(n - 1)n^3$ times, or approximately n^4 times. In Warshall's algorithm, we use conjunction once for each of the n^2 entries in W_k. Thus to find $W_1, \ldots, W_n$, we use conjunction only n^3 times.

| Section 7.2 | Exercises |

Exercises 1 through 8 all refer to the digraph D given in the figure that follows.

Figure for Exercises 1–8

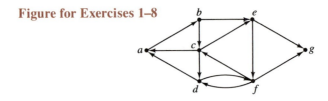

1. Find all the simple directed paths from a to f.

2. Find all the simple directed circuits that contain the edge $e = (c, a)$.

3. Find all the simple directed circuits in D.

4. Find the shortest directed path from b to e.

5. Find the longest simple directed path from a to g. Is there a longest path from a to g if we do not require the path to be simple?

6. Find the longest simple path from d to c.

7. Find two nodes x and y in D such that there is a directed path from x to y but not from y to x.

8. Find all the edges in D that do not lie on any directed circuit.

9. Determine which of the digraphs shown in the accompanying figure is strongly connected.

Figure for Exercise 9

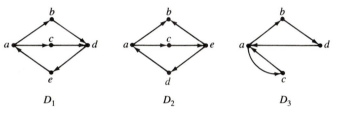

D_1 D_2 D_3

10. The adjacency matrices for two digraphs, D_1 and D_2, follow. Determine which of these digraphs is strongly connected.

D_1:

	a	b	c	d	e	f
a	0	1	0	0	0	0
b	0	0	1	1	0	0
c	0	0	0	1	0	0
d	0	0	0	0	1	1
e	0	0	1	0	0	1
f	1	1	0	0	0	1

D_2:

	1	2	3	4	5
1	0	1	1	1	0
2	0	0	0	1	0
3	0	0	0	0	0
4	0	0	0	0	1
5	0	0	0	1	0

11. Suppose that D is a digraph with at least two nodes. Prove that if the adjacency matrix of D has either a column or a row of 0's, then the digraph cannot be strongly connected. Is the converse also true?

12. Determine which of the digraphs shown in the accompanying figure has (a) a directed Euler circuit or (b) a simple directed Euler circuit.

Figure for Exercise 12

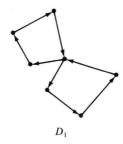
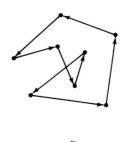

D_1 D_2

13. Suppose that a strongly connected digraph D has n nodes and at least one edge. Suppose also that D has a simple directed Euler circuit. What are the maximal and the minimal numbers of edges in D?

14. Find a different solution to the teleprinter's problem for $n = 3$ by finding a different Euler circuit.

15. Consider the digraph D with the following adjacency matrix M:

	a_1	a_2	a_3	a_4	a_5	a_6
a_1	1	1	0	0	0	0
a_2	0	0	1	1	1	0
a_3	0	0	0	1	0	0
a_4	0	0	0	0	0	0
a_5	0	0	0	0	0	1
a_6	0	0	1	0	0	0

(a) Compute M^3. For which pairs of nodes x, y is there a directed path of length 3 from x to y?

(b) Find the reachability matrix of D by finding the powers of M.

(c) Use the reachability matrix to find nodes x and y for which there is no directed path from x to y.

16. Consider the digraph D given in Exercise 15.

(a) Use Warshall's algorithm to find W_3.

(b) Verify that if the (j, k)th position of W_3 is 1, then there is a directed path in D from a_j to a_k in which the interior nodes are in the set $\{a_1, a_2, a_3\}$.

(c) Use Warshall's algorithm to find the reachability matrix of D. Verify that your results agree with the results of Exercise 15.

| **Weighted Digraphs**

A digraph D is **weighted** if each directed edge (x, y) in D is assigned a positive number n called its **weight** and denoted by $w(x, y)$. The digraph shown in Figure 7.15 is weighted and, for instance, $w(b, c) = 3$. If P is a directed path of length n in D, then the weight of P is the sum of the weights of the n edges occurring in P. Thus the weight of the path $P = (a, b, d, c)$ in the digraph of Figure 7.15 is 7 and the weight of the nonsimple path $P' = (a, b, d, a, e, c)$ is 11.

Figure 7.15

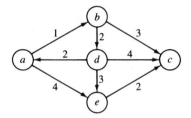

Example 1. A weighted digraph with n nodes $\{a_1, a_2, \ldots, a_n\}$ can be represented by an $n \times n$ **weighted adjacency matrix** W in which we enter the weight $w(a_j, a_k)$ of the edge from a_j to a_k in the (j, k)th position of W. If there is no edge from a_j to a_k, we enter the symbol ∞ in the (j, k)th position. The matrix W that represents the weighted digraph shown in Figure 7.16 follows.

	a	b	c	d	e
a	∞	2	∞	5	∞
b	3	∞	3	∞	∞
c	∞	3	∞	5	∞
d	∞	1	∞	∞	2
e	∞	∞	2	∞	∞

Figure 7.16

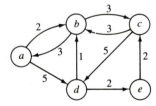

Example 2. A delivery truck transports goods from a warehouse W in the town of Uneuclid to a store located at S. The map shown in Figure 7.17 gives both the

Figure 7.17

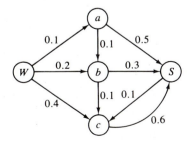

lengths (in miles) and the directions of the streets of the town. The truck must obey the one-way signs. In order to economize on fuel, the driver must find the shortest route from W to S. The legal routes from W to S are as follows:

1. (W, a, S): length 0.6
2. (W, a, b, S): length 0.5
3. (W, a, b, c, S): length 0.9

4. (W, b, S): length 0.5
5. (W, b, c, S): length 0.9
6. (W, c, S): length 1.0

Clearly, the driver should take route 2 or route 4. (A pedestrian who can violate the one-way signs can find a shorter route.)

In Example 2 we found the path of least weight, or the **shortest path,** between two specified nodes in a digraph by listing all the directed paths between those two nodes and comparing their weights. This is not efficient for graphs with large numbers of edges and nodes. For example, consider a digraph with n nodes in which there is a weighted, directed edge from each node x to every other node y. There are

$$1 + (n - 2) + (n - 2)(n - 3) + (n - 2)(n - 3)(n - 4) + \cdots + (n - 2)!$$

different simple paths from x to y. (To see this, note that there is exactly one path of length 1. There are $n - 2$ paths of length 2, $(n - 2)(n - 3)$ paths of length 3, etc.) On such a graph with merely 8 nodes, we would need to compare the lengths of 1,957 different paths.

For the rest of this section we shall investigate a more efficient algorithm (Dykstra's algorithm) by which we can determine the path of least weight from node a to node z in a digraph D. We shall call such a path the "shortest path" from a to z. In a general, unweighted digraph, the length of a path is the number of edges in that path. However, in many applications the weight of an edge in a weighted digraph is interpreted as the length of that edge, as we interpreted it in Example 2. So it has become standard to call the path of least weight between two nodes the "shortest path" and the path of greatest weight the "longest path." We shall continue using this convention unless there is danger of confusion.

During each iteration on Dykstra's algorithm, we shall assign labels to each of the nodes in D—some permanent, but some temporary and subject to change during subsequent iterations of the algorithm. We shall denote the permanent label of a node x by $PL(x)$ and a temporary label by $TL(x)$. The algorithm terminates when the node z acquires a permanent label. That label, $PL(z)$, will be the length of the shortest directed path from a to z.

Dykstra's Algorithm.   To find the shortest path from a to z:

Step 1. (The initial assignment of labels) Assign node a the permanent label 0 so that $PL(a) = 0$. Assign every other node in D the temporary label ∞ so that $TL(x) = \infty$ for each node $x \neq a$. Set $V = a$. (During each iteration of this algorithm, we shall use V to denote the node most recently assigned a permanent label.)

Step 2. (The assignment of new temporary labels) To each node x without a permanent label, compute a new temporary label by setting $TL(x)$ equal to whichever of the following is smaller: the value of its old temporary label or the sum $PL(V) + w(V, x)$. That is, set

$$TL(x) = \min\{\text{old temporary label of } x, \ PL(V) + w(V, x)\}$$

(Note that for any real number n, $\min\{n, \infty\} = n$ and also that $\min\{\infty, \infty\} = \infty$.)

Step 3. (The assignment of a permanent label) Let q be a node with the smallest temporary label that is not ∞. If no q can be found, stop: there is no directed path from a to z. Otherwise, permanently label q with its temporary label; that is, set $PL(q) = TL(q)$. If $q = z$, stop: $PL(z)$ is the length of the shortest path from a to z. Otherwise, set $V = q$ and return to the beginning of step 2.

Example 3. We shall apply Dykstra's algorithm to the digraph shown in Figure 7.18 and find the shortest path from a to z. The label of each node will be enclosed in square brackets.

Step 1. Set $PL(a) = 0$. Label every other node x with $TL(x) = \infty$. Set $V = a$. (We shall doubly circle permanently labeled nodes. See Figure 7.19.)

Figure 7.18

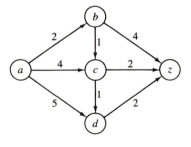

Figure 7.19

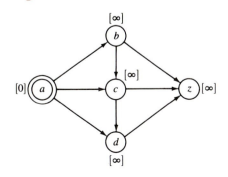

Iteration 1.

Step 2.

$$TL(b) = \min\{\infty, 0 + 2\} = 2$$
$$TL(c) = \min\{\infty, 0 + 4\} = 4$$
$$TL(d) = \min\{\infty, 0 + 5\} = 5$$
$$TL(z) = \min\{\infty, \infty\} = \infty$$

Step 3.   Node b has the smallest new temporary label. Set $PL(b) = 2$.
(See Figure 7.20.) Because $b \neq z$, set $V = b$ and return to step 2.

Figure 7.20

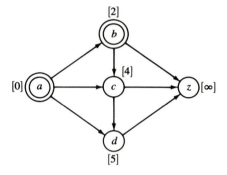

Iteration 2.

Step 2.

$$TL(c) = \min\{4, 2 + 1\} = 3$$
$$TL(d) = \min\{5, \infty\} = 5$$
$$TL(z) = \min\{\infty, 2 + 4\} = 6$$

Step 3.   Node c has the smallest new temporary label. Set $PL(c) = 3$.
(See Figure 7.21.) Because $c \neq z$, set $V = c$ and return to step 2.

Figure 7.21

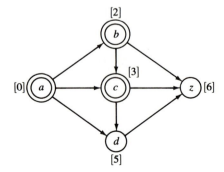

Iteration 3.

Step 2.

$$TL(d) = \min\{5, 3 + 1\} = 4$$
$$TL(z) = \min\{6, 3 + 2\} = 5$$

Step 3. Node d has the smallest new temporary label. Set $PL(d) = 4$. (See Figure 7.22.) Because $d \neq z$, set $V = d$ and return to step 2.

Figure 7.22

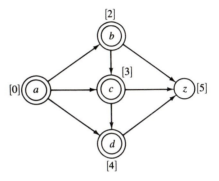

Iteration 4.

Step 2.

$$TL(z) = \min\{5, 4 + 2\} = 5$$

Step 3. Set $PL(z) = 5$ and stop. The weight of the shortest path from a to z is 5. (See Figure 7.23.)

Figure 7.23

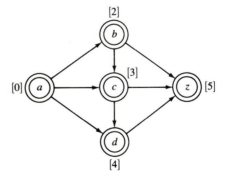

The algorithm as presented gives only the weight of the shortest path from a to z, not the path itself. We can modify steps 1 and 2 of the algorithm so that at each iteration we also label node x with the shortest path $SP(x)$ from a to x that passes through only permanently labeled nodes (except perhaps for x itself). We shall use the following notation. If no such path exists, we shall set $SP(x) = \emptyset$. If P is a path $(a, v_1, v_2, \ldots, V)$ and P' is the path $(a, v_1, v_2, \ldots, V, x)$ obtained by adjoining the edge (V, x) to P, we shall denote P' by (P, x). We modify the algorithm by adding the following instructions to step 1 and step 2.

Step 1.   Also label node a with $SP(a) = (a)$. Also label every other node x in D with $SP(x) = \emptyset$.

Step 2.   (We change $SP(x)$ only when we change the temporary label of x.) If the old label of x is not equal to its new label, set

$$SP(x) = (SP(V), x)$$

Example 3 (continued). In Figures 7.24 through 7.28, we apply the modified algorithm to the graph shown in Figure 7.18. The path from a to x is listed before the label $TL(x)$ or $PL(x)$ in the square brackets.

Step 1.

Figure 7.24

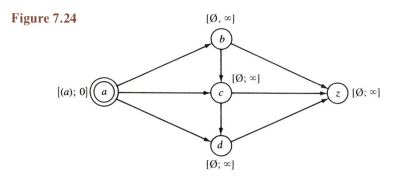

Iteration 1.

Figure 7.25

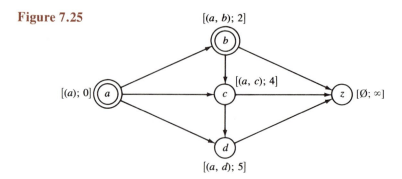

Iteration 2.

Figure 7.26

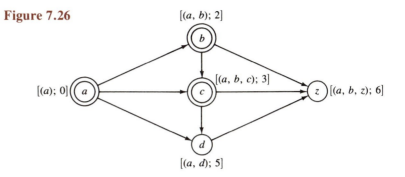

Iteration 3.

Figure 7.27

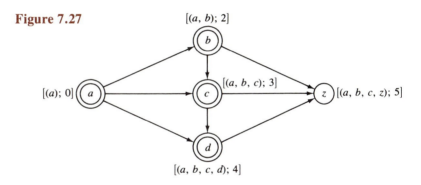

Iteration 4.

Figure 7.28

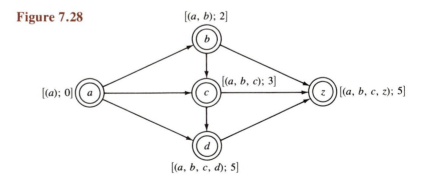

Thus at the termination of the algorithm, we indeed find z labeled with the shortest path from a to z. We may verify that every other permanently labeled node x is labeled with the shortest path from a to x.

Let us now see why Dykstra's algorithm produces the weight of the shortest directed path from a to z in D. Suppose that the algorithm permanently labeled the nodes in D in the following order: $a, x_1, x_2, \ldots, x_{n-1}, z$ so that x_i was labeled after the ith iteration of the algorithm—that is, after the ith application of step 3. We shall prove by induction on i that, after its ith iteration, the algorithm has accomplished the tasks described in the following three statements:

(i) If $j < i$, then $PL(x_j) \leq PL(x_i)$.

(ii) If x has a temporary label, then $TL(x)$ is the shortest path from a to x through the set $Q = \{a, x_1, \ldots, x_{i-1}\}$.

(iii) $PL(x_i)$ is the weight of the shortest path from a to x_i overall.

[Note that statement (iii) is the purpose of the algorithm when $x_i = z$.] The three statements are certainly true when $i = 0$ and the only permanently labeled node is node a with $PL(a) = 0$. Assume now that we have completed the ith application of step 3 and that the three statements hold so far. Recall that we have set $V = x_i$, the last permanently labeled node. Suppose that x is a node that is temporarily labeled with the weight of the shortest path from a to x through the set $Q = \{a, x_1, \ldots, x_{i-1}\}$. We change the temporary label on x in the $(i + 1)$th application of step 2 only if there is a shorter path from a to x that passes through $Q' = Q \cup \{V\}$. To see this, first note that any path from a to x through V that does not visit V just before x can be replaced with a shorter path whose interior nodes do not include V. This is because all the nodes in Q were labeled previous to V, and their distances to a are thus shorter. So if the shortest path from a to x through the nodes in Q' passes through V, it must visit V just before ending at x, and its weight would be $PL(V) + w(V, x)$. The term $\min\{$old $TL(x), PL(V) + w(V, x)\}$ compares this weight with the weight of the shortest path from a to x through Q and chooses the smaller number. We have thus shown that statement (ii) holds after the $(i + 1)$th iteration of the algorithm.

The ith application of step 3 selected $V = x_i$ for permanent labeling because its temporary label was the smallest. Thus the temporary labels on the remaining nodes are all at least as great as $PL(V)$. We replace the temporary labels in the $(i + 1)$th application of step 2 only if the weight of the path through V is smaller. But since the weights of the edges are all positive, the weight of a path from a to x through V must be greater than the weight of the shortest path from a to V. After the $(i + 1)$th application of step 2, the new temporary labels are still at least as big as $PL(V)$. Thus when we apply step 3 and permanently label x_{i+1} with its temporary label, we have that $PL(x_i) \leq PL(x_{i+1})$. Hence statement (i) holds after the $(i + 1)$th completion of step 3.

We shall now prove that statement (iii) holds after $i + 1$ iterations by contradiction. Suppose that $SP(x_{i+1})$ is the shortest path from a to x_{i+1} through Q', as guaranteed by statement (ii), but that it is not the shortest path overall. Then we could find another path $P = \{a, z_1, \ldots, z_j, y_1, \ldots, y_m, x_{i+1}\}$ with the smallest weight overall such that each of the z_i's was permanently labeled but such that none of the y_i's was permanently labeled. The length of the first part of the path $\{a, z_1,$

. . . , z_j, y_1} is $TL(y_1)$. Because x_{i+1} was chosen for permanent labeling instead of y_1, we know that $PL(x_{i+1}) \leq TL(y_1)$. Thus the weight of the entire path P is greater than that of $SP(x_{i+1})$, a contradiction. So statement (iii) holds after the $(i + 1)$th iteration of the algorithm, and, when we have permanently labeled z, we have found the weight of the shortest path from a to z.

| **Section 7.3** | **Exercises** |

Exercises 1 through 6 and Exercises 12 and 13 refer to the weighted digraph shown in Figure 7.29.

Figure 7.29

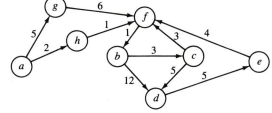

1. Find the weight of the path $P = (f, b, c, d)$.

2. Find the weight of the cycle $C = (b, d, e, f, b)$.

3. Find the weights of all the simple directed paths from a to e.

4. Find the directed path of least weight from a to e.

5. Find the weight of the simple directed path of greatest weight from a to e. Is there a directed path, not necessarily simple, of greatest weight from a to e?

6. Find the path of greatest weight from a to f. How does this situation differ from that given in Exercise 5?

Exercises 7 through 11 refer to the weighted digraph with the weighted adjacency matrix W given in the following table.

	a	b	c	d	e	f
a	∞	1	∞	1	1	∞
b	∞	∞	2	4	∞	∞
c	∞	∞	∞	3	∞	∞
d	∞	∞	2	∞	∞	1
e	∞	∞	∞	2	∞	1
f	∞	∞	3	∞	∞	∞

7. Find the weights of the following edges: (a, b), (d, c), and (c, d).

8. Find the weights of the following paths:

(a) $P_1 = (a, b, d, c)$ (b) $P_2 = (a, e, d, f, c)$

9. Find the edge of least weight that ends at d.

10. Find all paths of length 2 that start at d, and then determine which is the path of least weight.

11. Find the path of least weight from a to f by finding all simple directed paths from a to f and comparing their weights.

12. Apply Dykstra's algorithm to the weighted digraph shown in Figure 7.29 to find the shortest path (the path of least weight) from node a to node e. After each iteration, list the nodes and their labels. Verify that after each iteration, each node x that is permanently labeled is labeled with the shortest path from a to x and that path passes through only nodes that have been labeled previous to x.

13. Apply Dykstra's algorithm again to the digraph shown in Figure 7.29 to find the shortest path (the path of least weight) from e to a.

14. Apply Dykstra's algorithm to the digraph whose adjacency matrix is given for Exercises 7 through 11 to find the shortest path (the path of least weight) from node a to node f.

Section 7.4   **Acyclic Digraphs**

A digraph is **acyclic** if it contains no proper directed circuits. Both digraph D_1 and digraph D_2 shown in Figure 7.30 are acyclic. The first is clearly so (its underlying undirected graph is a tree), but digraph D_2 is also acyclic. We can verify that every

Figure 7.30

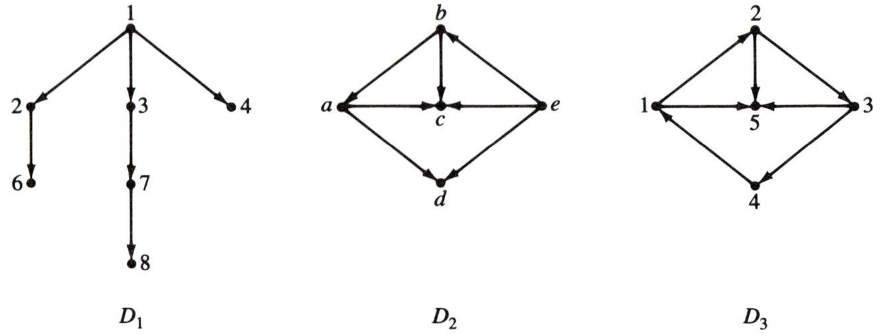

D_1 D_2 D_3

undirected circuit violates the orientation of at least one edge. For instance, the circuit (a, c, e, b, a) violates the orientation of the edge (e, c). Digraph D_3 is not acyclic because $(1, 2, 3, 4, 1)$ is a directed circuit.

A digraph is acyclic if and only if every diagonal entry in its reachability matrix is 0. We can verify this by inspecting the reachability matrix of D_2, which is given in Table 7.2.

Table 7.2

	a	b	c	d	e
a	0	0	1	1	0
b	1	0	1	1	0
c	0	0	0	0	0
d	0	0	0	0	0
e	1	1	1	1	0

In any digraph, a node with in-degree 0 is called a **source** and a node with out-degree 0 is called a **sink**. For example, node 5 in D_3 is a sink and node e in D_2 is a source. Any path that includes a source must start at that source, and any path that includes a sink must end at that sink. In general, a digraph need not have a source or a sink (D_3 has no source), but every acyclic digraph must have at least one of each. Thus node e is a source in D_3 and node c is a sink. In D_1, node 1 is a source and nodes 4, 6, and 8 are all sinks.

Theorem 1. Let D be an acyclic digraph. Then D has at least one source and one sink.

PROOF: Let x be any node in D. Starting at x, construct the longest path possible and call it P. (Such a path must exist because D is acyclic and has only a finite number of nodes.) Suppose that $P = (x, v_1, v_2, \ldots, v_n)$. Then the last node in P, v_n, is a sink (or else we could extend P). Similarly, to find a source in D, find the longest path possible that terminates at x. If $P' = (v_0, v_1, \ldots, x)$ is such a path, then the first node of P', v_0, is a source.

In the adjacency matrix of a digraph, the column labeled with a source contains only 0's and the row labeled with a sink contains only 0's.

We can use Theorem 1 to prove that the vertices in an acyclic digraph D may be numbered in such a way that any edge leads from a lower number to a higher number, and hence the nodes on any directed path in D are visited in ascending order. Such an enumeration is called a **topological sorting** of the nodes of D. The nodes of D_1 given in Figure 7.30 are topologically sorted.

Theorem 2. The nodes of an acyclic digraph D can be topologically sorted.

PROOF: We prove this by induction on the number of nodes in D. If D has only 1 node x, any number assigned to x gives a topological sorting and so the theorem holds. Assume that the theorem holds for acyclic digraphs with n nodes, and assume that D has $n + 1$ nodes. By Theorem 1 we know that D has a source we shall call s. Label s with 0. Then remove s and all its incident edges from D. What remains is an acyclic digraph with n nodes—call it D'. By the induction hypothesis we may topologically sort the nodes of D'. Start the numbering of such a sorting at 1. This topological sorting of D', together with the assignment of 0 to s, yields a topological sorting of all the nodes of D.

The proof of Theorem 2 gives us a method for obtaining a topological sorting: Find and label a source; then remove it and repeat the process. We shall use this method to find a topological sorting of the nodes of the acyclic digraph shown in Figure 7.31(a). At each stage we shall circle and label the source. We begin in

Figure 7.31

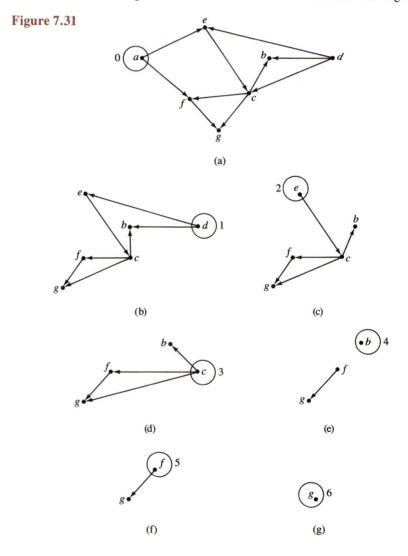

(a)

(b)

(c)

(d)

(e)

(f)

(g)

Figure 7.31(a) and continue through Figure 7.31(g). The graph, with its nodes topologically sorted (and renamed accordingly), is shown in Figure 7.32.

Figure 7.32

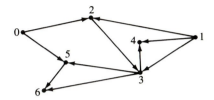

Example 1. A student in a certain liberal arts college must complete all the courses on the following list in order to fulfill the requirements of a major in mathematics. Table 7.3 gives each course, its abbreviation, and its prerequisites.

Table 7.3

Course	Prerequisite
Calculus 1 (CLC 1)	none
Calculus 2 (CLC 2)	CLC 1
Calculus 3 (CLC 3)	CLC 2
Linear algebra (LA)	CLC 1
Ordinary differential equations (ODE)	CLC 3, LA
Partial differential equations (PDE)	ODE
Discrete mathematics (DM)	CLC 1
Modern algebra 1 (MA 1)	LA, DM
Modern algebra 2 (MA 2)	MA 1
Advanced calculus 1 (AC 1)	CLC 3, DM
Advanced calculus 2 (AC 2)	AC 1
Numerical analysis (NA)	PDE

The math department wants to number these courses so that the prerequisite for any given course has a lower number than the course itself. An ambitious math student wishes to determine the least number of semesters needed to complete the major. We can solve both problems by constructing a digraph in which each of the courses is represented by a node and in which we have placed a directed edge from course x to course y whenever x is listed as a prerequisite for y. The resulting digraph is shown in Figure 7.33. The topological sorting of the nodes numbers the courses.

The reader may verify that this digraph is acyclic. It must be: a circuit would require a course to be a prerequisite for itself—an insurmountable difficulty at even the best of schools. We have topologically sorted the courses. Such a numbering of the courses meets the requirement that each course have a higher number than any of its prerequisites. A student will need at least 6 semesters to complete this major. To see this, we find the longest directed path in the digraph and count its nodes. That path is (101, 102, 103, 206, 307, 312), and it will take a student at least 6 semesters to complete the major.

Figure 7.33

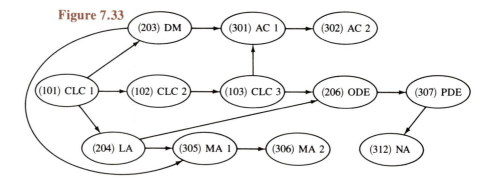

A directed graph that is not acyclic does not necessarily have a longest path between any two given nodes. For instance, if $(a, \ldots, x, \ldots, b)$ is a path from a to b and if there is a circuit at x, then one can obtain a path of arbitrarily large length by looping around the circuit at x as often as needed. This cannot be done in an acyclic graph, and so there is a longest path (if there is any at all) between any two nodes in such a graph. Similarly, there is a path of maximal weight between any two nodes in a weighted acyclic digraph.

Example 2. One rushed evening the operator at a computer center finds he must process 10 programs labeled $A, B, C, \ldots, J$. He cannot batch-process the entire set because some of the programs require the results of others in order to run. The programs, their requirements, and their running times are given in Table 7.4. The operator wishes to determine the earliest time at which he can start running each of the programs after starting the set and also the minimal amount of time required to process the entire set.

Table 7.4

Program	Requirements	Time (in hours)
A	none	0.5
B	none	0.7
C	A	0.3
D	B	0.4
E	C, D	0.6
F	B	0.3
G	A	0.5
H	B	0.6
I	H	0.4
J	G, E, F	0.5

We shall model this problem with a weighted acyclic digraph. The programs and their times will be represented by weighted edges. Each node will stand for the starting of the programs represented by the edges leaving it. If program x is a requirement for program y, then edge x will enter the node that represents the start of program y. The digraph is shown in Figure 7.34, wherein we have also topologically sorted and labeled the nodes.

Figure 7.34

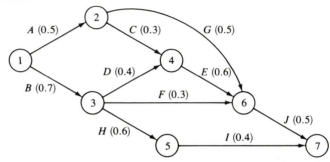

No program can be started until all the programs entering its starting node are completed. Thus the earliest time any program x can be started is the length of the longest path from node 1 to the node at which x begins. For instance, the earliest time at which program E can be started is 1.1 hours after the beginning of the job, after B and D are completed. (Programs A and C must also be completed before E can begin, but these programs require only 0.8 hour.) We shall denote the weight of the longest path from node 1 to node x by $L(1, x)$. These lengths are listed in column 2 of Table 7.5. We shall denote the longest path from node x to node 7 by $L(x, 7)$. These lengths are listed in column 3 of Table 7.5. (We shall use column 3 shortly.) The earliest times at which each of the programs can be started are given in Table 7.6.

Table 7.5

Node x	L(1, x)	L(x, 7)
1	0	2.2
2	0.5	1.4
3	0.7	1.5
4	1.1	1.1
5	1.3	0.4
6	1.7	0.5
7	2.2	0

Node 7 represents completion of all the programs. The earliest time of completion is the weight of the longest path from 1 to 0.7. It is 2.2 hours along the path (1, 3, 4, 6, 7). The edges on this path represent programs B, D, E, and J, respectively. Any delay in the starting of these programs or any time overruns in their execution will delay the completion of the entire set of programs. However, not all the programs need to be started at the earliest possible time. Our operator can relax

Table 7.6

Program	Starting Node	Earliest Time	Latest Time
A	1	0	0.3
B	1	0	0
C	2	0.5	0.8
D	3	0.7	0.7
E	4	1.1	1.1
F	3	0.7	1.4
G	2	0.5	1.2
H	3	0.7	1.2
I	5	1.3	1.8
J	6	1.7	1.7

a bit and, for instance, start program H as late as 1.2 hours after the initial set (as opposed to starting after 0.7 hour) and still finish the entire set in 2.2 hours. To see this, we note that H is required for I, and together these programs require 1 hour. Thus the entire set of programs can be completed on time if H is started no later than 1 hour before the projected completion time. More generally, we can determine the latest possible starting time for each program x by finding the longest path from the terminal node of the edge representing x to node 7 and then subtracting the sum of this and the weight of x from 2.2. For example, to determine the latest starting time of program C, we find that the longest path from node 4 (the terminal node of C) to node 7 is 1.1. The weight of C is 0.3. Thus the latest starting time of C is $2.2 - (1.1 + 0.3) = 0.8$. The latest starting time for each of the programs is given in column 4 of Table 7.6. The operator will note that the earliest and latest times at which programs B, D, E, and J can be started are identical, reconfirming our observation that any delay in *these* programs will delay completion of the entire set.

In this example we have analyzed a very simple scheduling problem in terms of a digraph. But the same reasoning can be used very successfully to obtain optimal schedules in activities comprising many events and complicated precedence relations. We urge the interested reader to pursue this subject further in an operations research course.

Section 7.4 | **Exercises**

1. Determine whether any of the digraphs shown in the figure at the top of page 327 are acyclic.

2. The adjacency matrices M_1 and M_2 of digraphs D_1 and D_2 are given in Table 7.7. Find the reachability matrices of each of these digraphs and determine which is acyclic.

Figure for Exercise 1

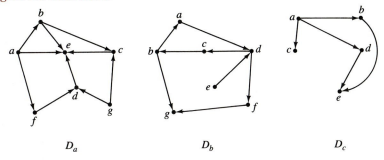

D_a D_b D_c

Table 7.7 D_1:

	a	b	c	d	e
a	0	1	1	0	0
b	1	0	1	1	0
c	0	0	0	0	0
d	1	0	1	1	0
e	1	0	0	1	0

D_2:

	1	2	3	4	5	6
1	0	1	1	1	1	1
2	0	0	1	0	0	0
3	0	0	0	0	0	0
4	0	0	1	0	0	0
5	0	0	0	1	0	1
6	0	0	0	0	0	0

3. Find all the sources and sinks in each of the digraphs given in the figure accompanying Exercise 1.

4. Find all the sources and sinks in D_1 and D_2 by inspecting the rows and columns of their adjacency matrices given in Table 7.7.

5. Topologically sort the nodes of the accompanying digraph. Start your numbering at the source node a.

Figure for Exercise 5

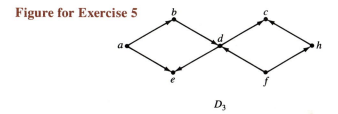

D_3

6. Find a different topological sorting of the nodes of the digraph that accompanies Exercise 5 by starting at the source node f.

7. Find the adjacency matrix of the digraph that accompanies Exercise 5. Label the rows and columns of the matrix in the order of the topological sorting you obtained in Exercise 5. Repeat the problem, labeling the rows and columns in the order of the topological sorting you obtained in Exercise 6.

8. Without drawing the digraph itself, topologically sort the nodes of the digraphs with the adjacency matrix given in Table 7.8. When you find a source node, cross out its row and column. Find the next source to label by inspecting the smaller matrix.

Table 7.8

	a	b	c	d	e	f	g	h	i	j
a	0	0	1	0	1	0	0	0	0	1
b	0	0	0	0	0	0	0	0	1	0
c	0	0	0	0	0	0	0	1	0	0
d	1	0	1	0	0	0	0	1	0	0
e	0	1	0	0	0	1	0	0	0	0
f	0	0	0	0	0	0	0	0	0	0
g	0	0	0	0	0	1	0	0	0	0
h	0	0	0	0	0	0	1	0	0	0
i	0	0	1	0	0	1	0	0	0	1
j	0	0	0	0	0	1	0	0	0	0

9. The offices of the administrative staff of a certain corporation are to be numbered in such a way as to reflect the corporation's power structure. If X must answer to Y, then Y must have the office with the higher number. The initial of each administrator is given in the first column in Table 7.9. That of his or her immediate superior is given in the second column. Model the power structure with the appropriate acyclic digraph, and number the offices accordingly.

Table 7.9

Administrator	Superior
L	N
I	M
H	M
F	O
G	O
A	C
C	E
B	C
E	H
D	E
M	N
J	L
K	L
O	H
N	no one

10. The computer operator discussed in Example 2 finds that he has the wrong information about the running times of programs B and D. Program D runs

0.7 hour and program *B* runs 0.3 hour. Re-solve his problem, taking this new information into account.

Finite State Machines

Throughout this text we have drawn our examples from formal languages: sets of words over an alphabet of symbols. In Chapter 2 we investigated ways of defining such languages; in Chapter 3 we used injective functions to encode symbol sets; in Chapter 4 we developed tools for counting the numbers of words in certain finite languages; in Chapter 5 we investigated relations, partial orders, and equivalence relations on languages. In this section we shall use digraphs to model the process of determining when a given word is in a language. The digraph, called a finite state machine, accepts (or rejects) a word entered as input and thus serves as a "language acceptor." (*Note*: In some of the digraphs that follow, we have allowed parallel directed edges—that is, two or more edges with the same orientation between a pair of nodes.)

Let $E = \{a, b\}$ be a symbol set, and consider the digraph G shown in Figure 7.35. Each node has exactly two out-going edges, one labeled a and the other labeled b. Two nodes have been specially designated, one with the label B (for "begin") and another with the label F (for "finish").

Figure 7.35

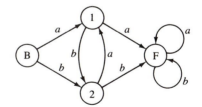

Let x be a word in E^* so that $x = x_1 x_2 \ldots x_n$, where each x_i is a symbol in E. The word x defines a path in G. Starting at the node labeled B, we simply follow the labeled edges in the same sequence as that in which the symbols appear in the word x. Thus the word $x = abbbab$ defines a path that visits the nodes of G in the following order: B, 1, 2, F, F, F, F. (We always begin at the node labeled B.) After the initial visit to F, we follow the self-loops and return there three times. We shall consider a word "accepted" by G if the path it defines ends at F. Thus *abbbab* is accepted. The path defined by the word *abab* ends at 2 rather than F. It is not accepted. We can verify that a word w defines a path that ends at F if and only if the word w contains either *aa* or *bb*. Thus the graph G accepts the language $L(G) = \{x : x \text{ is a word in } E^* \text{ in which either } aa \text{ or } bb \text{ appears}\}$.

We formalize the notions presented in this discussion in the following definition.

> **DEFINITION.** Let $E = \{a_1, a_2, \ldots, a_m\}$ be a finite set of symbols. A directed graph G is called a **finite state machine** for E if each of the following conditions hold.
>
> > (a) Each node of G has exactly m out-going edges, each labeled with a different symbol of E.
> >
> > (b) There is exactly one node labeled B (for "begin").
> >
> > (c) At least one of the nodes of G is labeled F (for "finish").
>
> The nodes of G are called **states** and the nodes labeled with F are called **accepting states.** The language $L(G)$ defined by G is the set of all words defining paths that, when they begin at node B, end at an accepting state labeled with F. We say that $L(G)$ is the language **accepted** by G.

Example 1. Let $E = \{a, b, c\}$ and let G be the finite state machine given by the graph shown in Figure 7.36. The word *aabc* defines the path (B, 2, F, F). The word

Figure 7.36

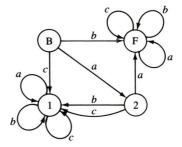

is accepted because the path ends at an accepting state. The word *abc* is rejected. If a word begins with c, with ab, or with ac, its path must visit node 1, a dead end that is not an accepting state. However, the path of a word that begins with b or with aa ends at F, an accepting state. Thus the language accepted by G is the set of all words over $\{a, b, c\}$ that begin with b or aa.

For the rest of the examples in this section, we shall take our symbol set to be the set $E = \{a, b\}$ unless we note otherwise explicitly.

Example 2.   Parity check. The finite state machine shown in Figure 7.37 accepts only words that contain an even number of a's. The initial state B is also an accepting state.

Figure 7.37

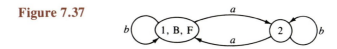

Example 3. The machine shown in Figure 7.38 accepts only words in which the pattern *abb* appears.

Figure 7.38

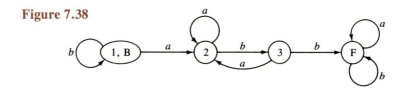

The term "states" used for the nodes of a finite state machine is quite suggestive of the modeling process involved in constructing a machine to accept a given formal language. For instance, consider the parity check machine given in Example 2 and consider the process of scanning a word from left to right in order to determine the parity of the number of *a*'s. At any position in the word, a scanner is in one of two possible states: the state of having detected an even number of *a*'s or the state of having detected an odd number of *a*'s. The scanner switches states whenever another *a* is encountered. The finite state machine also has two states. A path is at state 1 if an even number of *a*'s have been encountered and at state 2 if not.

Suppose we wish to design a finite state machine that accepts a word *w* if *aa* appears somewhere in *w* and *bb* appears somewhere later in the word *w*. (Thus *baababb* and *bbaabb* are accepted, whereas *bbaa* is rejected.) Let us analyze the states encountered as we inspect a word from left to right. (Use the word *bbabaababbaaa*, for instance.)

1. No *a* is encountered.
2. One *a* is encountered and we are looking for another to follow immediately.
3. The subword *aa* is encountered and we are looking for *b*.
4. The subword *aa* is encountered and *b* is encountered later; we are looking for another *b* to follow immediately.
5. Both *aa* and *bb* have been detected and the word is accepted.

A scanner will change from state 1 to state 2 when it has found the first *a* in the word under investigation. If that *a* is immediately followed by another, then the first part of the task (finding *aa*) is accomplished and it enters state 3. But if that *a* is followed immediately by *b*, then the search for the first *a* of the pair *aa* must begin again. So the scanner is back in state 1. The states listed above and the transitions between them are reflected in the states of the finite state machine shown in Figure 7.39.

Figure 7.39

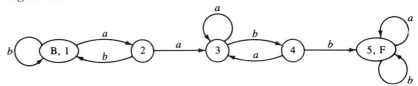

Example 4. Suppose we wish to design a finite state machine that will accept words over {*a*, *b*} in which the number of *a*'s is congruent to 1 modulo 3 and the number of *b*'s is congruent to 2 modulo 3. As we scan a word from left to right, there are 9 possible states: the number of *a*'s can be congruent to 0, 1, or 2, and the number of *b*'s can be congruent to 0, 1, or 2. We can represent each state by an ordered pair (*n*, *m*), where *n* is the number of *a*'s (mod 3) and *m* is the number of *b*'s (mod 3). For example, if 7 *a*'s and 5 *b*'s have been encountered, then we are in state (1, 2). If the next letter encountered is an *a*, then we enter state (2, 2). If it is *b*, then we enter state (1, 0). We start with no *a*'s or *b*'s encountered, in (0, 0). A word is accepted if its final state is (1, 2). The finite state machine is shown in Figure 7.40.

Figure 7.40

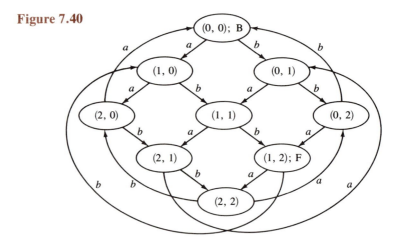

Let *G* be the graph of a finite state machine with symbol set *E* and state set *S*. We can use the directed edges of *G* to define a function $f: S \times E \rightarrow S$ as follows: Let *x* be a symbol in *E* and let *s* be a state in *S*. Let $f(s, x) = z$ if there is a directed edge in *G* from state *s* to state *z* labeled with *x*. The function *f* is called the **transition function** of the finite state machine.

Example 5. Let *G* be the finite state machine defined in Example 3. The transition function for this machine is given by the following diagram:

$$(1, a) \longrightarrow 2$$
$$(1, b) \longrightarrow 1$$
$$(2, a) \longrightarrow 2$$
$$(2, b) \longrightarrow 3$$
$$(3, a) \longrightarrow 2$$
$$(3, b) \longrightarrow 4$$
$$(4, a) \longrightarrow 4$$
$$(4, b) \longrightarrow 4$$

If we know the transition function for a finite state machine with symbol set E and state set S, we can reconstruct its digraph or **state diagram**. We take S to be the set of nodes and place a directed edge labeled with the symbol x from state s to state z if and only if $f(s, x) = z$.

Example 6. Let $E = \{a, b\}$ and let S be the state set $\{1, 2, 3, 4\}$. Suppose that 1 is the beginning state and 4 the accepting state. Let f be the transition function defined by the following diagram:

$$(1, a) \longrightarrow 2$$
$$(1, b) \longrightarrow 3$$
$$(2, a) \longrightarrow 4$$
$$(2, b) \longrightarrow 3$$
$$(3, a) \longrightarrow 2$$
$$(3, b) \longrightarrow 4$$
$$(4, a) \longrightarrow 4$$
$$(4, b) \longrightarrow 4$$

The state diagram for f is shown in Figure 7.41. The language accepted by this finite state machine is the set of all words over $\{a, b\}$ in which either aa appears or bb appears.

Figure 7.41

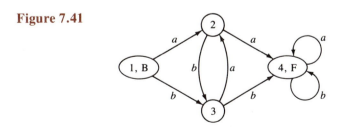

These considerations lead us to the following equivalent but nongraphical definition of a finite state machine. (Any finite state machine can be described by either definition.)

> **DEFINITION.** A finite state machine M over a finite alphabet E consists of a finite set of states S and a transition function $f : S \times E \to S$. Exactly one state in S is designated as a starting state, and a finite subset of S is designated as the set of accepting states.

Example 7. Define the finite state machine M over the alphabet $E = \{a, b\}$ as follows. Let $S = \{1, 2, 3, 4, 5\}$, where 1 is the starting state and 4 and 5 are the accepting states of M. Let $f : S \times E \to S$ be defined by the following diagram:

$$(1, a) \longrightarrow 2 \qquad (3, b) \longrightarrow 5$$
$$(1, b) \longrightarrow 3 \qquad (4, a) \longrightarrow 4$$
$$(2, a) \longrightarrow 4 \qquad (4, b) \longrightarrow 2$$
$$(2, b) \longrightarrow 2 \qquad (5, a) \longrightarrow 3$$
$$(3, a) \longrightarrow 3 \qquad (5, b) \longrightarrow 5$$

The state diagram for M is shown in Figure 7.42. The language accepted by M is the language in which all words that begin with a also end with a and all words that begin with b also end with b.

Figure 7.42

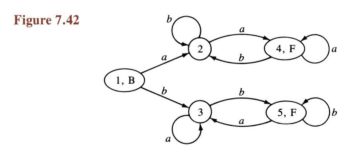

In a nongraphical description of a finite state machine, the notion of a path in the state diagram is replaced by the notion of a production. A **production** is a finite sequence of states defined by a word x of E^*. Let $x = a_1 a_2 \ldots a_n$ be a word in E^*. Then x defines a production $s_1 s_2 \ldots s_n$ inductively as follows: s_1 is the starting state of the machine. If s_i is the ith state of the production and a_i is the ith symbol in the word x, then we set s_{i+1} equal to $f(s_i, a_i)$. A word is in the language accepted by the machine if and only if the production it defines ends with an accepting state.

Example 8. Let M be the machine defined in Example 7. We obtain the production defined by the word *abaaba* as follows: $(a, 1) \to 2, (b, 2) \to 2, (a, 2) \to 4, (a, 4) \to 4, (b, 4) \to 2, (a, 2) \to 4$. Thus we have the production $\{1, 2, 2, 4, 4, 2, 4\}$. Because 4 is an accepting state, the word *abaaba* must be in the language accepted by this machine.

Example 9. Suppose that G is a finite state machine over the symbol set $E = \{a_1, a_2, \ldots, a_n\}$ and that the states of G are the members of the set $S = \{s_1, s_2, \ldots, s_q\}$. The transition function $f : S \times E \to S$ can be readily transformed into a program that accepts the words in $L(G)$. The statement $f(s_i, a_k) = s_j$ becomes "If (state $= s_i$) and (symbol $= a_k$) then state $:= s_j$."

The Pascal program "Acceptword," which follows, accepts words over $\{a, b\}$ that contain an even number of a's and an odd number of b's and words that contain an odd number of a's and an even number of b's. The finite state machine for this language and its transition function are given in Figure 7.43.

Figure 7.43

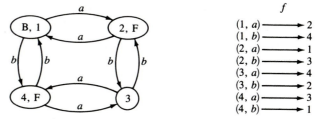

Program Acceptword(Input,Output);
```
{Scans input word composed solely of lowercase letters  a  and  b ,  }
{and accepts the word if and only if it is composed of an even number}
{of  a's  and an odd number of  b's  or an odd number of  a's  and  }
{an even number of  b's.                                             }
const
  maxchar=50;
var
  wordlen,letindex,counter,state:integer;
  word:array[1..maxchar] of char;
  invalidword:boolean;
{---------------------------------------------------------------------}
procedure state1(cursymbol:char;
                 var state:integer);
begin {state1}
  if cursymbol = 'a' then
    state:=2
  else
    state:=4;
end; {state1}
{---------------------------------------------------------------------}
procedure state2(cursymbol:char;
                 var state:integer);
begin {state2}
  if cursymbol = 'a' then
    state:=1
  else
    state:=3;
end; {state2}
{---------------------------------------------------------------------}
```

```
procedure state3(cursymbol:char;
                 var state:integer);
begin {state3}
  if cursymbol = 'a' then
    state:=4
  else
    state:=2;
end; {state3}
{-------------------------------------------------------------------}
procedure state4(cursymbol:char;
                 var state:integer);
begin {state4}
  if cursymbol = 'a' then
    state:=3
  else
    state:=1;
end; {state4}
{===================================================================}
begin {main program}
  invalidword:=false;
  write('Number of letters in word?  (maximum is ',maxchar:1,'): ');
  readln(wordlen);
  write('Ready for a ',wordlen:1,'-letter word (a''s  b''s only): ');
  for letindex:=1 to wordlen do
    read(word[letindex]);
  writeln;
  write('The word received is  ');
  for letindex:=1 to wordlen do
    write(word[letindex]);
  writeln;
  state:=1; {Start in State 1}
  for counter:=1 to wordlen do
    if word[counter] in ['a','b'] then
      case state of
        1:state1(word[counter],state);
        2:state2(word[counter],state);
        3:state3(word[counter],state);
        4:state4(word[counter],state);
        end {case state}
    else
      invalidword:=true; {Current symbol is not in alphabet [a,b]}
  if invalidword then
    writeln('Invalid word (contains symbol not in alphabet [a,b] )')
  else
    if (state = 2) or (state = 4) then
      writeln('Word accepted')
    else
      writeln('Word rejected')
end. {program acceptword}
```

Not all languages can be accepted by finite state machines. For example, consider the language $Q = \{w : w = a^n b^n, n > 0\}$. Q contains all words over $\{a, b\}$ in which a string of a's is followed by exactly the same number of b's. The language Q is not accepted by any finite state machine. As we can intuitively appreciate, a machine that could accept exactly the words in Q would need a state to remember the number of a's in a word so as to match the number of b's. Thus it would need a state for each possible value of n, and we could not make do with a finite number of states. The following theorem enables us to prove that Q cannot be accepted by any finite state machine.

Theorem 1.   The Pumping Theorem. Let M be a finite state machine and let L be the language accepted by M. Suppose that L is an infinite set. Then we can find words x, y, and z (with y not the empty word) such that xy^nz is an element of L for all values of $n > 0$.

PROOF: Suppose that M has exactly m states. Let $w = a_1 \ldots a_n$ be a word in L with length n, and suppose that $n > m$. Let $s_1, \ldots , s_n$ be the production defined by w. Because $n > m$, one of the states in the production must be repeated. So let us suppose that $s_i = s_j$ and that $i < j$. Let y be the word $a_i \ldots a_{j-1}$, let x be the prefix of y in w, and let z be the suffix of y in w. Then any word of the form xy^nz will define a production in which the subsequence $s_i, \ldots , s_{j-1}$ is repeated n times. This production also ends in an accepting state and so the word xy^nz must be in L.

Using Theorem 1, we can show that the language Q defined above cannot be accepted by any finite state machine. Suppose the word xy^nz is in Q for all values of n. If y is a word in which both a's and b's appear, then xy^nz will be a word in which some b's precede some a's, which is impossible in Q. If y is a string of a's, then the number of a's will exceed the number of b's in xy^nz. Similarly, if we take y to be a string of b's, the number of b's will exceed the number of a's. Both these situations are also impossible in Q. So y must be the empty word, contradicting Theorem 1. Thus no finite state machine accepts the language Q.

Finite state machines can be modified to perform other tasks besides language acceptance. We can, for instance, ask that certain states of the machine print (or otherwise output) certain symbols each time those states are visited. The result of processing a word with such a machine will then be a sequence of output symbols, not just an "acceptance" or "rejection."

Example 10. The finite state machine shown in Figure 7.44 prints "1" every time the sequence abb occurs in a word being processed. We consider all states to be

Figure 7.44

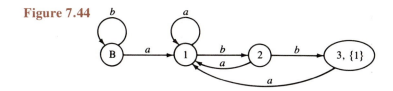

accepting states. State 3 has an extra label enclosed in curly brackets, {1}. We use this notation to indicate that the machine should print "1" each time state 3 is encountered. The path defined by the word *babbbabb* has node sequence (B, B, 1, 2, 3, B, 1, 2, 3). State 3 is encountered twice, so the machine prints the sequence "1, 1." The number of 1's that the machine prints is the same as the number of times the sequence *abb* occurs in the word being processed. Thus the machine serves as a counter for the number of times this pattern occurs.

Example 11. In this example we shall design a finite state machine that will print out the results of adding two binary numbers, x and y. We let $x = x_n x_{n-1} \ldots x_1$ and $y = y_n y_{n-1} \ldots y_1$, where each x_k and y_k is either 1 or 0. If, for instance, we let $x = 01011$ and $y = 01111$, we shall want the machine to print $z = 11010$. The machine will read the digits of x and y in pairs (x_k, y_k) from right to left. There are three possibilities: $(0, 0)$, $(1, 1)$, and $(0, 1)$. [We shall regard $(0, 1)$ and $(1, 0)$ as equal for purposes of addition.] The sum of two digits x_k and y_k depends on their values and on the value of the digit carried from the previous computation. So if $x_k = 1$, $y_k = 0$, and the digit carried from the addition of x_{k-1} and y_{k-1} is 1, then the machine must print "0" and carry 1 to the next computation. (We shall also print from right to left.) A finite state machine that accomplishes this task is shown in Figure 7.45. The symbol to be printed at each state is given at the top of its label and enclosed in curly brackets. The symbol to be carried is below it. For simplicity, we shall assume that the binary representations of both x and y have the same number of digits and that both representations start with 0. (Otherwise we would have to make special arrangements to print the last carried digit.) Every state except B is an accepting state.

Figure 7.45

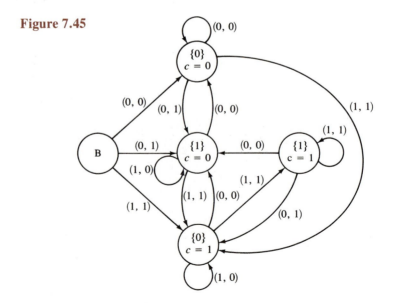

Section 7.5 | Exercises

1. The accompanying digraph is a finite state machine over $\{a, b\}$.

Figure for Exercise 1

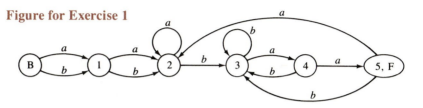

(a) Determine the path (beginning at B) defined by each of the following words: *abbaaababa, bbaaa, aabbabb, ababbaa, bbbbaaa.*

(b) Which of the words given in part (a) is accepted by the machine?

(c) What is the length of the shortest words accepted by the machine?

(d) Can a word ending in a be accepted by the machine?

(e) Find at least five different words accepted by the machine. Describe the language accepted by the machine.

2. Consider the finite state machine over $\{a, b\}$ shown in the accompanying figure.

Figure for Exercise 2

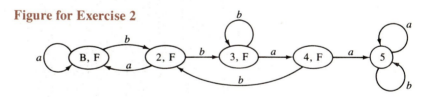

(a) Determine which of the following words are accepted: *babbaba, aaaaabbbb, abaabaaa, baabaabb, abababaabbbbb*?

(b) Describe the language defined by this machine in terms of the words it does not accept.

For Exercises 3 through 6, describe the language accepted by the given finite state machine. All are over the symbol set $\{a, b\}$.

3.

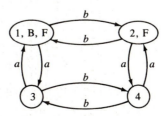

4.

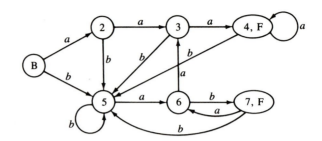

5.

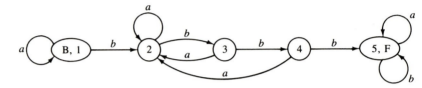

6.

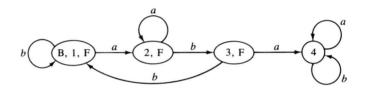

For Exercises 7 through 11, design a finite state machine that accepts the language described. All are over the symbol set {a, b}.

7. The number of *a*'s in each word is a multiple of 3.

8. The number of *a*'s and *b*'s in each word is odd.

9. The pattern *abba* appears in each word.

10. If the pattern *bab* appears in a word, then the pattern *abb* appears somewhere later in the word.

11. All words end in *bbb* or *aba*.

12. Find the transition functions of each of the machines given in Exercises 3 through 6.

13. Suppose that F is a finite state machine with 6 states labeled 1 through 6. Suppose that the starting state is 1 and that 4 and 6 are the accepting states. The transition function is as follows:

$$(1, a) \longrightarrow 2 \qquad (3, a) \longrightarrow 5 \qquad (5, a) \longrightarrow 4$$
$$(1, b) \longrightarrow 3 \qquad (3, b) \longrightarrow 3 \qquad (5, b) \longrightarrow 6$$
$$(2, a) \longrightarrow 4 \qquad (4, a) \longrightarrow 4 \qquad (6, a) \longrightarrow 5$$
$$(2, b) \longrightarrow 3 \qquad (4, b) \longrightarrow 3 \qquad (6, b) \longrightarrow 3$$

(a) Find the productions defined by each of the following words and determine which are accepted by the machine: *abbaabb*, *ababaa*, *ababab*, *bbbbaaaa*, *aaaaaabab*.

(b) Draw the state diagram of the machine.

(c) Characterize the language accepted by the machine.

14. The machine given in the accompanying figure is over the symbol set {0, 1}. It prints the symbol "0," "1," "2," or "3" when states 3, 4, 5, and 6 are visited, respectively. All states are accepting states. Suppose the machine both reads and prints from right to left. Find the output of the machine for the following words: 10001011, 100100, 111111. (Suppose we interpret the words over {0, 1} as binary numbers and insist that all numbers contain an even number of digits by adding a 0 to the beginning of a number where necessary. The machine converts the binary number that it reads to the base-4 expression of that number.)

Figure for Exercise 14

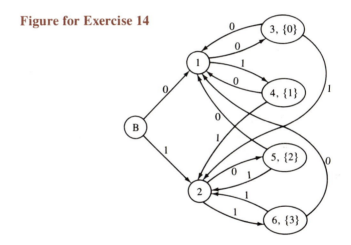

15. Design a machine that prints "HI" every time the sequence *aabb* is encountered in a word over {a, b}.

Bridge to Computer Science

7.6 Kleene's Theorem (optional)

Inductive procedures are an important tool in mathematics and computer science. We have defined sets—especially formal languages—inductively throughout this text, and we have seen how readily inductive procedures are implemented on the computer. We now use this bridge to link those languages accepted by our computational modeling device (the finite state machine) and a special class of inductively defined languages called **regular languages**. The result is Kleene's theorem.

First we define (inductively) a set of expressions called **regular expressions**. (These are like well-formed algebraic expressions.) Then we show how each regular expression defines a regular language. Finally, we show that a language is regular if and only if it is accepted by a finite state machine. This is Kleene's theorem. It provides us with a completely mathematical characterization of the capability of the finite state machine. (This time, we cross a bridge from computer science to mathematics.) We begin with the definition of the regular expression.

DEFINITION. Let $E = \{a_1, a_2, \ldots, a_n\}$, a finite alphabet. The set of **regular expressions** over E is defined inductively as follows:

(1) The empty set $\emptyset$ is a regular expression. Any word w in E^* is also a regular expression.

(2) If R and S are regular expressions, then so are

   (i) $R + S$

   (ii) $(R)(S)$

   (iii) $(R)^*$

(3) Only expressions formed by a finite number of iterations of steps 1 and 2 are regular expressions.

For example, if $E = \{a, b\}$, then $R = \{(aa)\,[(a + b)^*]\}\,(bb)$ is a regular expression over E. If there is no danger of ambiguity, we shall omit some of the "fences" from such expressions and, for instance, rewrite R as $(aa)\,(a + b)^*(bb)$. The following are also regular expressions: Λ, $(aba + bbb)^* + (b)^*$ and $(a + b)^*$ $(bb)\,(a + b)^*$.

Let R be a regular expression over the symbol set E. We now show how to obtain $L(R)$, the language defined by R.

DEFINITION.

(a) If $R = \emptyset$, then $L(R)$ is the empty set. If $R = w$, and w is a word in E^*, then $L(R)$ is the singleton set $\{w\}$.

> (b) If R and S are regular expressions and $L(R)$ and $L(S)$ are the languages they define, then
>
> (i) $L(R + S) = L(R) \cup L(S)$
>
> (ii) $L[(R)(S)] = \{x : x = yz,$ where y is in $L(R)$ and z is in $L(S)\}$
>
> (iii) $L[(R)^*] = \{x : x = \Lambda$ or $x = x_1 x_2 \ldots x_m,$ where each x_i is a word in $R\}$

Thus $L[(R)(S)]$ is the set of words obtained by concatenating words in R with words in S and $L[(R)^*]$ is the set of words formed by concatenating any finite set of words of R in any order. In terms of the notation of Chapter 2, $L[(R(S)] = L(R)L(S)$, and $L[(R)^*] = [L(R)]^*$.

Example 1. Let $E = \{a, b\}$.

(a) If $R = (a)^*$, then $L(R) = \{\Lambda, a, aa, aaa, \ldots\}$.

(b) If $R = (aa + bb)^*$, then $L(R) = \{\Lambda, aa, aaaa, \ldots, bbaa, bbbb, aabb, \ldots\}$, the set of all words in which a's and b's always occur in pairs.

(c) If $R = (a + b)^*$, then $L(R) = E^*$.

(d) If $R = (aa)(a + b)^*(bb)$, then $L(R)$ is the set of all words that begin with aa and end with bb.

Example 2.

(a) Suppose we wish to find a regular expression for the set L of words in which the sequence abb appears. This sequence can be preceded by any configuration of letters and followed by any configuration of letters. Thus the regular expression that defines L is $(a + b)^* (abb) (a + b)^*$.

(b) Let Q be the language of words in which either a appears exactly once or b appears exactly twice. The regular expression for this language is $(b)^*(a)(b)^* + (a)^*(b)(a)^*(b)(a)^*$.

Kleene's Theorem. A language L is regular if and only if L is a language accepted by some finite state machine.

We will prove Kleene's theorem with two constructions. In the first construction, we will show how to obtain the regular expression that defines the same language as a given finite state machine. In the second construction, we proceed in the opposite manner and construct a finite state machine from a regular expression. To perform these constructions, we need the notions of the transition graph and the generalized transition graph, variations on the idea of a state diagram.

> **DEFINITION.** A **transition graph** T over an alphabet E is a directed graph in which each directed edge is labeled with a word in E^*. One node is designated as the starting state of T and labeled with B. A nonempty subset of the nodes of T is designated as the set of accepting states of T. These nodes will be labeled with F.

Every finite state machine is also a transition graph. The graphs shown in Figure 7.46 are transition graphs that are not finite state machines.

Figure 7.46

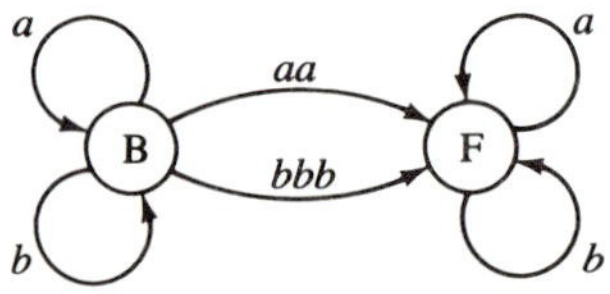

(a) 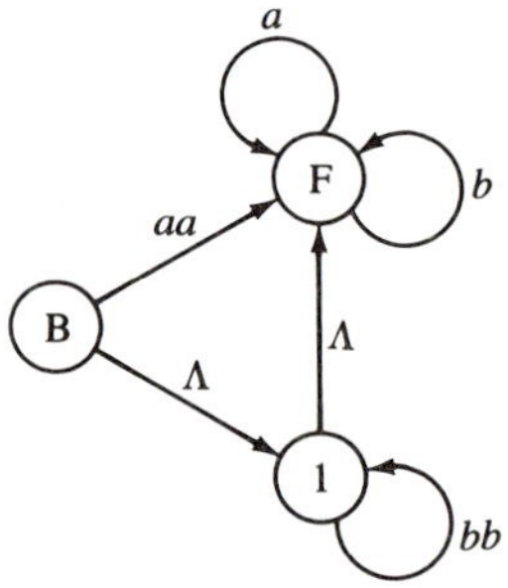   (b)

A word x in E^* is accepted by a transition graph T if both of the following conditions are met:

(i) There is a path P in T that begins at the starting state of T and ends at an accepting state of T.

(ii) The concatenation (in order) of the labels of the edges traversed along the path P spells x.

For example, the word *bbaaab* is accepted by the graph shown in Figure 7.46(a), whereas the word *abab* is not. We note that there are several paths in this graph that correspond to the word *bbaaab*. One path, for instance, never leaves the starting node. The word *bbaaab* is accepted by the machine because at least one of the paths it defines ends at an accepting state. In Figure 7.46(b) there is no path that corresponds to the word *baba*, and so it is not accepted.

Example 3.

(a) A transition graph for the language defined by the regular expression $(a)^*(bb)(a + b)^*$ is shown in Figure 7.47.

Figure 7.47

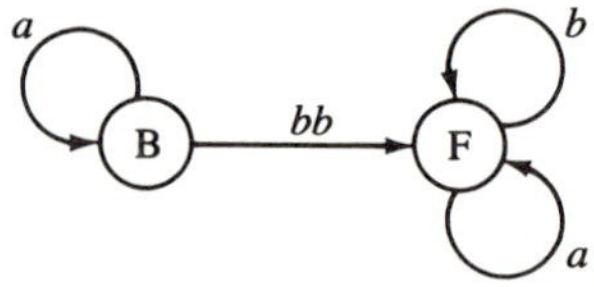

(b) A transition graph for the language $(a)(a + b)*(bb) + (b)(a + b)*(aa)$ is shown in Figure 7.48.

Figure 7.48

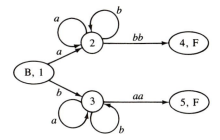

In a **generalized transition graph** we label each edge with a regular expression over an alphabet E. A word x is accepted by a generalized transition graph T if there is a path P in T that begins at the starting state of T, ends at an accepting state, and is such that there are words in the languages defined on the edges of the path P, the concatenation of which spells x.

Example 4.

(a) Figure 7.49 shows a generalized transition graph over $E = \{a, b\}$. This graph accepts such words as *aaabb* and *bbbbaa*—words that are either a string of *a*'s followed by two *b*'s, or a string of *b*'s followed by two *a*'s.

Figure 7.49

(b) The generalized transition graph shown in Figure 7.50 defines the set of words that have length at least 4 and begin and end with *aa*, or begin and end with *bb*; or have length at least 5 and begin with *bb* and end with *baa*.

Figure 7.50

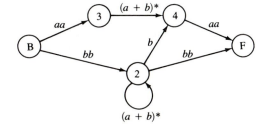

Construction I. Let G be a finite state machine that accepts the language L. We will construct a regular expression that also defines L. To do this we will eliminate the nodes of the graph of G one by one until all we are left with is a starting node and one accepting node. When we eliminate a node x, we shall replace all paths through x by edges labeled with the regular expressions that define the same words as the paths they replace. Eventually we shall obtain the configuration shown in Figure 7.51. The regular expression that accepts the same language as this generalized transition graph is $R_1 + R_2 + \cdots + R_m$.

Figure 7.51

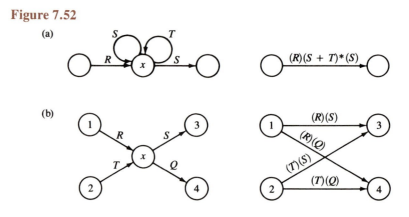

Figure 7.52 shows how we eliminate a node x from a generalized transition graph. In both cases, the configuration on the left may be replaced by that on the right without changing the language defined by the graph.

Figure 7.52

(a)

(b)

To begin our elimination process on a finite state machine G, we first add two new nodes A and Z, where A will serve as our new starting state and Z as our new sole accepting state. We add a directed edge from A to B (the beginning node of G) and directed edges from each of the accepting states of G) to Z. We label each of these new edges with the empty word. We then proceed to eliminate all other nodes, using the replacements indicated in Figure 7.52 until all that remain are nodes A and Z. At each replacement we obtain a generalized transition graph that accepts the same language as G. When we have reached the configuration shown in Figure 7.51, we obtain a regular expression that also defines this language. We now demonstrate with some examples.

Example 5. Let F be the finite state machine given in Figure 7.53. Since no word that defines a path through state 5 is accepted, we simply eliminate it and its incident edges.

Figure 7.53

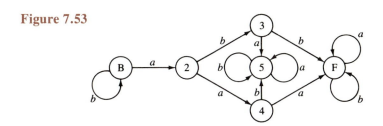

Step 1. We add the new starting state A and the new sole accepting state Z and obtain Figure 7.54.

Figure 7.54

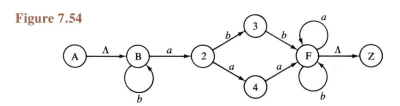

Step 2. We replace B and obtain Figure 7.55.

Figure 7.55

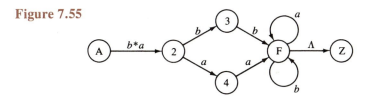

Step 3. We eliminate nodes 3 and 4 and obtain Figure 7.56.

Figure 7.56

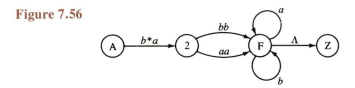

Step 4. We eliminate node 2 by replacing both the upper path and the lower path through 2. We obtain Figure 7.57.

Figure 7.57

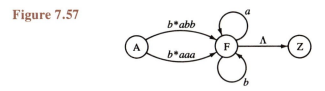

Step 5. Finally, we replace node F and obtain Figure 7.58. We now obtain the regular expression $(b)*(abb)(a + b)* + (b)*(aaa)(a + b)*$, which defines the same language as the finite state machine shown in Figure 7.53.

Figure 7.58

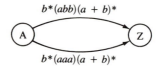

Example 6. Consider the finite state machine shown in Figure 7.59. Node 5 and its incident edges can be eliminated and *not* replaced, because no path through node 5 leads to an accepted word. (It is a dead end.) Thus we obtain Figure 7.60. The regular expression it defines is $(a)(b)*(a) + (b)(a)*$.

Figure 7.59

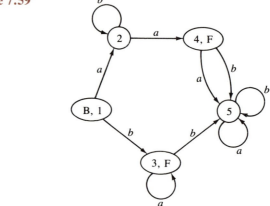

Figure 7.60

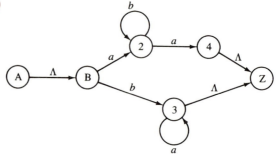

Construction II. Let R be a regular expression. We now show how to construct a finite state machine that accepts the language $L(R)$ defined by R. (We shall assume that R is an expression over $E = \{a, b\}$.) First we construct a transition graph that accepts $L(R)$. Each edge is labeled with a, b, or the empty word. To do this we note that in a generalized transition graph:

(a) Figure 7.61(a) corresponds to the regular expression $(R)(S)$.

(b) Figure 7.61(b) corresponds to $(R + S)*$.

(c) Figure 7.61(c) corresponds to $R + S$.

Figure 7.61

(a)

(b)

(c)

If R is a regular expression, we start with the generalized transition graph (Figure 7.62) and replace its edges until we have a (simple) transition graph.

Figure 7.62

Example 7. A transition graph that accepts the language defined by the expression $(a + b)*(bb) + (bb + aa)*(aa)$ is obtained as shown in Figure 7.63.

Now suppose that we have constructed a transition graph T in which each of the edges is labeled with a, b, or the empty word. Suppose that the starting state of T is A and that the accepting states are the members of the set $C = \{C_1, C_2, \ldots, C_n\}$. We will construct a finite state machine F that accepts the same language as T by the **subset** method: each of the nodes of F will be labeled with a subset of the nodes of T. The starting state of F is labeled with the set containing the starting state of T, together with all nodes of T that are reachable from A by paths in which all the edges are labeled with the empty word. Call this set B_F. We find the sets that will label the remainder of the nodes of F by using a **transition table**. The transition table contains three columns labeled M, M_a, and M_b, respectively. The starting node of F—namely the set B_F—is the first entry in the first column M. We fill in the rest of the table inductively as follows. Suppose that Q is a set of states appearing in column M. Then, next to it in column M_a, we place the set Q_a. The set Q_a contains all states of T that can be reached from a state

Figure 7.63   (a)

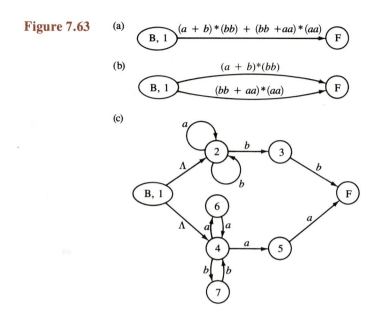

$(a + b)*(bb) + (bb +aa)*(aa)$

B, 1 — F

(b)

$(a + b)*(bb)$

B, 1 — $(bb + aa)*(aa)$ — F

(c)

in Q by a path that contains exactly one edge labeled with an a, and the remainder of its edges labeled with the empty word. Similarly, in column M_b we place Q_b, the set of states that can be reached from a state in Q by a path containing exactly one edge labeled with b, and the remainder of its edges labeled with the empty word. Now we list the sets Q_a and Q_b back in the first column (unless they have already appeared there). We fill in columns M_a and M_b for each set appearing in column M until no new sets are generated.

Once we have our transition table, we construct our finite state machine F as follows. The sets appearing in column M are the nodes of the finite state machine F under construction. The starting state of F is B_F, the first entry in column M, and the accepting states of F are those sets in column M that contain accepting states of T. We place a directed edge labeled with a from a node labeled with set Q to the node labeled with set Q_a when Q is in column M and Q_a is next to Q in column M_a. Similarly, we place a directed edge labeled with b from Q to Q_b. We illustrate with an example.

Example 8. A transition graph T for the regular expression $(a)*(ba)$ is shown in Figure 7.64.

Figure 7.64

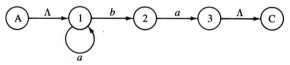

We now construct a transition table and finite state machine according to the instructions given in Construction II. The starting state for a finite state machine that accepts the same language is $B_F = \{A, 1\}$. (The set B_F contains A, together with the set of all states reachable from A in T via paths labeled with the empty word.) We place the set B_F in column M. From the states in B_F, we can reach only the state 1 by paths containing only one edge labeled with a, so we place the set $\{1\}$ next to B_F in column M_a. Similarly, we can reach only state 2 from the nodes in B_F by paths containing only one edge labeled with b. We place the set $\{2\}$ in column M_b. We now list the sets $\{1\}$ and $\{2\}$ back in column M and repeat the procedure. The transition table is as follows:

M	M_a	M_b
$\{1, A\}$	$\{1\}$	$\{2\}$
$\{1\}$	$\{1\}$	$\{2\}$
$\{2\}$	$\{3, C\}$	Ø
Ø	Ø	Ø
$\{3, C\}$	Ø	Ø

Thus our resulting finite state machine has five states connected as shown in Figure 7.65.

Figure 7.65

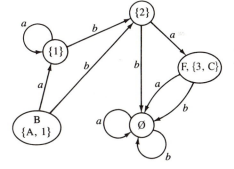

Example 9. We find a finite state machine that accepts the language defined by the regular expression $(ab + a)*(ab)$. A transition graph T that accepts this language is shown in Figure 7.66.

Figure 7.66

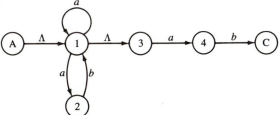

The transition table is as follows:

	M	M_a	M_b
B	{A, 1, 2, 3}	{1, 2, 3, 4}	Ø
1	{1, 2, 3, 4}	{1, 2, 3, 4}	{1, C}
2	Ø	Ø	Ø
F	{1, C}	{1, 2, 3, 4}	Ø

We relabel the sets with B, 1, 2, and F to obtain the finite state machine shown in Figure 7.67.

Figure 7.67

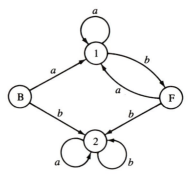

Section 7.6   **Exercises**

---

1. Characterize the languages defined by each of the following regular expressions.

   (a) $(aa + bb)* (b)*$
   (b) $(abb)(a + b)* (bba)$
   (c) $(a)* + (abb + b)*$
   (d) $(b)* + [(b)* (a)(b)* (a)(b)*]*$

2. Find the regular expression that defines each of the following languages over $\{a, b\}$.

   (a) All words that begin and end with the same letter
   (b) All words in which the pattern *abb* appears
   (c) All words in which the number of a's is a multiple of 3 or the number of b's is even
   (d) All words in which every b is followed immediately by at least 3 a's

3. For each of the regular expressions given in Exercise 1, find a transition graph that accepts the language it defines.

4. Suppose that L_1 and L_2 are regular languages accepted by the transition graphs

T_1 and T_2. Show how to assemble or otherwise modify these graphs so that the resulting graph accepts each of the following languages.

(a) $L_1 \cup L_2$ (b) $L_1 \cap L_2$ (c) $(L_1)'$

5. Characterize the languages accepted by the following transition and generalized transition graphs.

Figure for Exercise 5

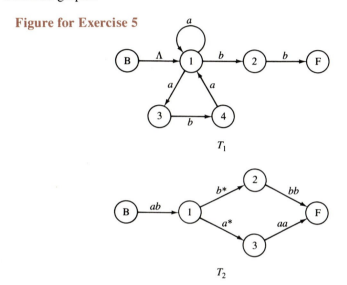

T_1

T_2

6. Use Construction I to find a regular expression that accepts the same language as each of the following finite state machines.

Figure for Exercise 6

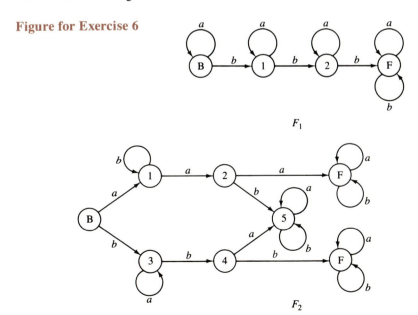

F_1

F_2

Figure for Exercise 6—*Cont.*

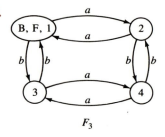

F_3

7. Use Construction II to find a finite state machine that accepts the language defined by each of the following regular expressions.

(a) $aa(aa + bb)*(bb)$ (b) $a(aa + b)*$ (c) $(a + b)*(abb)(a + b)*$

8. Modify Construction II so that you can construct a finite state machine that accepts the same language as defined by a regular expression over an alphabet with n symbols, $E = \{a_1, a_2, \ldots, a_n\}$. Use your modified construction to construct a machine that accepts the language defined by $(a + b)* (bb)(a + c)*$.

key concepts

7.1 Digraphs	digraph
	directed edge $e = (a, b)$ starting at a and ending at b
	adjacency matrix
	in-degree, out-degree, and total degree of a node
7.2 Paths and Connectivity	directed path
	directed circuit
	connected digraph
	strongly connected digraph
	Euler's theorem for digraphs
	reachability matrix
	Warshall's algorithm
7.3 Weighted Digraphs	weighted digraph
	$w(a, b)$, the weight of the edge (a, b)
	weight of a path
	weighted adjacency matrix
	shortest path from a to z (the path of least weight)
	longest path from a to z (the path of greatest weight)
	Dykstra's algorithm
7.4 Acyclic Digraphs	acyclic digraph
	sources and sinks
	topological sorting
7.5 Finite State Machines	finite state machine
	language acceptor
	states; beginning and accepting states
	path defined by a word
	language defined or accepted by a finite state machine
	transition function
	production defined by word
	state diagram
	pumping theorem
	machines that print (or have output)

Chapter 7

Exercises

1. Suppose that D is a digraph with nodes $\{a, b, c, d, e, f, g\}$ and that the directed edges of D are $\{(a, d), (b, a), (c, d), (d, e), (b, g), (g, a), (e, f), (f, g), (g, c), (b, c), (g, b)\}$.

   (a) Draw D.

   (b) List the in-degree and the out-degree of each node.

   (c) Find all the simple directed paths in D from a to d.

   (d) Find all the simple paths in D from g to d.

   (e) Find all the simple circuits in D that include the edge (f, g).

2. For each of the digraphs depicted in the accompanying figure:

   (a) Find a directed path from a to b if possible.

   (b) Determine whether the digraph is strongly connected.

   (c) Find the digraph's adjacency matrix and reachability matrix.

   Figure for Exercise 2

   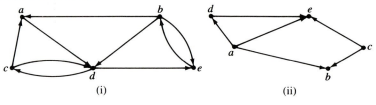

   (i)   (ii)

3. Find the number of edges in a digraph with 5 nodes, 2 of which have out-degree 2 and 3 of which have out-degree 3. Draw such a digraph.

4. Is it possible to construct a digraph such that each node has in-degree 2 and out-degree 1?

5. A digraph D has 7 nodes and 13 edges. Each node has out-degree either 1 or 3. Determine the number of nodes of each degree, and draw at least one digraph that satisfies these conditions.

6. A digraph is to have 6 nodes and 11 edges. Each node must have in-degree 1, 2, or 3. Determine all solutions to the problem of how many nodes there are of each in-degree.

7. Six teams will play in a soccer tournament. Each team will play every other team exactly once. Draw a digraph that models one possible outcome of such a tournament. How many different outcomes are possible if each game must continue until a winner is determined? What if ties are allowed?

8. The 7 nodes of a digraph D are numbered 1 through 7. The adjacency matrix of the digraph is given in Table 7.10.

Table 7.10

	1	2	3	4	5	6	7
1	0	1	1	0	1	0	0
2	0	0	1	0	0	1	1
3	0	0	0	1	1	0	0
4	0	0	0	0	0	0	1
5	0	0	0	0	0	0	0
6	1	0	0	0	0	0	0
7	0	0	1	0	0	0	0

(a) Use Warshall's algorithm to find the reachability matrix of digraph D.

(b) Draw D and use it to verify that there is a 1 in the (i, j)th position of W_5 if and only if there is a directed path P in D from node i to node j such that the interior nodes of P are from the set $\{1, 2, 3, 4, 5\}$.

9. Prove that, if W_k is the kth matrix obtained using Warshall's algorithm, then the (m, n)th of W_k is equal to 1 if and only if there is a path P from a_m to a_n whose interior nodes are in the set $\{a_1, a_2, \ldots, a_k\}$.

10. Determine which of the digraphs in the accompanying figure admits a directed Euler circuit. Find such a circuit for those that do.

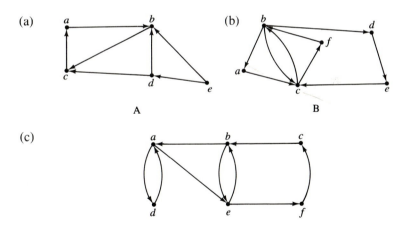

11. Prove Euler's theorem for digraphs.

12. Find the most general conditions under which a digraph has a directed Euler path—that is, a directed path that traverses each edge exactly once. (It need not be closed.) Prove your results.

13. Suppose that the nodes of D are the strings over $\{0, 1\}$ of length 3, and suppose there is a directed edge from node x to node y if and only if x and y differ in exactly one position (so there is an edge from 011 to 010 but not from 011 to 100). Show that any directed Hamiltonian path in D that is not closed can be used to represent the set of numbers from 0 to 7 with 3-bit strings in such a way that successive numbers differ in exactly one place. Find two such representations. Repeat the process for the numbers 0 through 15, using strings of 0's and 1's of length 4. (Such codes are called Gray codes.)

14. Consider the weighted digraph given in the accompanying figure.

   (a) Find the weight of all simple paths from a to z.

   (b) Find the path of least weight from a to z.

   (c) Find the path of least weight from z to a.

Figure for Exercise 14

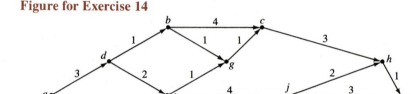

15. Use Dykstra's algorithm on the graph given in the figure accompanying Exercise 14 to find the shortest path from d to h.

16. The weighted adjacency matrix of a weighted digraph D is given in Table 7.11. Use Dykstra's algorithm to find the weight of the shortest path from node 1 to node 6.

Table 7.11

	1	2	3	4	5	6	7
1	∞	3	5	∞	1	∞	1
2	∞	∞	1	1	∞	∞	1
3	1	2	∞	∞	1	2	3
4	∞	3	∞	∞	3	1	3
5	∞	2	1	∞	∞	1	∞
6	∞	∞	∞	∞	∞	∞	6
7	∞	2	∞	4	∞	∞	∞

17. Suppose that W is the weighted adjacency matrix of a weighted digraph with n nodes, $a_1, a_2, \ldots, a_n$. The following modification of Warshall's algorithm finds the weight of the shortest path from a_i to a_j.

Warshall's Algorithm (modified).

(1) Let $W_0 = W$.

(2) Suppose that $0 < k < n$ and that W_k has been constructed. Construct W_{k+1} as follows:

Let $x_{i,j} = \min(w_{i,j}, w_{i,k+1} + w_{k+1,j})$, where $w_{i,j}$, $w_{i,k+1}$, and $w_{k+1,j}$ are entries in W_k. Enter $x_{i,j}$ in the (i, j)th position of W_{k+1}. The value of the (i, j)th entry of W_n is the weight of the shortest path from a_i to a_j. (If there is no path, that value will be ∞.)

Apply this modification of Warshall's algorithm to Table 7.11. Verify that the entry in the first row, sixth column of W_7 is the same as the value you found in Exercise 16 for the weight of the shortest path from node 1 to node 6.

18. (We refer again to Exercise 17.) Show that the (i, j)th entry in W_k is the weight of the shortest path (the path of least weight) from node a_i to node a_j with interior nodes from the set $\{a_1, a_2, a_3, \ldots, a_k\}$. Argue that this proves that the algorithm works as stated.

19. A certain airline operates in 6 cities, which we shall call A, B, C, D, E, and F. The cost of flying from city X to city Y appears in the (X, Y) position in Table 7.12. (If there is no flight between a pair of cities, we have entered the symbol ∞.) As a result of various economic factors, a direct flight is not always the least expensive. Use Warshall's algorithm as developed in Exercise 17 to obtain a matrix that yields the least expensive trip possible between any two cities.

Table 7.12

	A	B	C	D	E	F
A	0	25	∞	40	∞	∞
B	25	0	∞	∞	25	75
C	∞	∞	0	20	20	20
D	40	∞	20	0	50	80
E	∞	25	20	50	0	20
F	∞	75	20	80	20	0

20. Determine which of the digraphs shown in the accompanying figure are acyclic. Topologically sort the nodes of those digraphs that are acyclic.

Figure for Exercise 20

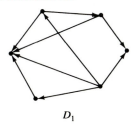

D_1

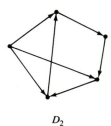

D_2

Figure for Exercise 20—*Cont.*

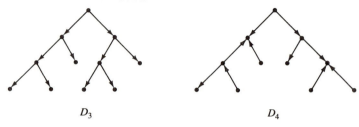

D_3 D_4

21. Prove that the digraph of an irreflexive transitive relation must be acyclic.

22. Either prove or give a counterexample: The digraph of an antisymmetric relation must be acyclic.

23. (a) Recall that a complete tournament is a digraph with no self-loops in which there is exactly one directed edge between every two distinct nodes. Draw one complete tournament on 5 nodes that is acyclic and draw one that is not.

   (b) Prove by induction on n that, for each value of n, there is at least one acyclic complete tournament on n nodes.

24. A matrix is said to be "upper triangular" if whenever $i > j$, then the entry in the ith row, jth column is 0. Prove that a digraph D is acyclic if and only if the nodes of D can be listed in some order such that the resulting adjacency matrix is upper triangular.

25. A certain manufacturing company assembles personal computers. The process requires 15 distinct tasks (such as installing the memory boards, wiring in the keyboard, and the like) that we shall call tasks $A_1, A_2, \ldots, A_{15}$. The time required for the completion of each task and the tasks that must precede any given task are given in Table 7.13.

   (a) Draw a digraph to represent the assembly process as follows. Represent each task by an edge. If task A_i must be completed before task A_j, then there must be a directed path from the initial node of A_i to the terminal node of A_j. All edges representing tasks with no prerequisites should start at the same node—call it B. All edges representing tasks that are not themselves prerequisites should end at the same node—call it E.

   (b) Using the completion time of a task as the weight of the edge representing it, find the shortest time in which a computer can be assembled by finding the longest path from B to E.

   (c) Find the earliest time at which each task can be started.

   (d) Find the latest time (assuming that all other tasks are not delayed) at which each task can be started without delaying the completion of the assembly.

Table 7.13	Task	Duration (min)	Prerequisite task
	A_1	15	none
	A_2	10	none
	A_3	20	none
	A_4	20	A_2
	A_5	30	A_1
	A_6	10	A_2
	A_7	15	A_3
	A_8	20	A_6, A_7
	A_9	20	A_4, A_5
	A_{10}	10	A_3
	A_{11}	40	A_8, A_9, A_{10}
	A_{12}	5	A_{11}
	A_{13}	10	A_{11}
	A_{14}	15	A_{12}
	A_{15}	15	A_{13}, A_{14}

26. (a) Determine which of the finite state machines shown in the accompanying figure accepts the word *aaababaabb*.

    (b) Characterize the language accepted by each of these finite state machines. All are over the symbol set $\{a, b\}$.

Figure for Exercise 26

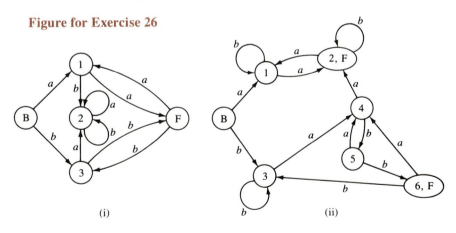

(i)   (ii)

27. Determine the language defined by each of the finite state machines shown in the accompanying figure. Each uses the symbol set $\{a, b\}$.

Figure for Exercise 27
(a)

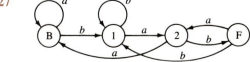

Figure for Exercise 27—Cont.

(b)

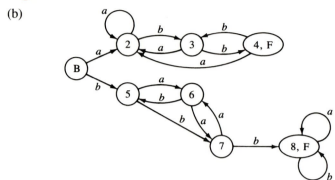

(c)

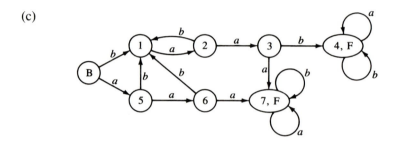

28. Construct a finite state machine that accepts each of the following languages over $\{a, b\}$.

   (a) All words in which the pattern *abbab* appears
   (b) All words ending in *abb*
   (c) All words in which neither *a* nor *b* appears three times in a row
   (d) All words that begin with *aa* must end with *bbb*

29. Find the transition functions of each of the machines shown in Exercise 27.

30. Find the state diagram for the finite state machines whose transition functions follow.

   (a) The machine over $\{a, b\}$ with state set $S = \{1, 2, 3, 4\}$, starting state 1, accepting state 3, and transition function

$$
\begin{array}{ll}
(1, a) \longrightarrow 2 & (3, a) \longrightarrow 4 \\
(1, b) \longrightarrow 3 & (3, b) \longrightarrow 3 \\
(2, a) \longrightarrow 2 & (4, a) \longrightarrow 4 \\
(2, b) \longrightarrow 3 & (4, b) \longrightarrow 4
\end{array}
$$

   (b) The machine over $\{a, b\}$ with state set $S = \{1, 2, 3\}$, state 1 as both the starting and accepting state, and transition function

$$(1, a) \longrightarrow 2 \qquad (2, a) \longrightarrow 3 \qquad (3, a) \longrightarrow 1$$
$$(1, b) \longrightarrow 2 \qquad (2, b) \longrightarrow 3 \qquad (3, b) \longrightarrow 1$$

31. For each of the machines given in Exercise 30, find the productions defined by the words *aaabba* and *babababb*. Determine whether these words are accepted.

32. For each of the machines given in Exercise 30, find words x, y, and z that satisfy the pumping theorem.

33. Let Q be the language of all words over $\{a, b\}$ consisting of a string of a's followed by twice as many b's. Use the pumping theorem to prove that Q is not a language accepted by any finite state machine.

34. (a) Re-prove the pumping theorem, using the digraph representation of a finite state machine. (*Hint*: Suppose that a finite state machine has n states and accepts a word w of length greater than n. Show the state diagram must have a proper directed circuit. Use it to find y. Also use the beginning of w to find x and the end of w to find z.)

(b) The proof you obtained in part (a) shows that *any* word of length greater than n can be written in the form xyz, where x and z are words such that the length of $xz \leq n$ and such that xy^iz is accepted for any i. Use this fact to prove that the language $L = \{x : x = ww^r$, where w is any word over $\{a, b\}$ and w^r is its reverse$\}$ is not accepted by any finite state machine.

35. (a) Let L be the set of all words in which the pattern *abbbaa* appears. Design a finite state machine to accept L. Using the same states and edges but changing your designation of accepting states, obtain a machine that accepts the complement of L.

(b) Show that if L is a language accepted by a finite state machine, then its complement L' is also accepted by a finite state machine.

36. Suppose that F_1 and F_2 are finite state machines that accept languages over a symbol set E. Suppose that the set of states of F_1 is S_1, that the language it accepts is L_1, and that its transition function is f_1. Similarly, assume that the set of states of F_2 is S_2, its language is L_2, and its transition function is f_2. A new machine—call it F—can be constructed from F_1 and F_2 as follows:

(i) Let the states of F be the set $S_1 \times S_2$.

(ii) Let the transition function f of F be the defined by

$$f((s, s'), x) = (f_1(s, x), f_2(s', x))$$

for each s in S_1, s' in S_2, and x in the symbol set E.

(iii) If B_1 and B_2 are the starting states of F_1 and F_2 respectively, let (B_1, B_2) be the starting state of F.

(iv) Let (s, s') be an accepting state of F if either s or s' is an accepting state.

(a) Carry out the preceding construction, letting F_1 be the machine that accepts only words over $\{a, b\}$ that end with aa, and letting F_2 be the machine over $\{a, b\}$ that accepts words with an odd number of b's. Determine the language accepted by this new machine.

(b) Prove that if L_1 and L_2 are languages over a symbol set E that are accepted by the finite state machines F_1 and F_2, then their union $L_1 \cup L_2$ is also accepted by a finite state machine.

37. (a) Modify the construction given in Exercise 36 so that the new machine accepts words in the intersection of L_1 and L_2.

(b) Design a machine that accepts words over $\{a, b\}$ that end in aa and have an odd number of b's.

(c) Design a machine that accepts words over $\{a, b\}$ in which the pattern $abbb$ appears and in which the number of a's is congruent to 1 mod 3.

Note: **Exercises 35, 36, and 37 show that the languages over a symbol set E that are accepted by some finite state machine are closed under union, intersection, and complementation.**

38. Consider the finite state machine given in the accompanying figure. If a state shows a symbol enclosed in braces, then the machine will print (from left to right) that symbol every time that state is encountered.

Figure for Exercise 38

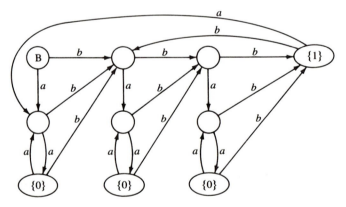

(a) Determine the sequences printed when the following words are read: *aaabbbaaabbabbbbbbabbaaa* and *bbbbbbbbbbbbbabba*.

(b) Describe the output of this machine.

39. (a) Design a machine that prints a "1" every time the pattern *abbbb* is encountered.

(b) Design a machine that prints the number of a's modulo 5 appearing in a word.

40. Design a machine that reads the binary representation of a number x and prints the octal representation of x. Assume that the number of places in the binary representation of x is a multiple of 3. (You may have to prefix the binary representation of x with one or two 0's.)

AN INTRODUCTION TO ALGEBRA

outline

8.1 Binary Operations, Semigroups, and Monoids

8.2 The Integers

8.3 Groups

8.4 Subgroups

8.5 Rings and Fields

8.6 Bridge to Computer Science: More About Coding Theory (optional)

Key Concepts

Exercises

Introduction

OUR mathematical studies are usually focused on sets equipped with one or more binary operations. For example, we have studied

(i) The set of integers under addition $+$ and multiplication $\cdot$

(ii) The set of functions $f : X \rightarrow X$ on a fixed set X under composition $\circ$

(iii) Logical propositions under conjunction $\wedge$ and disjunction $\vee$

Such binary operations often exhibit similarities in the ways in which they can be manipulated. For instance, we have a distributive law for arithmetic and a distributive law for the propositional calculus:

(i) $x(y + z) = xy + xz$

(ii) $P \wedge (Q \vee R) = (P \wedge Q) \vee (P \wedge R)$

Algebra is the study of such similarities in an abstract setting. It is a two-way street. On the one hand, the knowledge we gain from our inquiries within abstract structures gives us new knowledge about our familiar, concrete structures. For instance, our study of abstract groups yields Fermat's theorem, a fact about prime powers of integers. Recently this theorem has been used as the basis of an important technique in coding theory. On the other hand, our knowledge of familiar computations leads us to look for what might be true in the general abstract situation. Thus the study of algebra looks closely at old-fashioned arithmetic as it investigates less familiar structures such as groups, rings, and fields.

| Section 8.1 | Binary Operations, Semigroups, and Monoids |

We begin our investigation of algebraic structures by looking at the most basic setting for mathematical computation, a set together with one *binary* operation. The familiar operations of addition and multiplication are the prototypical binary operations, but other examples abound. Subtraction, composition, and concatenation are just a few. What they have in common is summarized in the following definition.

> **DEFINITION.** A **binary operation** $*$ on a set A is a function $f_* : A \times A \rightarrow A$.

367

To see that addition on the integers satisfies this definition, note that we may define $f_+ : \mathbf{Z} \times \mathbf{Z} \to \mathbf{Z}$ by $f_+(m, n) = m + n$. As with addition, we shall usually denote $f_*(a, b)$ by the *in-fix* notation $a * b$.

Example 1.

(a) Let Q be the set of rational numbers. Addition, multiplication, and sub-traction all satisfy the preceding definition. Division is not a binary operation because it is not defined for any pair (x, y) in $Q \times Q$ in which $y = 0$.

(b) Let X be a nonempty set, and let $C(X)$ be the set of functions on X of the form $f : X \to X$. (If $X = \{a, b, c\}$, for instance, there are 27 such functions and so the set $C(X)$ has 27 elements.) Composition is a binary operation on $C(X)$ because for any pair of functions f and g in $C(X)$, their compo-sition is yet another element in $C(X)$.

(c) Let E be a finite alphabet and E^* the set of all words over E. Concatena-tion is a binary operation on E^*.

(d) Let WWF be the set of propositional forms that can be written using the logical variables P, Q, R, S, and T. Then disjunction $\vee$ and conjunction $\wedge$ are binary operations on WWF.

Example 2. The binary operations discussed in Example 1 are all very familiar. But binary operations can arise in quite abstract settings as well. Table 8.1 defines a binary operation $*$ on the set $\{a, b, c\}$. To obtain $x * y$, we find the entry in the row labeled x and the column labeled y. Thus $b * c = a$. Such a table is called a **Cayley table** for the operation $*$.

Table 8.1

$*$	a	b	c
a	a	b	c
b	b	c	a
c	c	a	b

The matrix given in Table 8.1 is symmetric; that is, the entry in the (x, y) position is the same as the entry in the (y, x) position. Thus for any pair (x, y), we have $x * y = y * x$. The operation is commutative. We also note that $a * (b * c) = (a * b) * c$. In fact, we may verify that for any of the 27 possible triples (x, y, z), we have $x * (y * z) = (x * y) * z$. The operation is associative.

> **DEFINITION.** A binary operation $*$ on a set A is said to be **associa-tive** if for every three elements a, b, and c in A, we have $(a * b) * c = a * (b * c)$. A binary operation $*$ on a set A is said to be **com-mutative** if for every pair of elements a and b in A, we have $a * b = b * a$.

Example 3.

(a) Let X be a nonempty set, and let $C(X)$ be the set of functions of the form $f : X \rightarrow X$. The operation composition is associative. To see this, let f, g, and h be any functions in $C(X)$, and let x be an element of X. Then we have $(f \circ g) \circ h(x) = f \circ g(h(x)) = f(g(h(x))) = f \circ (g \circ h)(x)$. Thus $(f \circ g) \circ h = f \circ (g \circ h)$. This proves associativity. However, composition is not commutative.

(b) Addition and multiplication on the integers are each both commutative and associative. But the operation of subtraction on the integers is neither associative nor commutative. To see that subtraction is not associative, note that $(2 - 3) - 1 = -2$, whereas $2 - (3 - 1) = 0$. To see that subtraction is not commutative, note that $2 - 3 \neq 3 - 2$.

(c) Let $\mathbf{R}$ be the set of real numbers, and define the binary operation $\#$ on $\mathbf{R}$ by $x \# y = |x - y|$. The operation $\#$ is commutative because $|x - y| = |y - x|$, but it is not associative. For instance, $2 \# (2 \# 3) = |2 - |2 - 3|| = 1$, whereas $(2 \# 2) \# 3 = ||2 - 2| - 3| = 3$.

The simplest algebraic structures are formed by a set together with an associative binary operation.

DEFINITION. Let A be a set and let $*$ be an associative binary operation on A. We call A a **semigroup** under $*$ and denote this algebraic structure by $(A, *)$.

Our familiar settings provide many examples of semigroups: $(\mathbf{Z}, +)$; $(\mathbf{Z}, \cdot)$; $(E^*, *)$, the set of all words over the alphabet E under concatenation; $(C(X), \circ)$; etc. The set of integers under subtraction is not a semigroup because associativity fails.

If we look again at the binary operation $*$ defined on the set $A = \{a, b, c\}$ by the Cayley table given in Table 8.1, we find that for each x in A, we have $a * x = x$. The same role is played by the number 1 on the integers $\mathbf{Z}$ under multiplication, because $1 \cdot x = x$ for each x in $\mathbf{Z}$. It is played by 0 for the integers under addition, because $0 + x = x$ for each x in $\mathbf{Z}$, and it is played by id_X for $C(X)$ under composition because $\text{id}_X \circ f = f$ for each function f in $C(X)$. The set of integers under subtraction has no element e such that $e - x = x$, but moving to the right, we find that $x - 0 = x$ for each x in $\mathbf{Z}$. What a, 0, 1, and id_X have in common is summarized in the following definition.

DEFINITION. Let $*$ be a binary operation on a set A. Let e be an element of A.

(a) We say that e is a **right identity** element if for each element x in A, we have $x * e = x$.

(b) We say that e is a **left identity** element if for each element x in A, we have $e * x = x$.

(c) We say that e is an **identity** element if for each x in A we have $e * x = x * e = x$. (So e is an identity element if it is both a right and a left identity element.)

Example 4.

(a) The element 0 is both a left and a right identity element for the addition operation on the integers. The element 1 is both a right and a left identity element for the multiplication operation on the integers. But 0 is only a right identity element for the integers under subtraction.

(b) Let X be a set and let id_X be the identity function on X. Then id_X is an identity for $C(X)$ under composition.

(c) The empty word Λ is both a right and a left identity element for the operation of concatenation, and hence it is an identity element.

(d) Consider the operation $*$ defined on the set $A = \{a, b, c\}$ by Table 8.2. Both a and c are left identities, but there is no right identity.

Table 8.2

	a	b	c
a	a	b	c
b	a	c	b
c	a	b	c

In our examples we have not found distinct right and left identity elements. This cannot be done.

Proposition 1. Suppose that $*$ is a binary operation on a set A and that e is a right identity element and that f is a left identity element. Then $e = f$.

PROOF: We have $f * e = e$ because f is a left identity element. Similarly, we have $f * e = f$ because e is a right identity element. Thus $e = f$.

In our examples we have not found more than one identity element. It follows as a simple corollary to Proposition 1 that this also cannot be done. (We leave the proof as an exercise.)

Corollary. Let $*$ be a binary operation on a set A. Then A has at most one identity element.

> **DEFINITION.** Let A be a set with an associative binary operation $*$ and an identity element e. We call A a **monoid** and denote the algebraic structure by $(A, *, e)$.

Thus most of our familiar examples—$(\mathbf{Z}, +, 0)$, $(\mathbf{Z}, \cdot, 1)$, $(C(X), \mathrm{id}_X, \circ)$—are monoids as well as semigroups.

Example 5. Let S be a set and consider its power set $\mathcal{P}(S)$ under the operation of intersection $\cap$. Since $\cap$ is associative, $(\mathcal{P}(S), \cap)$ is a semigroup. For any subset A of S, we have $A \cap S = S \cap A = A$. Thus the set S is the identity element for $\mathcal{P}(S)$ under $\cap$, and we see that $(\mathcal{P}(S), \cap, S)$ is a monoid.

Let $*$ be a binary operation on a set A, and let B be a subset of A. We say that B is **closed** under $*$ if for every pair of elements (x, y) in $B \times B$, we have that $x * y$ is also in B. The set E of even integers is closed under addition. Since 0 is an even integer, the subset E of $\mathbf{Z}$ is itself a monoid. The set of odd integers is not closed under addition because, for instance, $3 + 5 = 8$ and 8 is not odd.

Example 6. Let X be a nonempty set. The set of bijections from X to X is closed under composition.

> **DEFINITION.** Let $(A, *, e)$ be a monoid, and let B be a subset of A. The set B is a **submonoid** of A if B is closed under $*$ and the identity element of A is an element of B.

The set of even integers E under addition is a submonoid of $\mathbf{Z}$. The set E is also closed under multiplication, but it is not a submonoid under this operation because 1, the identity element for multiplication, is not a member of E. The set of odd integers is a submonoid of $\mathbf{Z}$ under multiplication.

Example 7. Let S be a nonempty set, and let A be a proper subset of S that is also not empty. Then $\mathcal{P}(A)$ is a subset of $\mathcal{P}(S)$. We know from Example 5 that $(\mathcal{P}(A), \cap, A)$ is monoid. However, it is not a submonoid on $(\mathcal{P}(S), \cap, S)$ because the identity element of $\mathcal{P}(S)$ is not an element of $\mathcal{P}(A)$. (To be called a submonoid, the subset in question must contain the identity element of the original monoid.)

Example 8. We wish to obtain the smallest submonoid S of $(\mathbf{Z}, +, 0)$ that contains the elements 6 and -4. The set S must contain 0 and must be closed under addition. Thus it must contain

$$0, \quad 6, \quad -4, \quad 6 + (-4) = 2, \quad (-4) + 2 = -2, \quad 2 + 2 = 4$$

and so forth. We obtain S inductively as follows:

(1) 0, 6, and -4 are in S.

(2) If x and y are in S, then $x + y$ is in S.

(3) Only elements obtained from Step (1) and iterations of Step (2) are in S.

Step (2) ensures that S is closed under addition, and Step (1) ensures that S contains the identity element of $\mathbf{Z}$, namely 0. Thus S is a submonoid of $\mathbf{Z}$. All the elements of S are even. Any submonoid of $\mathbf{Z}$ that contains 6 and -4 must contain 2 and hence all the other even integers. Thus E must be the smallest submonoid of $\mathbf{Z}$ that contains both -4 and 6.

In Example 8 we began with a subset B of a monoid A and used an inductive definition to construct the smallest submonoid of A that contained all the elements of B. Using Example 8 as a model, we generalize this procedure to arbitrary monoids.

DEFINITION. Let $(A, *, e)$ be a monoid. Let S be a subset of A. The monoid **generated** by S, denoted $\langle S \rangle$, is defined as follows:

(1) The identity element e of A is in $\langle S \rangle$ and, if x is in S, then x is also in $\langle S \rangle$.

(2) If x and y are elements of $\langle S \rangle$, then $x * y$ is also in $\langle S \rangle$.

(3) Only elements obtained from Step (1) and iterations of Step (2) are in $\langle S \rangle$.

We verify that $\langle S \rangle$ is indeed a submonoid of S by first noting that step (1) directly establishes that e is a member of S. To show that $\langle S \rangle$ is closed, we let x and y be elements of $\langle S \rangle$. Suppose that x was obtained in the ith iteration of step (2) and y in the jth iteration. Suppose $j \leq i$. Then both $x * y$ and $y * x$ are obtained in the $(i + 1)$th iteration of step (2). Associativity holds in $\langle S \rangle$ because it holds in A. (If $*$ is commutative on A, it is also commutative on $\langle S \rangle$.) Thus $\langle S \rangle$ is indeed a monoid under its inherited operation $*$.

Example 9.

(a) Let $E = \{a, b\}$ and $S = \{aa, bbb\}$. The submonoid of E^* generated by S is $\langle S \rangle = \{\Lambda, aa, bbb, aabbb, bbbaa, aaaa, \ldots\}$. In terms of the notation given in Section 2.8, $\langle S \rangle = S^*$.

(b) Let $X = \{a, b, c, d\}$, and let $f : X \rightarrow X$ be defined by the following diagram:

The submonoid of $(C(X), \circ, \mathrm{id}_X)$ generated by f—namely $\langle\{f\}\rangle$—is the set $\{\mathrm{id}_X, f, f^2, f^3\}$.

Throughout this section, most of our examples have come from very familiar settings. We now turn our attention to a quite different way of obtaining semigroups: the **semigroups of finite state machines.**

Let F be a finite state machine over the alphabet E, and suppose that S is the set of states of F. Let x be any word in E^*. We use x to define a function $g_x : S \rightarrow S$ as follows: let $g_x(s) = t$ if t is the state reached from state s by following the path defined by x. If F is the machine shown in Figure 8.1 and $x = abab$, then g_x is the function diagrammed in Figure 8.1. Let $A_F = \{g_x : x \in E^*\}$. We now define a binary operation $*$ on A_F and show that $(A_F, *)$ is a semigroup (and actually a monoid) under $*$. Let x and y be words in E^*. Define $g_x * g_y$ to be $g_y \circ g_x$. (Note the reversed order of x and y.) Because composition is associative, so is $*$. Letting x be the empty word, we find that $g_\Lambda = \mathrm{id}_S$. Thus $(A_F, *, g_\Lambda)$ is a monoid; it is usually called the **semigroup** of the finite state machine F.

Figure 8.1

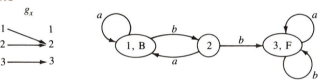

For F as given in Figure 8.1, the members of A_F are as shown in Figure 8.2. We note that $g_{aa} = g_a$, that $g_{aba} = g_a$, and that $g_{abb} = g_{bba} = g_{bbb} = g_{bb}$. Also, $g_{bab} = g_b$. Hence there are no functions in A_F other than the six diagrammed in Figure 8.2. The Cayley table for the monoid A_F is given as Table 8.3.

Figure 8.2

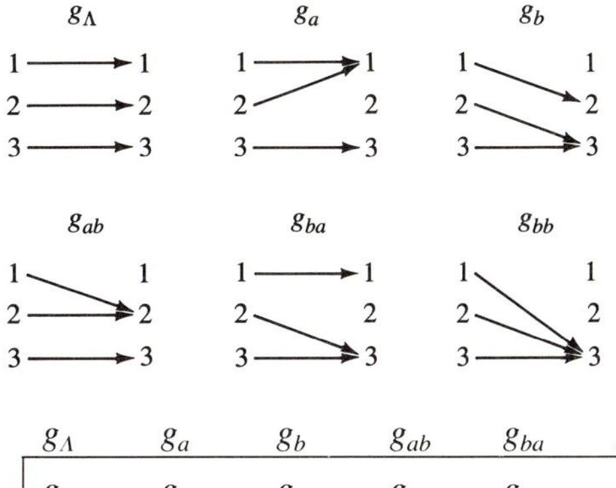

g_Λ	g_a	g_b
1 → 1	1 → 1	1 → 1
2 → 2	2 → 2	2 → 2
3 → 3	3 → 3	3 → 3

g_{ab}	g_{ba}	g_{bb}
1 → 1	1 → 1	1 → 1
2 → 2	2 → 2	2 → 2
3 → 3	3 → 3	3 → 3

Table 8.3

$*$	g_Λ	g_a	g_b	g_{ab}	g_{ba}	g_{bb}
g_Λ	g_Λ	g_a	g_b	g_{ab}	g_{ba}	g_{bb}
g_a	g_a	g_a	g_{ab}	g_{ab}	g_a	g_{bb}
g_b	g_b	g_{ba}	g_{bb}	g_b	g_{bb}	g_{bb}
g_{ab}	g_{ab}	g_a	g_{bb}	g_{ab}	g_{bb}	g_{bb}
g_{ba}	g_{ba}	g_{ba}	g_b	g_b	g_{ba}	g_{bb}
g_{bb}	g_{bb}	g_{bb}	g_{bb}	g_{bb}	g_{bb}	g_{bb}

| Section 8.1 | Exercises |

1. Let $A = \{a, b, c\}$, and let $*$ be the binary operation defined in the accompanying table. Determine the associativity and the commutativity of $*$. Determine also whether $*$ has a right or a left identity element.

$*$	a	b	c
a	b	c	a
b	c	a	b
c	a	b	c

2. Repeat Exercise 1 for the operation $\#$ defined in the accompanying table.

$\#$	a	b	c
a	a	b	c
b	b	b	a
c	a	c	b

3. Determine the commutativity and the associativity of each of the following binary operations defined on $\mathbf{Z}$.

   (a) $f_*(m, n) = 2mn$
   (b) $f_\#(m, n) = 2m + n$

4. Let X be a set. Determine the commutativity and associativity of the operation $A \backslash B$ on $\mathcal{P}(X)$. Does this operation have a right or a left identity element?

5. Determine whether the operations given in Exercise 3 have identity elements.

6. Let n be an integer, and let B^n be the set of strings of 0's and 1's of length n. For strings $x = x_1 x_2 \ldots x_n$ and $y = y_1 y_2 \ldots y_n$ in B_n, define $x \wedge y$ to be the string $z = z_1 z_2 \ldots z_n$, where $z_i = 1$ if and only if both x_i and y_i are 1. Show that B^n is a monoid under $\wedge$. Write down the Cayley table for B^2.

7. Let $X = \{a, b, c\}$, and let $B(X)$ be the set of bijections from X to X. Write down the Cayley table for $B(X)$ under the operation of composition. Show that the operation is not commutative.

8. Is the set of prime numbers closed under multiplication? Under addition?

9. Show that the set of integers of the form $3m + 1$ is closed under multiplication. Is this set a submonoid of $(\mathbf{Z}, \cdot, 1)$?

10. Let $A = \{a, b, c, d\}$, and let S be the collection of subsets of A that contain b. Show that S is closed under $\cap$.

11. (a) Determine the submonoid of $(\mathbf{Z}, +, 0)$ generated by 6 and 9.
    (b) Determine the submonoid of $(\mathbf{Z}, +, 0)$ generated by -3 and 5.

12. Find the semigroup defined by each of the following finite state machines.

(a)

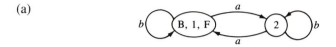

(b)

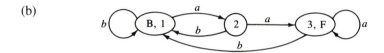

The Integers

In this section we take a closer look at the algebraic structure of the set of integers. We focus our attention on the notion of divisibility. Our results are basic to everyday arithmetic and are of great interest and importance in themselves. (The classical study of number theory probes this topic far more deeply.) Even so, a clear statement of the facts of arithmetic is necessary to solve problems that arise in the more abstract settings of groups, rings, and fields—algebraic structures that we look at in later sections.

Suppose that a and b are integers. We say that a **divides** b if there is an integer q such that $b = qa$. In this case we write $a|b$. Thus $2|6$ but $6 \nmid 2$. (If $a|b$, then we also say that b is a **multiple** of a and that a is a **factor** of b.) If p is an integer greater than 1 and if the only factors of p are ± 1 and $\pm p$, then we say that p is a **prime** number. Some primes are 2, 3, 5, 7, . . . , 101, ... Later in this section we shall show that there is an infinite number of prime numbers. We also show that every integer m can be factored into a product of primes in exactly one way. (We use this fact every time we reduce a fraction to lowest terms.) We begin with some facts about divisibility.

Proposition 1. Suppose that $a|b$ and $a|c$. Then $a|(mb + nc)$ for any pair of integers m and n.

PROOF: We can find integers s and t such that $b = sa$ and $c = ta$. Thus the sum $mb + nc$ can be written as $a(ms + nt)$.

If a and b are integers and a is not divisible by b, then a has a remainder r upon division by b. Theorem 1 states that we can always find r such that $0 < r < b$. In the proof, we use the fact that the set of whole numbers $\{0, 1, 2, 3, 4, . . .\}$

is well ordered; that is, every nonempty subset S of W has a smallest element. (Well-ordered sets were discussed in Chapter 5.)

Theorem 1.   The Division Theorem. Let a and b be integers and suppose that $b > 0$. Then we can find integers q and r such that

$$a = bq + r \qquad \text{and} \qquad 0 \le r < b$$

Further, r and q are unique.

For example, if $a = 19$ and $b = 5$, then, letting $q = 3$ and $r = 4$, we have $19 = 5 \cdot 3 + 4$. Letting $a = -19$ and $b = 5$, we must take $q = -4$ and $r = 1$. Then $-19 = 5 \cdot (-4) + 1$.

PROOF: Let $S = \{m : m \ge 0 \text{ and } m = a - bq \text{ for some integer } q\}$. The set S is not empty. Suppose $a \ge 0$. Then, taking $q = 0$, we see that a itself is in S. If $a < 0$, we can find a positive value of q sufficiently large that $a - b(-q) \ge 0$. Then the number $a - b(-q)$ is in S. Because S is not empty, it has a smallest element. Call this number r. Because r is an element of S, $a = bq + r$ for some integer q and $r \ge 0$. We must now prove that $r < b$. We do this by contradiction. Suppose that $r \ge b$ and that $r = a - bq$. We can subtract b from both sides of this equation and rewrite it as $r - b = a - b(q + 1)$. Thus $r - b$ is an element of S, but $r - b < r$, contradicting our choice of r as the smallest element of S. So $r < b$.

We must now show that r and q are unique. That is, if r_0 and q_0 satisfy the theorem, then $r = r_0$ and $q = q_0$. Let us assume that r_0 and q_0 also satisfy the theorem, so that $r = a - bq$ and $r_0 = a - bq_0$. Solving for a in both expressions and rearranging, we find that

$$(q_0 - q)b = (r - r_0)$$

Let us assume that $0 \le r_0 < r < b$. Then $0 < (r - r_0) < b$. Since $b > 0$, we must have that $(q_0 - q) > 0$, and since q and q_0 are integers, $(q_0 - q) \ge 1$. But then $(r - r_0) \ge b$, a contradiction. Thus $r = r_0$ and $q = q_0$. Our theorem is proved.

When we put the rational number $x = 12/42$ into lowest terms, we factor out the greatest common divisor of 12 and 42—namely 6—from the numerator and denominator to find that $x = 2/7$. When we try to do the same for $s/t = 10237/423141$, experience tells us that s and t have a greatest common divisor, but how to find it is not obvious. In what follows, we define what it means to be the greatest common divisor of two integers s and t, and we show how to find it in terms of s and t. We will need the following theorem.

Theorem 2.   Suppose that s and t are integers that are both greater than 0. Let S be the nonempty set $\{x : x > 0 \text{ and } x = ms + nt \text{ for some integers } m \text{ and } n\}$. Let d be the smallest element of S. Then $d|s$ and $d|t$. Further, if c is any other divisor of both s and t, then $c|d$.

PROOF: Since d is an element of S, we can find integers m and n such that $d = ms + nt$; d is the smallest number greater than 0 that can be so expressed. Now suppose that s is not divisible by d. By the division theorem, we can find integers q and r such that $s = dq + r$ and $0 < r < d$. Substituting $ms + nt$ for d and solving for r, we find that $r = (1 - qm)s - (qn)t$. Thus r is also an element of S, but $r < d$, a contradiction. Thus s must be divisible by d. Similarly, t must be divisible by d. Now let c be any other common divisor of s and t. Since $d = ms + nt$, d is also divisible by c and our theorem is proved.

> **DEFINITION.** Let a and b be integers. We say that d is the **greatest common divisor** of a and b if d divides both a and b, and further, if c is any other common divisor of a and b, then $c|d$. The greatest common divisor (gcd) of a and b is denoted by (a, b).

Let $a = 30$ and let $b = 72$. The set of common divisors of a and b is the set $\{1, -1, 2, -2, 3, -3, 6, -6\}$. Of this set, both 6 and -6 satisfy the definition of the greatest common divisor of 30 and 72; they are both divisible by every other common divisor. (Note that when we speak of "the" greatest common divisor, we shall usually choose the positive alternative.)

Theorem 2 guarantees that every pair of integers a and b has a greatest common divisor—namely d as found in the theorem—because d divides both a and b and is divisible by every other divisor. Further, it says that (a, b) can be written as a linear combination of a and b. So, for instance, $(30, 72) = 6$ and $6 = 5 \cdot 30 - 2 \cdot 72$. The theorem tells us more: it says that the positive greatest common divisor of two integers a and b is the *smallest* positive number that can be written as $ma + nb$. So, for instance, we can see that 453 and 227 share no common factors greater than 1 because $1 = 2 \cdot 227 + (-1) \cdot 453$. Because 1 must indeed be the smallest positive such expression, it must be the greatest common divisor of 227 and 453. However, the theorem does not show us how to find the gcd of two integers. That is accomplished by Euclid's algorithm, which we shall present shortly. We first give two propositions about gcd's that are necessary to see that Euclid's algorithm works.

Proposition 2.   Suppose that a and b are integers and that $b|a$. Then $b = (a, b)$.

PROOF: Since b divides both a and b, and since every common divisor of both a and b divides b, b satisfies the definition of the greatest common divisor of a and b.

Proposition 3. Suppose that a and b are integers such that $b > 0$, and that we have found r according to the division theorem: $a = bq + r$ where $0 < r < b$. Then $(a, b) = (b, r)$.

PROOF: Let $d = (a, b)$. Since $r = a - qb$, and d divides both a and b, d must divide r. Let c be any common divisor of r and b. Since $a = qb + r$, c also divides a. Thus c is also a common divisor of a and b. Since $d = (a, b)$, d is divisible by c. Since d is a common divisor of r and b and is divisible by any other common divisor of r and b, we have that $d = (b, r)$.

For example, let $a = 72$ and $b = 30$. Then $a = 2b + 12$ and we have that $(72, 30) = (30, 12) = 6$. We can divide again to find that we have $30 = 2 \cdot 12 + 6$ and that $(30, 12) = (12, 6) = 6$. That 6 is the common divisor of 12 and 6 follows from Proposition 2, because $6|12$. Iterating Proposition 3 as we have just done is the basis of Euclid's algorithm, a method by which we can determine the gcd of two integers a and b. We now present Euclid's algorithm.

Euclid's Algorithm. Let a and b be integers and assume that $0 < b < a$.

(1) Let $y = a$ and let $x = b$.

(2) Find r so that $y = qx + r$, $0 \leq r < x$. If $r = 0$, stop. The gcd of a and b is x. If $r \neq 0$, let y assume the value of x and let x assume the value of r. Return to the beginning of step 2.

Example 1. We apply Euclid's algorithm to find the gcd of 15 and 24.

1. $y = 24$, $x = 15$, and $24 = 1 \cdot 15 + 9$. So $r = 9$.
2. $y = 15$, $x = 9$, and $15 = 1 \cdot 9 + 6$. So $r = 6$.
3. $y = 9$, $x = 6$, and $9 = 1 \cdot 6 + 3$. So $r = 3$.
4. $y = 6$, $x = 3$, and $6 = 2 \cdot 3 + 0$. Since $r = 0$, stop: $3 = (24, 15)$.

Note: The gcd of a and b is the last nonzero remainder.

Example 2. We apply Euclid's algorithm to $a = 105$ and $b = 22$.

1. $105 = 4 \cdot 22 + 17$
2. $22 = 1 \cdot 17 + 5$
3. $17 = 3 \cdot 5 + 2$
4. $5 = 2 \cdot 2 + 1$
5. $2 = 2 \cdot 1 + 0$

Thus $(105, 22) = 1$, and we see that 105 and 22 share no common factors other than 1 and -1.

> **DEFINITION.** We say that two integers a and b are **relatively prime** if $(a, b) = 1$—that is, if a and b share no common factors other than 1 and -1.

Thus 15 and 4 are relatively prime even though neither number is itself a prime number.

Proposition 4. Suppose that a and b are relatively prime and suppose that a divides the product bc. Then a divides c.

> PROOF: Since a and b are relatively prime, $(a, b) = 1$. By Theorem 2, we know that we can find integers m and n such that $1 = ma + nb$. Multiplying through by c, we find that $c = cma + cnb$. Since a divides both cma and cnb, it divides its sum, namely c.

The proposition fails if a and b are not relatively prime. For instance, if $c = 4$, $b = 15$, and $a = 6$, we have that $6 \mid 60$, but $6 \nmid 4$ and $6 \nmid 15$.

> **Corollary.** If p is prime and $p \mid ab$, then either p divides a or p divides b.

Proposition 5. Let a and b be integers and let $d = (a, b)$. Then $(a/d, b/d) = 1$.

> PROOF: By Theorem 2, we can find integers m and n such that $d = ma + nb$. Dividing both sides by d, we obtain that $1 = m(a/d) + n(b/d)$.

To illustrate Proposition 5, we let $a = 12$ and $b = 15$. Then $(12, 15) = 3$ and $(12/3, 15/3) = (4, 5) = 1$.

In Chapter 5 we looked for the solution sets of congruences of the form $ax \equiv b \pmod{m}$. For instance, the solution set of the congruence $3x \equiv 2 \pmod 5$ is exactly the equivalence class of 4, namely $\{\dots, -1, 4, 9, \dots\}$. The solution set of the congruence $4x \equiv 2 \pmod 6$ is the union of the equivalence classes of 2 and 5. The solution set of the congruence $2x \equiv 3 \pmod 4$ is empty. These three congruences are representative of the general situation: the solution set of a congruence of the form $ax \equiv b \pmod{m}$ is exactly one congruence class $\pmod m$, the union of equivalence classes $\pmod m$, or empty. We can use our results about greatest common divisors to determine which situation obtains. We summarize our results in the following proposition.

Proposition 6. The congruence $ax \equiv b \pmod{m}$ has a solution if and only if $(a, m) \mid b$. Further, x and y are both solutions if and only if
$x \equiv y \;[\mathrm{mod}\;(m/(a, m))]$.

Before we prove Proposition 6, let us see what it says for various congruences. Consider $5x \equiv 2 \pmod 6$. Since $(5, 6) = 1$ and $1|2$, we are assured that the congruence has a solution. We find that $x = 4$ is a solution. Since y is also a solution if and only if $y \equiv 4$ [mod $(6/1)$], we know that the solution set is exactly the equivalence class of 4. Now $4x \equiv 5 \pmod 6$ has no solution since $(4, 6) = 2$ and $2 \nmid 5$. However, $4x \equiv 2 \pmod 6$ does have a solution. We find that $x = 2$ is a solution. We know that y is also a solution if and only if $y \equiv 2 \pmod 3$. Thus the solution set is the set $\{. . . , -1, 2, 5, . . .\}$, which is the union of the congruence classes of 2 and 5, modulo 6.

PROOF OF PROPOSITION 6. Suppose that $d = (a, m)$ and that $d|b$. We show that congruence $ax \equiv b \pmod m$ has a solution. By Theorem 2, we can find integers s and t such that $d = sm + ta$. Since $d|b$, we can find q such that $b = qd = qsm + qta$. Then $x = qt$ is a solution to the congruence because $ax - b = (-qs)m$. Conversely, we show that if $ax \equiv b \pmod m$ has a solution, then $d|b$. Let t be a solution so that $at - b = qm$ for some integer q. Solving for b, we see that $b = at - qm$. Since both a and m are divisible by their greatest common divisor d, the number b is also.

Now assume that x is a solution to the congruence. We prove that y is a solution if and only if $x \equiv y$ [mod $(m/(a, m))$]. Assume first that y is a solution so that $ay \equiv b \pmod m$. Thus we have $ax \equiv ay \pmod m$—that is, $ax - ay = qm$ for some integer q. Dividing both sides of the latter expression by (a, m), we have $[a/(a, m)] (x - y) = q[m/(a, m)]$. Recall that $a/(a, m)$ and $m/(a, m)$ are relatively prime. So $(x - y)$ is divisible by $m/(a, m)$. That is, $x \equiv y$ [mod $(m/(a, m))$]. Conversely, assume that $x \equiv y$ [mod $(m/(a, m))$]. We show that y is also a solution to $ax \equiv b \pmod m$. Let $d = (a, m)$. We know that $x - y = qm/d$ for some integer q or, multiplying through by d, that $dx - dy = qm$. Multiplying through by a/d, we find that $ax \equiv ay \pmod m$, so that $ay \equiv b \pmod m$. Thus y is also a solution to the congruence.

Corollary to Proposition 6. Let p be a prime number and let a and b be integers. Then every congruence of the form $ax \equiv b \pmod p$ has a unique solution modulo p. (That is, the solution set is exactly one equivalence class modulo p.)

We now turn our attention to prime factorization. Our experience with arithmetic tells us that a number like 12 can be factored into primes in exactly one way, $(12 = 2 \cdot 2 \cdot 3)$. Similarly, $27{,}600{,}573 = 3 \cdot 7 \cdot 7 \cdot 11 \cdot 13 \cdot 13 \cdot 101$, but neither this factorization nor its uniqueness is so obvious as that of 12. In the following two propositions, we prove that every natural number greater than 1 can be factored as the product of primes in a unique way. Thus every non-zero integer can be factored as ± 1 and the product of primes in a unique way.

Proposition 7. Let m be an integer that is greater than or equal to 2. Then m can be factored into the product of primes.

PROOF: We prove this by induction on m. If $m = 2$, the statement is true since 2 itself is a prime. Now we assume that if n is an integer such that $2 \leq n \leq m$, then n can be factored into primes. We show that the proposition holds for $m + 1$. If $m + 1$ is a prime, then we are finished. If not, then $m + 1$ can be properly factored as $m + 1 = xy$, where both x and y satisfy our induction hypothesis. Thus both x and y can be factored as the products of primes and, multiplying the factorizations of x and y, we conclude that $m + 1$ can be factored as the product of primes.

Proposition 8. Let m be an integer, $m \geq 2$. Then m can be factored into primes in exactly one way.

PROOF: Again we proceed by induction. The proposition is true for $m = 2$. Let us assume that it is true for every integer $s < m$. We show that it holds for m, so suppose that $m = p_1 p_2 \ldots p_k$ and also that $m = q_1 q_2 \ldots q_j$. (Because we have not used exponents, a prime may occur more than once in the factorization.) Since $p_1 | m$, we know that p_1 must divide and hence be equal to one of the q_i's. Let us assume that we have listed the q_i's in such a way that $p_1 = q_1$. Thus $p_2 p_3 \ldots p_k = q_2 q_3 \ldots q_j = m/p_1$. Now m/p_1 satisfies our induction hypothesis so that, except for the order in which they appear, the list of primes in the product $p_2 p_3 \ldots p_k$ must be identical to the list of primes appearing in the product $q_2 q_3 \ldots q_j$. Our factorizations of m must be identical, and we have proved the proposition.

The results of Propositions 7 and 8 are summarized in Theorem 3.

Theorem 3. Unique Factorization. Any nonzero integer m can be factored into the product of ± 1 and the product of primes in exactly one way.

The existence of the factorization of any integer into the product of primes is guaranteed. However, determining how to obtain that factorization is a difficult and time-consuming task even in the age of the supercomputer. Later in this chapter we shall see an application of the unique factorization theorem to a coding problem. The relative unbreakability of the codes obtained depends on the enormous amount of time required to factor numbers with many digits into a product of primes.

We conclude this section by showing that there are an infinite number of primes.

Proposition 9. There are an infinite number of prime numbers.

PROOF: (We prove this by contradiction.) Suppose that there are only a finite number of primes and that the complete set is $S = \{p_1, p_2, \ldots, p_n\}$. Let $q = (p_1 p_2 \cdots p_n) + 1$. Since q is larger than any number in S, it cannot be a prime. Thus it must be factorable into the product of primes and hence divisible by one of the primes in S—say p_i. Now p_i divides q, and it divides the product $p_1 p_2 \cdots p_n$. Thus it must divide the difference $q - p_1 p_2 \cdots p_n = 1$. But the number 1 is not divisible by p_i, and we have a contradiction. Thus q must be prime, and the finite set S cannot be complete.

<table>
<tr><td>**Section 8.2**</td><td>**Exercises**</td></tr>
</table>

1. For $b = 19$, obtain the remainder r and q as guaranteed by the division theorem for $a = 40$, -10, -10001, and 400592.

2. For each of the following pairs of numbers s and t, find the greatest common divisor (s, t) and express it as $ms + nt$.
   (a) 9 and 15   (b) 7 and 5   (c) 24 and 33

3. Apply Euclid's algorithm to obtain the greatest common divisor of each of the following pairs of numbers.
   (a) 45 and 33   (b) 1573 and 36   (c) 70 and 91

4. Determine which of the following pairs of numbers are relatively prime.
   (a) 1155 and 247   (b) 153 and 4   (c) 221 and 1111

5. Give several counterexamples to Proposition 4 in the case that $(a, b) \neq 1$.

6. Solve the following congruences. In each case, express your answer as the union of equivalence classes.
   (a) $2x \equiv 7 \pmod{13}$ (b) $3x \equiv 1 \pmod{5}$
   (c) $2x \equiv 6 \pmod{12}$ (d) $3x \equiv 1 \pmod{9}$

7. Use Proposition 6 to determine whether the solution set to each of the following congruences is one equivalence class, the union of two or more equivalence classes, or empty.
   (a) $3x \equiv 5 \pmod{17}$ (b) $2x \equiv 4 \pmod{24}$ (c) $6x \equiv 5 \pmod{18}$

8. Solve $ax \equiv 1 \pmod{p}$ for $p = 5$ and $1 \leq a \leq p - 1$.

9. Repeat Exercise 8 for $p = 3$ and 7.

10. Prove that if $(a, b) = 1$ and $(a, c) = 1$, then $(a, bc) = 1$.

11. Prove that the congruence $ax \equiv 1 \pmod{m}$ has a nonempty solution set if and only if a and m are relatively prime.

12. Prove that the solution set to the congruence $ax \equiv 1 \pmod{m}$ is either empty or exactly one equivalence class modulo m.

13. Factor each of the following as the product of primes: 243, 1800, 1155, 5831, 9797.

14. List the first 20 prime numbers.

15. Prove that if $m > 1$ is not prime, then m has a prime factor p such that $p \leq m^{1/2}$.

16. (Least Common Multiples) Let x and y be two nonzero integers. The **least common multiple** of x and y is an integer m such that m is divisible by both x and y and, if z is any other integer divisible by both x and y, then z is divisible by m also.

   (a) Find the least common multiple of each of the following pairs of integers: 5 and 7, 15 and 10, 15 and 14.
   (b) Characterize the least common multiple of two integers x and y in terms of the prime factorizations of x and y.

   (*Note*: In the chapter exercises, we shall ask for a proof of the fact that the least common multiple of any two nonzero integers can always be found.)

Section 8.3	Groups

Within the structure of a monoid $(A, *, e)$, we are not guaranteed that we can solve for x in even the simplest equation of the form $a * x = b$, where a and b are members of A. For example, there is no integer solution to $3x = 5$ in $(\mathbf{Z}, \cdot, 1)$ and there is no word x over $E = \{a, b\}$ for which $ababx = baba$. In $(\mathbf{Z}, +, 0)$, however, we can always solve the equation $a + x = b$ for exactly one value of x. We simply add $(-a)$ to both sides of the equation, where $(-a)$ is the unique solution to the equation $a + x = 0$. In this section we examine monoids like $(\mathbf{Z}, +, 0)$, in which the equation $a \cdot x = e$ always has a solution. Such structures are called **groups.**

> **DEFINITION.** Let $*$ be a binary operation on the set A, and let e be the identity element for $*$. Let a and b be elements of A. We say that b is the **inverse** of a if $a * b = b * a = e$. In this case we write $b = a^{-1}$.

Note: If we use additive notation and denote the operation on A by $+$ rather than by $*$, we denote the inverse of an element a by $-a$.

Every element m in $(\mathbf{Z}, +, 0)$ has an inverse; it is $(-m)$ because $(-m) + m$

= 0. Every nonzero rational number x in the monoid $(Q, \cdot, 1)$ has an inverse: if $x = s/t$, then $x^{-1} = t/s$ since $s/t \cdot t/s = 1$. No elements other than 1 and -1 have inverses in $(\mathbf{Z}, \cdot, 1)$.

DEFINITION. Let A be a set and let $*$ be an associative binary operation on A with identity element e. We call $(A, *, e)$ a **group** if every element x in A has an inverse.

A group is a monoid in which every element has an inverse. In a group we can always solve the equation $a * x = b$ for x by letting $x = a^{-1} * b$. Substituting for x, we have $a * (a^{-1} * b) = (a * a^{-1}) * b = e * b = b$. If $ax = ay$, we can multiply both sides on the left by a^{-1} to see that $x = y$. Thus the equation $ax = b$ has exactly one solution.

DEFINITION. If $(A, *, e)$ is a group and the operation $*$ is commutative, we call A an **abelian group.**

Example 1. $(\mathbf{Z}, +, 0)$ is an abelian group. The set of rational numbers Q under multiplication is not a group because the element 0 does not have an inverse. However $(Q^+, \cdot, 1)$, the set of rational numbers greater than zero under multiplication, is an abelian group.

DEFINITION. Let $(A, *, e)$ be a group. The **order of A** is the cardinality of A. It is denoted by $|A|$.

The groups $(\mathbf{Z}, +, 0)$ and $(Q^+, \cdot, 1)$ are groups of infinite order. In the following examples, we investigate some finite groups.

Example 2. Let A be the set $\{a, b, c, d\}$, and let $*$ be the operation on A defined by the Cayley table given as Table 8.4. The operation is both associative and commutative. Inspecting both the column and the row labeled with a, we see that $a * x = x * a = x$ for each x in A. Thus a is the identity element in A under $*$. Further, the identity element a appears exactly once in each row and each column of the table. This says that for each x in A, there is a solution to the equation $b * x = a$ that also solves $x * b = a$. Thus every element in A has an inverse. The operation $*$ is commutative, and so we conclude that $(A, *, a)$ is an abelian group of order 4.

Table 8.4

*	a	b	c	d
a	a	b	c	d
b	b	c	d	a
c	c	d	a	b
d	d	a	b	c

Example 3. Let $X_n = \{1, 2, 3, \ldots, n\}$. Recall that a permutation of X_n is a bijection $f : X_n \to X_n$. Let S_n be the set of all permutations of X_n. There are $n!$ elements in S_n. We noted in Section 8.1 that $(S_n, \circ, \mathrm{id}_{X_n})$ is a monoid. Each bijection f in S_n has an inverse function $f^{-1} : X_n \to X_n$ such that $f \circ f^{-1} = f^{-1} \circ f = \mathrm{id}_{X_n}$. The inverse function f^{-1} is the inverse of f with respect to the binary operation of composition. Thus $(S_n, \circ, \mathrm{id}_{X_n})$ is a group. It is called the **nth symmetric group.**

In Figure 8.3 the $3! = 6$ permutations in S_3 are listed. Table 8.5 is the Cayley table for the group $(S_3, \circ, \mathrm{id}_{X_3})$. We can see that $f_3 \circ f_1 \neq f_1 \circ f_3$, so that S_3 is not abelian. (Composition in general is not commutative.) Let us solve for g in the following two equations: $f_2 \circ g = f_3$ and $g \circ f_2 = f_3$. The inverse of f_2 is f_1. To solve the first equation, we multiply both sides on the *left* by f_1 and find that $g = f_1 \circ f_3 = f_5$. To solve the second equation for g, we multiply both sides on the *right* by f_1 and find that $g = f_3 \circ f_1 = f_4$. Note that the solutions for g are different.

Figure 8.3

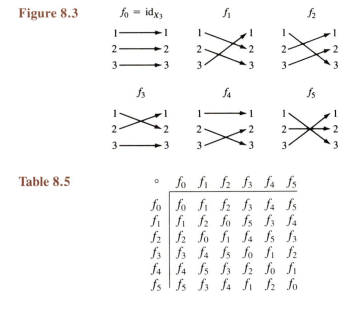

Table 8.5

$\circ$	f_0	f_1	f_2	f_3	f_4	f_5
f_0	f_0	f_1	f_2	f_3	f_4	f_5
f_1	f_1	f_2	f_0	f_5	f_3	f_4
f_2	f_2	f_0	f_1	f_4	f_5	f_3
f_3	f_3	f_4	f_5	f_0	f_1	f_2
f_4	f_4	f_5	f_3	f_2	f_0	f_1
f_5	f_5	f_3	f_4	f_1	f_2	f_0

We now turn to geometry to find an example of a non-abelian group of order 8.

Example 4. **The Symmetries of the Square.** Consider the rigid motions of the square (see Figure 8.4) that take the square onto itself.

1. We can rotate the square 0, 90, 180, or 270 degrees in a clockwise direction. We shall call these motions I, R_1, R_2, and R_3, respectively.

2. We can reflect about the diagonal axis through a and c (D_1) or through the diagonal axis through b and d (D_2).

3. We can reflect about the vertical axis (V) or through the horizontal axis (H).

Figure 8.4

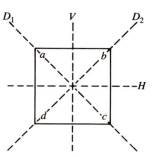

The set $S = \{I, R_1, R_2, R_3, D_1, D_2, V, H\}$ is a complete list of possible rigid motions. We equip S with a binary operation $*$ as follows: if x and y are motions in S, then $x * y$ denotes the motion y followed by the motion x. Note that we perform the motion y first in $x * y$. Thus $D_2 * R_1$ denotes the motion R_1 followed by the motion D_2. We have $D_2 * R_1 = V$. But $R_1 * D_2 = H$, and so $*$ is not commutative. The complete table for the operation $*$ on S appears as Table 8.6. An easy way to perform the operation $*$ is to cut out a square, label the corners, and execute the motions by hand.

The binary operation $*$ is an associative operation, but it is not commutative. The motion I is the identity element. Thus $(S, *, I)$ is a monoid, but $(S, *, I)$ has a further structural property: given any motion x in S, we can solve the equation $y * x = I$. Geometrically, this means that if we move the corners of the square via the motion x, there is always another motion y that returns the corners of the square

Table 8.6

	I	R_1	R_2	R_3	V	H	D_1	D_2
I	I	R_1	R_2	R_3	V	H	D_1	D_2
R_1	R_1	R_2	R_3	I	D_2	D_1	V	H
R_2	R_2	R_3	I	R_1	H	V	D_2	D_1
R_3	R_3	I	R_1	R_2	D_1	D_2	H	V
V	V	D_1	H	D_2	I	R_2	R_1	R_3
H	H	D_2	V	D_1	R_2	I	R_3	R_1
D_1	D_1	H	D_2	V	R_3	R_1	I	R_2
D_2	D_2	V	D_1	H	R_1	R_3	R_2	I

to their original positions. For instance, $R_1 * R_3 = I$ and $V * V = I$. Thus $(S, *, I)$ is a group. We solve the equation $R_3 * x = V$ in S. To do this, we operate on the left on both sides of the equation by the inverse of R_3, which is R_1. We obtain $x = R_1 * V = D_2$. Checking our table, we find that indeed $R_3 * D_2 = V$. (Note that because S is not abelian, we must be sure to multiply both sides of the equation on the left. Had we not done this, we would have obtained the wrong answer because $V * R_1 = D_1$.)

DEFINITION. Let $(A, *, e)$ be a group, and suppose that a is an element of A and that $a \neq e$. The **order of a** is the smallest positive value of n such that $a * a * a * \cdots * a = a^n = e$. If no such value of n can be found, we say that a is of infinite order. We denote the order of an element a by $o(a)$.

Note: If the binary operation $+$ is used for a group $(G, +, e)$, we use the convention that $a + a + \cdots + a = na$. In this case, $o(a)$ is the smallest value of n for which $na = e$.

In $(\mathbf{Z}, +, 0)$ there are no elements of finite order because for any integer $a \neq 0$, $a + a + a + \cdots + a = na \neq 0$. However, if we look at the element c in the group $(A, *, a)$ given in Example 2, we see that $c * c = a$ and so $o(c) = 2$. The order of the element d is 4 since $d * d * d * d = d^4 = a$, but $d * d \neq a$ and $d * d * d \neq a$.

Example 5. Let $B = \{e, s, t, u\}$. The binary operation $+$ defined on B in Table 8.7 is associative and commutative. The element e is the identity element, and each element has an inverse. Thus $(B, +, e)$ is an abelian group of order 4. Each of the nonidentity elements in B has order 2.

Table 8.7

$+$	e	s	t	u
e	e	s	t	u
s	s	e	u	t
t	t	u	e	s
u	u	t	s	e

Table 8.8 is the Cayley table for the group $(K_4, *, i)$. It is called the **Klein 4 group.** Its similarity to the group $(B, +, e)$ defined in Example 5 is obvious: both are groups of order 4 in which each nonidentity element has order 2. Further, we can find a bijection $f : B \rightarrow K_4$ (given in Figure 8.5) that is **operation-preserving.** That is, for each pair of elements x and y in B, we have $f(x + y) = f(x) * f(y)$. The groups $(B, +, e)$ and $(K_4, *, i)$ are said to be **isomorphic.** They are identical except for the names we give their elements and operations.

Table 8.8

K_4	i	a	b	c
i	i	a	b	c
a	a	i	c	b
b	b	c	i	a
c	c	b	a	i

Figure 8.5

f

$e \longrightarrow i$

$s \longrightarrow a$

$t \longrightarrow b$

$u \longrightarrow c$

DEFINITION. The groups $(G, *, e)$ and $(H, \#, i)$ are **isomorphic** if there is a bijection $f : G \rightarrow H$ that is **operation-preserving.** That is, for any pair of elements x and y in G, $f(x * y) = f(x) \# f(y)$. The bijection f is called an **isomorphism.**

Example 6.

(a) The groups $(\mathbf{Z}, +, 0)$ and $(E, +, 0)$ are isomorphic. The function f : $\mathbf{Z} \rightarrow E$ defined by $f(x) = 2x$ is an isomorphism. To see this, we note first that f is a bijection. Now let x and y be two integers. Then $f(x + y) = 2(x + y) = 2x + 2y = f(x) + f(y)$, so f is operation-preserving and hence an isomorphism.

(b) The groups $(G_1, *, e)$ and $(\mathbf{Z}_4, +, 0)$ with the Cayley tables shown as Tables 8.9 and 8.10, respectively, are isomorphic. The function g : $G_1 \rightarrow \mathbf{Z}_4$ shown in Figure 8.6 is an isomorphism.

Table 8.9

G_1:	$*$	e	a	b	c
	e	e	a	b	c
	a	a	b	c	e
	b	b	c	e	a
	c	c	e	a	b

Table 8.10

$\mathbf{Z}_4$:	$+$	0	1	2	3
	0	0	1	2	3
	1	1	2	3	0
	2	2	3	0	1
	3	3	0	1	2

Figure 8.6

g

$e \longrightarrow 0$

$a \longrightarrow 1$

$b \longrightarrow 2$

$c \longrightarrow 3$

We can see that wherever an element x appears in Table 8.9, $g(x)$ appears in the same position in Table 8.10. The bijection g is thus operation-preserving and hence an isomorphism.

Example 7. All groups of order 3 are isomorphic. To prove this, we first note that in the Cayley table for any finite group, each element of the group appears exactly once in each row and column. (Do Exercise 5 at the end of this section.) Now we show that, given a set of three elements $A = \{a, b, c\}$, there is only one way to define a group operation $*$ on A if we agree that a will be the identity element. We must determine the values of $b * c$, $c * b$, $b * b$, and $c * c$. The value of $b * c$

cannot be c since $a * c = c$, and it cannot be b since $b * a = b$. Thus $b * c = a$ and, similarly, $c * b = a$. The value of $b * b = c$ since $b * a = b$ and $b * c = a$. Similarly, $c * c = b$. The Cayley table must be as shown in Table 8.11.

Table 8.11

*	a	b	c
a	a	b	c
b	b	c	a
c	c	a	b

We can see that A is a group under $*$. The operation is also commutative, and so all groups of order 3 are isomorphic to the abelian group $\{A, *, a\}$.

Proposition 1. Let $(G, *, e)$ and $(H, \circ, i)$ be groups, and suppose that $f : G \to H$ is an isomorphism. Let x be an element of G. Then:

(a) $f(e) = i$
(b) $f(x^k) = (f(x))^k$ for each k in $\mathbf{N}$
(c) $o(x) = o(f(x))$
(d) $f(x^{-1}) = (f(x))^{-1}$

PROOF: We shall prove part (a) and leave the proofs of the remaining parts as exercises. For each element x in G, we have $f(x) = f(e * x)$. Since f is operation-preserving, we have $f(x) = f(e) \circ f(x)$. Multiplying both sides of the last equation on the right by $(f(x))^{-1}$ in H, we have $i = f(e)$.

We can use part (c) of Proposition 1 to show that the group of order 4 defined in Example 2 is not isomorphic to the group of order 4 defined in Example 5. The first has an element of order 4, whereas the second does not.

> **DEFINITION.** Let $(G, *, e)$ and $(H, \circ, i)$ be groups, and let $f : G \to H$ be a function that is operation-preserving. We call f a **homomorphism**.

A homomorphism is an operation-preserving function that is not necessarily bijective. Surjectivity, injectivity, or both may fail. Every isomorphism is also a homomorphism, but not conversely.

Example 8. The function $f : \mathbf{Z} \to E$ defined by $f(x) = 4x$ is a homomorphism from the group $\{\mathbf{Z}, +, 0\}$ to the group of even numbers $\{E, +, 0\}$. It is not surjective but it is operation-preserving. For any pair of integers x and y, we have $f(x + y) = 4(x + y) = 4x + 4y = f(x) + f(y)$.

For the rest of this section we turn our attention to the principal example of finite abelian groups, the **modular groups** $(\mathbf{Z}_n, +, [0]_n)$. In Chapter 5 we investigated the equivalence relation $x \equiv y \pmod{n}$ on the integers $\mathbf{Z}$. Recall that we write $x \equiv y \pmod{n}$ if and only if we can find an integer q such that $x - y = qn$. We denote the equivalence class of x modulo n by $[x]_n$. We shall denote the set of equivalence classes modulo n by $\mathbf{Z}_n$. Thus $\mathbf{Z}_5 = \{[0]_5, [1]_5, [2]_5, [3]_5, [4]_5\}$ and $\mathbf{Z}_3 = \{[0]_3, [1]_3, [2]_3\}$. (We shall usually delete the subscript on $[x]_n$ if the value of n is clear.)

We now define the operations $+$ and $\cdot$ on $\mathbf{Z}_n$. We will show that $\mathbf{Z}_n$ is a group under the operation $+$. (We will investigate $\mathbf{Z}_n$ under the pair of operations $+$ and $\cdot$ in Section 8.5.) Let $[x]$ and $[y]$ be equivalence classes in $\mathbf{Z}_n$, and let x and y be elements of $[x]$ and $[y]$, respectively. We define $[x] + [y]$ to be $[x + y]$ and $[x] \cdot [y]$ to be $[x \cdot y]$. Thus if $n = 5$, we have

$$[3] + [4] = [3 + 4] = [7] = [2]$$
$$[2] \cdot [4] = [2 \cdot 4] = [8] = [3]$$

If $n = 6$, we have

$$[3] + [4] = [7] = [1]$$
$$[2] \cdot [4] = [8] = [2]$$

We must show that our operations $+$ and $\cdot$ are "well defined." That is, if s is another representative of $[x]$ and t another representative of $[y]$, then $[s + t] = [x + y]$ and $[s \cdot t] = [x \cdot y]$. This ensures that the resulting equivalence class is independent of the representatives we choose from $[x]$ and $[y]$ to perform our computations. (Indeed, in $\mathbf{Z}_7$ we have $[3] + [2] = [24] + [51] = [5]$.) So suppose that $s \equiv x \pmod{n}$ and $t \equiv y \pmod{n}$. Then $(x - s)$ and $(y - t)$ are both divisible by n. Thus the sum $(x - s) + (y - t) = (x + y) - (s + t)$ is also divisible by n, and we have that $[x + y] = [s + t]$. This shows that addition is well defined. To show that $\cdot$ is well defined, we note that we can find integers q and r such that $s = x + qn$ and $t = y + rn$. Thus $st = xy + (qy + rx + rq)n$, and we see that $st \equiv xy \pmod{n}$ or $[xy] = [st]$. Thus the operation $\cdot$ is also well defined.

Tables 8.12 and 8.13 are the Cayley tables for $+$ and $\cdot$, respectively, on $\mathbf{Z}_6$. We see, for instance, that both operations are associative and commutative. The identity element for $+$ is $[0]$, and the identity element for $\cdot$ is $[1]$. Every element has an additive inverse (for instance, $[2] + [4] = [0]$ and so $[4] = -[2]$), but there is no multiplicative inverse for $[0]$ because $[0] \cdot [x] = [0]$ for all x. Thus $(\mathbf{Z}_6, +, [0])$ is an abelian group, but $\mathbf{Z}_6$ is not a group under multiplication, $\cdot$. The structure of $\mathbf{Z}_6$ reflects the more general situation: $(\mathbf{Z}_n, +, [0]_n)$ is an abelian group.

Proposition 2. For each positive integer n, $(\mathbf{Z}_n, +, [0])$ is an abelian additive group. Also the operation $\cdot$ is associative and commutative.

Table 8.12

+	[0]	[1]	[2]	[3]	[4]	[5]
[0]	[0]	[1]	[2]	[3]	[4]	[5]
[1]	[1]	[2]	[3]	[4]	[5]	[0]
[2]	[2]	[3]	[4]	[5]	[0]	[1]
[3]	[3]	[4]	[5]	[0]	[1]	[2]
[4]	[4]	[5]	[0]	[1]	[2]	[3]
[5]	[5]	[0]	[1]	[2]	[3]	[4]

Table 8.13

$\cdot$	[0]	[1]	[2]	[3]	[4]	[5]
[0]	[0]	[0]	[0]	[0]	[0]	[0]
[1]	[0]	[1]	[2]	[3]	[4]	[5]
[2]	[0]	[2]	[4]	[0]	[2]	[4]
[3]	[0]	[3]	[0]	[3]	[0]	[3]
[4]	[0]	[4]	[2]	[0]	[4]	[2]
[5]	[0]	[5]	[4]	[3]	[2]	[5]

PROOF: The operations $+$ and $\cdot$ are associative because, for each triple of integers x, y, and z, we have

$$([x] + [y]) + [z] = [x + y] + [z] = [(x + y) + z] = [x + (y + z)]$$
$$= [x] + [y + z] = [x] + ([y] + [z])$$

Also we have

$$([x] \cdot [y]) \cdot [z] = [xy] \cdot [z] = [(xy)z] = [x(yz)] = [x] \cdot [yz] = [x] \cdot ([y] \cdot [z])$$

The operations $+$ and $\cdot$ are commutative because, for each pair of integers x and y, we have

$$[x] + [y] = [x + y] = [y + x] = [y] + [x]$$

Also we have

$$[x] \cdot [y] = [xy] = [yx] = [y] \cdot [x]$$

The identity element of $\mathbf{Z}_n$ for $+$ is [0] because $[0] + [x] = [0 + x]$ $= [x]$ for each equivalence class $[x]$ in $\mathbf{Z}_n$. The identity element for $\cdot$ is [1] because $[1] \cdot [x] = [1 \cdot x] = [x]$ for each $[x]$.

Because $[x] + [n - x] = [n + (x - x)] = [n] = [0]$, we see that each element $[x]$ has an inverse, namely $[n - x]$. Thus $(\mathbf{Z}_n, +, [0])$ is an abelian group and we conclude our proof.

Example 9. In this example we investigate the order of the elements of $\mathbf{Z}_n$ for various values of n. Since we are using additive notation, we need to determine the smallest value of m such that $[x] + [x] + \cdots + [x] = [mx] = [0]$.

(a) In each of the groups $\mathbf{Z}_n$, the order of $[1]_n$ is n, because $[n] = [0]$ and, if $0 < m < n$, then $[m] \neq [0]$.

(b) The order of $[3]_5$ in $\mathbf{Z}_5$ is 5, because $[5 \cdot 3]_5 = [15]_5 = [0]_5$ and no smaller multiple of 3 is congruent to 0 modulo 5.

(c) The order of $[3]_6$ is 2.

(d) The order of $[15]_{23}$ is 23, because no smaller multiple of 15 is divisible by 23.

(e) The order of $[15]_{18}$ in $\mathbf{Z}_{18}$ is 6, because $6 \cdot 15 = 90$, and 90 is the smallest multiple of 15 that is also divisible by 18.

The order of an element $[x]_n$ in $(\mathbf{Z}_n, +, [0])$ is related to the greatest common divisor of x and n as follows:

Proposition 3. Let $[x]$ be an element in $\mathbf{Z}_n$, $[x] \neq [0]$. Suppose that x is a representative of the equivalence class $[x]$. Then the order of $[x]$ in $\mathbf{Z}_n$ is equal to $n/(x, n)$, where (x, n) is the greatest common divisor of x and n.

We leave the proof as an exercise. We verify that, for instance, $(4, 6) = 2$ and that the order of $[4]$ in $\mathbf{Z}_6$ is $6/2 = 3$. Similarly, $(15, 18) = 3$ and the order of $[15]$ in $\mathbf{Z}_{18}$ is $18/3 = 6$.

Corollary. The order of $[a] = n$ in $\mathbf{Z}_n$ if and only if a and n are relatively prime.

Section 8.3   Exercises

1. Each of the following tables determines a binary operation on the set $B = \{0, 1\}$. Under which of the operations is B a group?

$*$	0	1
0	0	0
1	0	1

$\#$	0	1
0	0	1
1	1	1

$@$	0	1
0	0	1
1	1	0

2. The set $A = \{a, b, c, d\}$ is a group under the operation $*$ defined in the accompanying table. Find the identity element of the group, and find the inverse of each element in the group. Solve the equation $b * x = d$ in the group.

$*$	a	b	c	d
a	c	d	a	b
b	d	a	b	c
c	a	b	c	d
d	b	c	d	a

3. In the group given in Exercise 2, find b^2, b^3, and b^4.

4. Find all the permutations of the set $\{1, 2\}$, and write down the Cayley table for the symmetric group $(S_2, \circ, \mathrm{id}_{x_2})$. Is this group abelian?

5. Let $(G, *, e)$ be a finite group. Prove that every element x of G appears exactly once in each row and column of the Cayley table for $*$.

6. Find the order of each of the elements in $(S_3, \circ, \mathrm{id}_{x_3})$. Solve each of the following equations in S_3 for g.

   (a) $f_3 \circ g = f_2$ (b) $g \circ f_4 = f_1$

7. Find two of the elements of $(S_4, \circ, \mathrm{id}_{x_4})$ (out of a possible 24), and find the order of each.

8. Give an example to show that if x and y are elements of a group $(G, *, e)$, then the inverse of the product $x * y$ is not in general $x^{-1} * y^{-1}$. (Your example must come from a nonabelian group.) Prove that $(x * y)^{-1} = y^{-1} * x^{-1}$.

9. Let $(\mathbf{Z}(3), +, 0)$ denote the group of all integer multiples of 3 under addition. Find an isomorphism from this group to $(E, +, 0)$, the group of even numbers under addition.

10. Find an isomorphism between the groups $(G, *, e)$ and $(\mathbf{Z}_5, +, [0])$. The Cayley table for G under $*$ is

$*$	e	a	b	c	d
e	e	a	b	c	d
a	a	c	d	b	e
b	b	d	a	e	c
c	c	b	e	d	a
d	d	e	c	a	b

11. Is $(S_3, \circ, \mathrm{id}_{x_3})$ isomorphic to $(\mathbf{Z}_6, +, [0])$?

12. Find a homomorphism from $(\mathbf{Z}, +, 0)$ to $(\mathbf{Z}_5, +, [0])$.

13. Let $f: G \to H$ be an isomorphism from $(G, *, e)$ to $(H, \circ, i)$. Prove that if G is abelian, then H is also abelian.

14. Give the Cayley table for the group $(\mathbf{Z}_8, +, [0])$.

15. Find the order and the inverse of each element in $(\mathbf{Z}_{12}, +, [0])$.

16. Solve the following equations in $(\mathbf{Z}_{12}, +, [0])$.

   (a) $[5] + x = [2]$ (b) $[7] + x = [5]$

17. Let x and y be elements of an abelian group $(G, *, e)$. Suppose that the order of x is m and the order of y is n. Let p be the least common multiple of n and m. Prove that $(x * y)^p = e$.

18. Show that if $n > 2$, then $(S_n, \circ, \mathrm{id}_{x_n})$ is not abelian.

Section 8.4	**Subgroups**

If we look at the subset $A = \{I, R_1, R_2, R_3\}$ of the group S of the symmetries of the square, we find that it is closed under the operation $*$, that it contains the identity element, and that the inverse of each element of A is also in A. The operation $*$ is associative on the larger set S, so it remains associative when restricted to the subset A. Thus $(A, *, I)$ is itself a group. It is a subgroup of S.

DEFINITION. Let $(G, *, e)$ be a group and let Q be a nonempty subset of G. Then Q is a **subgroup** of G if both of the following conditions hold.

    (a) Q is closed under $*$.
    (b) Whenever x is in Q, then x^{-1} is also in Q.

We must check that conditions (a) and (b) are enough to ensure that the subset Q is indeed a group under the inherited operation $*$. The operation certainly remains associative when restricted to a subset. Since Q is nonempty, we are guaranteed the existence of at least one element x in Q. Condition (b) guarantees that its inverse is also in Q. Condition (a) guarantees that Q is closed under $*$ and, since $x * x^{-1} = e$, it also guarantees that the identity element e is in Q. So $(Q, *, e)$ is a group.

Example 1. The set of even integers E is a subgroup of $(\mathbf{Z}, +, 0)$. Since the sum of even integers is again even, E is closed under addition. Since the negative of any even integer is also even, the inverse of any element in E is also in E. Conditions (a) and (b) are thus satisfied.

Example 2. The set $\{[0], [2]\}$ is a subgroup of $(\mathbf{Z}_4, +, [0])$. Since $[2] + [2] = [0]$, the set is closed under $+$ and contains the inverse of each of its elements.

In any group $(G, *, e)$, the singleton set $\{e\}$ containing only the identity element is a subgroup. The entire set G is also a subgroup. We call a subgroup A **proper** if it is neither the singleton set $\{e\}$ nor the entire group G.

Example 3. In this example we list all the proper subgroups of the group S of the symmetries of the square. The subgroups of order 2 are

$$\{I, R_2\} \qquad \{I, H\} \qquad \{I, V\} \qquad \{I, D_1\} \qquad \{I, D_2\}$$

The subgroups of order 4 are

$$\{I, R_1, R_2, R_3\} \qquad \{I, D_1, D_2, R_2\} \qquad \{I, V, H, R_2\}$$

There are no subgroups of order 3, 5, 6, or 7. (We shall see why later in this section.) Thus the preceding list is a complete list of proper subgroups of S. We see that S has two subgroups isomorphic to the Klein 4 group and one subgroup isomorphic to $(\mathbf{Z}_4, +, [0])$.

In the following inductive definition, we show how to generate a subgroup of a group $(G, *, e)$ starting with a subset S of G.

DEFINITION. Let $(G, *, e)$ be a group, and let S be a nonempty subset of G. Then the **subgroup generated by S,** denoted $\langle S \rangle$, is defined as follows:

    (1) If x is an element of S, then x is also an element of $\langle S \rangle$.

    (2) (i) If x is in $\langle S \rangle$, then x^{-1} is also in $\langle S \rangle$.

        (ii) If x and y are in $\langle S \rangle$, then $x * y$ is also in $\langle S \rangle$.

    (3) Only elements obtained by a finite number of iterations of (1) and (2) are in $\langle S \rangle$.

Step (2) guarantees that $\langle S \rangle$ is a subgroup of G, and Step (1) guarantees that the set S is contained in $\langle S \rangle$.

Example 4.

    (a) The subgroup generated by the element 2 in $(\mathbf{Z}, +, 0)$ is the set of even numbers E. Thus $E = \langle\{2\}\rangle$. When we wish to indicate the subgroup generated by a finite set of elements $S = \{a_1, a_2, \ldots, a_n\}$, we omit the braces and simply write $\langle S \rangle = \langle a_1, a_2, \ldots, a_n \rangle$. Thus $E = \langle 2 \rangle$.

    (b) To find the subgroup generated by [2] in $\mathbf{Z}_6$, we first find the inverse of [2], namely [4]. We then see that the set $\{[0], [2], [4]\}$ is closed under addition and that the inverse of each element is included in the set. Thus $\langle [2] \rangle = \{[0], [2], [4]\}$.

    (c) We now find the subgroup of $\mathbf{Z}_5$ generated by [2]. Its inverse, [3], must be in $\langle [2] \rangle$. Also, $[2] + [3] = [0]$, $[2] + [2] = [4]$, and $[3] + [3] = [1]$ must all be in $\langle [2] \rangle$. Thus all the elements of $\mathbf{Z}_5$ are in $\langle [2] \rangle$. (So $\langle [2] \rangle$ is not proper.) We say that [2] is a **generator** of the group $(\mathbf{Z}_5, +, [0])$ because $\mathbf{Z}_5 = \langle [2] \rangle$.

    (d) The subgroup generated by the set $\{V, H\}$ in S is the subgroup $\{I, V, H, R_2\}$. The set $\{V, R_1\}$ generates the entire group S.

(e) The subgroup of the integers generated by the set $\{4, 6\}$ is the group of all even integers, but the set $\{3, 5\}$ generates the entire set of integers. (The number 2 is in $\langle 3, 5 \rangle$ since $2 = 5 - 3$. So the number $1 = 3 - 2$ is also in $\langle 3, 5 \rangle$. Hence every integer is in $\langle 3, 5 \rangle$.)

DEFINITION. A group $(G, *, e)$ is **cyclic** if there is an element a in G such that the set $G = \langle a \rangle$. In this case, the set G can be written as $\{ \ldots, a^{-2}, a^{-1}, e, a, a^2, a^3, \ldots \}$.

The subgroup $\{I, R_1, R_2, R_3\}$ is cyclic and is generated by R_1. To see this, we note that $R_2 = R_1^2$ and $R_3 = R_1^3$. The group of integers is an infinite cyclic group since $(\mathbf{Z}, +, 0) = \langle 1 \rangle$.

The group of symmetries of the square is not cyclic. There is no element of order 8, and each element in S generates a cyclic subgroup of order at most 4.

For each $n \in \mathbf{N}$, the group $(\mathbf{Z}_n, +, [0])$ is a finite cyclic group. In each, the equivalence class $[1]_n$ is a generator. There may be more than one element that serves as a generator for a given cyclic group. We saw that $\mathbf{Z}_5 = \langle [2] \rangle$. Whereas $[2]$, $[3]$, and $[4]$ all generate proper cyclic subgroups in $\mathbf{Z}_6$, the order of the element $[5]$ in $\mathbf{Z}_6$ is 6 and thus $\mathbf{Z}_6 = \langle [5] \rangle$.

The following proposition tells us how to determine the generators of the group $(\mathbf{Z}_n, +, [0])$. Again, we use a fact about greatest common divisors.

Proposition 1. Let x be an integer. The equivalence class $[x]_n$ generates the cyclic group $(\mathbf{Z}_n, +, [0])$ if and only if x and n are relatively prime.

We leave the proof as an exercise.

Any natural number n generates a cyclic subgroup of $(\mathbf{Z}, +, 0)$—namely, $\mathbf{Z}(n) = \{x : x = nq \text{ for some integer } q\}$. Thus $\mathbf{Z}(2)$ is the subgroup of even numbers, and $\mathbf{Z}(10)$ is the subgroup of all multiples of 10. We may express each of the equivalence classes of the modular group $\mathbf{Z}_n$ in terms of $\mathbf{Z}(n)$ as follows:

$$[x]_n = \{m : m = x + y \text{ for some element } y \text{ in } \mathbf{Z}(n)\}$$

In other words, to find an integer m in the equivalence class of x, we simply add a multiple of n to x. We saw in Chapter 5 that the equivalence classes of the modulo n relation partition the set of integers into disjoint subsets. In what follows, we shall see how to obtain a similar partition on a group $(G, *, e)$ from a subgroup Q.

DEFINITION. Let $(G, *, e)$ be a group, and let Q be a subgroup of G. Let a be an element of G. The **right coset of Q with respect to a,** written Qa, is the set $\{x : x = s * a \text{ for some } s \text{ in } Q\}$. Similarly, the **left coset of Q with respect to a,** written aQ, is the set $\{x : x = a * s \text{ for some } s \text{ in } Q\}$.

If we use additive notation, then $aQ = \{x : x = a + s$ for some s in $Q\}$ and $Qa = \{x : x = s + a$ for some s in $Q\}$. Thus if we consider the group $(\mathbf{Z}, +, 0)$ and its subgroup $\mathbf{Z}(5)$, then $2\mathbf{Z}(5) = \{x : x = 2 + 5q\} = [2]_5$. More generally, $a\mathbf{Z}(n) = \{x : x = a + nq\} = [a]_n$. The cosets obtained from the subgroups $\mathbf{Z}(n)$ are exactly the equivalence classes of the modulo n relation.

Example 5. Let $Q = \{[0]_6, [3]_6\}$. The left cosets of Q in $\mathbf{Z}_6$ are

$$[0]Q = \{[0] + [0], [0] + [3]\} = \{[0], [3]\}$$

$$[1]Q = \{[1] + [0], [1] + [3]\} = \{[1], [4]\}$$

$$[2]Q = \{[2] + [0], [2] + [3]\} = \{[2], [5]\}$$

We note that $[3]Q = [0]Q$, $[4]Q = [2]Q$, and $[5]Q = [1]Q$, etc. Since $\mathbf{Z}_6$ is abelian, left and right cosets with respect to an element $[x]$ are identical.

Example 6. Let $(S, *, I)$ be the group of the symmetries of the square, and let $Q = \{I, V\}$. Let $a = R_1$. Then the left coset $R_1Q = \{R_1, D_2\}$, whereas the right coset $QR_1 = \{R_1, D_1\}$. When a group is not abelian, right and left cosets are different in general.

Proposition 2. Let $(G, *, e)$ be a group, and let Q be a subgroup of G. Then there is a bijection between any two right cosets of Q. Similarly, there is a bijection between any two left cosets.

PROOF: Let aQ and bQ be two left cosets. We shall show that the map that sends an element $a * x$ to $b * x$ is a bijection. (The argument is similar for right cosets.) We first note that an element z in bQ can be expressed in exactly one way as $b * x$. This is because, if $b * x = b * y$, we can multiply both sides on the left by b^{-1} to find that $x = y$. Now define $f : aQ \rightarrow bQ$ by $f(ax) = bx$. The map f is injective since if $b * x_1 = b * x_2$, then $x_1 = x_2$. The map is surjective since any element in bQ is of the form $b * x$ for some x in Q, and $b * x = f(a * x)$.

Proposition 2 assures us that any two right (or left) cosets are of the same cardinality: the cardinality of Q itself. (*Note:* $Q = eQ$.) Proposition 3 says that, like the equivalence classes of the modulo n relation, cosets are either disjoint or identical.

Proposition 3. Let Q be a subgroup of $(G, *, e)$, and let a and b be elements of Q. Then either $aQ = bQ$ or their intersection is empty. Similarly, either $Qa = Qb$ or their intersection is empty.

PROOF: Suppose that the intersection of aQ and bQ is not empty and that x is in the intersection. Then $x = a * s = b * t$ for some s and t in Q. Solving for a, we have $a = b * t * s^{-1}$. Let y be an element of aQ. Then $y = a * w$. Substituting for a, we have $y = b * t * s^{-1} * w$. Thus y is an element of bQ and we have that aQ is contained in bQ. A similar argument shows that bQ is contained in aQ. Thus the cosets aQ and bQ are identical if they are not disjoint. (The proof is similar for right cosets.)

LaGrange's Theorem.   Let $(G, *, e)$ be a finite group, and let Q be a subgroup of G. Then the order of Q divides the order of G.

PROOF: The subgroup Q partitions G into a finite set of disjoint left cosets, each with the cardinality of Q. Thus the order of G is a multiple of the order of Q.

LaGrange's theorem tells us that we cannot find groups of order 3, 5, 6, or 7 in S, the group of the 8 symmetries of the square, because 8 is not divisible by these numbers. It also tells us that if p is a prime number, then $(\mathbf{Z}_p, +, [0])$ has no proper subgroups because p is divisible only by 1 and p.

In any finite group $(G, *, e)$, the order of an element a in G is the order of the cyclic subgroup that it generates. That is, $o(a) = |\langle a \rangle|$. Thus we have the following corollaries to LaGrange's theorem.

Corollary 1.   The order of any element a in a finite group $(G, *, e)$ divides the order of the group.

Corollary 2.   Let n be the order of a finite group $(G, *, e)$, and let a be an element of G. Then $a^n = e$.

Thus all the elements of $\mathbf{Z}_6$ must be of order 1, 2, 3, or 6. (There are indeed no elements in $\mathbf{Z}_6$ of order 4 or 5.) Similarly, we cannot have an element in the symmetric group $(S_3, \circ, \mathrm{id}_{X_3})$ of order 5, because 5 does not divide $4! = 24$. LaGrange's theorem does not preclude the existence of an element of order 12 in S_3, but in fact each element in S_3 is of order 4 or less. (The converse of LaGrange's theorem is not in general true.)

In the bridge section of this chapter, we shall see an application of La-Grange's theorem to coding theory.

Section 8.4	**Exercises**

1. In each case, determine whether the given set is a subgroup of the indicated group.

   (a) $\{[0], [3], [5]\}$ in $(\mathbf{Z}_6, +, [0])$

(b) $\{m : m = 5n$ for some integer $n\}$ in $(\mathbf{Z}, +, 0)$

(c) $\{m : m = 5 + n$ for some integer $n\}$ in $(\mathbf{Z}, +, 0)$

(d) $\{f_0, f_3\}$ in $(S_3, \circ, id_{X_3})$

(e) $\{f_1, f_1^2\}$ in $(S_3, \circ, id_{X_3})$

2. Find all the proper subgroups of $(S_3, \circ, id_{X_3})$.

3. Find all the proper subgroups of $(\mathbf{Z}_{12}, +, [0])$.

4. Find a subgroup of $(\mathbf{Z}_{24}, +, [0])$ of order 3.

5. Find all the proper subgroups of the group $(G, *, e)$ with the following Cayley table:

*	a	b	c	d
a	a	b	c	d
b	b	a	d	c
c	c	d	a	b
d	d	c	b	a

6. Find the subgroup of $(\mathbf{Z}, +, 0)$ generated by the set $\{6, 9\}$.

7. Find the subgroup of $(\mathbf{Z}_{12}, +, [0])$ generated by the set $\{[6], [9]\}$.

8. Find the subgroup of $(\mathbf{Z}_{11}, +, [0])$ generated by the set $\{[6], [9]\}$.

9. Find the subgroup of the group given in Exercise 5 generated by the set $\{b, c\}$. Is G cyclic?

10. Find the subgroup of $(S_3, \circ, id_{X_3})$ generated by the set $\{f_1, f_3\}$.

11. Which elements of $(\mathbf{Z}_{12}, +, [0])$ generate proper cyclic subgroups?

12. Which elements of $(S_3, \circ, id_{X_3})$ generate proper cyclic subgroups? Which pairs of elements generate the entire group?

13. Find the left cosets of the subgroup $\{[0], [3], [6], [9]\}$ of $(\mathbf{Z}_{12}, +, [0])$.

14. The set $A = \{f_0, f_3\}$ is a subgroup of $(S_3, \circ, id_{X_3})$.

    (a) Find the left cosets of A.

    (b) Find the right cosets of A.

15. Prove: If n is divisible by d, then $(\mathbf{Z}_n, +, [0])$ has a subgroup of order d. (This exercise shows that the converse of LaGrange's theorem holds for the modular groups. It does not hold in general.)

16. Prove Proposition 1 of this section.

17. Let $f : G \to H$ be a homomorphism from the group $(G, *, e)$ to the group $(H, \circ, i)$. The **kernel** of f, denoted K_f, is the set $\{x : f(x) = i\}$; that is, $K_f = f^{-1}(\{i\})$.

    (a) Find the kernel of the homomorphism $f : \mathbf{Z} \to \mathbf{Z}_n$ defined by $f(x) = [x]_n$ for each integer x.

    (b) Prove that for every homomorphism $f: G \to H$, K_f is a subgroup of G.

18. (a) Let $f: \mathbf{Z} \to \mathbf{Z}_8$ be the homomorphism defined by $f(x) = [2x]_8$. Find the direct image $f(\mathbf{Z})$ in $\mathbf{Z}_8$.

    (b) Prove that if $f: G \to H$ is a homomorphism from the group $(G, *, e)$ to the group $(H, \circ, i)$, then the direct image $f(G)$ is a subgroup of H.

Section 8.5 | Rings and Fields

The set of integers comes equipped with two algebraic operations, addition and multiplication, that are related through the distributive law $x(y + z) = xy + xz$. We now turn our attention to similarly equipped algebraic structures.

> **DEFINITION.** Let R be a set with two binary operations $+$ and $*$. Then $(R, +, *)$ is a **ring** if all of the following conditions hold.
>
>   (a) R is an abelian additive group under $+$ with identity element denoted by 0.
>
>   (b) The operation $*$ is associative and has an identity element denoted by 1.
>
>   (c) The operations $+$ and $*$ satisfy the following distributive law: for each triple of elements x, y, and z in R, we have $x * (y + z) = x * y + x * z$.

The set of integers $\mathbf{Z}$ under addition and multiplication is a ring. The set of real numbers and the set of rational numbers are also rings under addition and multiplication. Although the set of even integers E is closed under both addition and multiplication, it is not a ring under our definition because there is no multiplicative identity element in E. (*Note:* Some authors do not require that a ring have a multiplicative identity. For them, E is a ring because it is an abelian group under addition, and the distributive law holds.)

Example 1. For each positive integer n, $(\mathbf{Z}_n, +, \cdot)$ is a ring. Recall that we defined the multiplication operation $\cdot$ in Section 8.3 by $[x]_n \cdot [y]_n = [x \cdot y]_n$ so that, for instance, in $\mathbf{Z}_5$ we have $[4]_5 \cdot [3]_5 = [12]_5 = [2]_5$. We showed that $\cdot$ was well defined and associative. The equivalence class $[1]_n$ is the multiplicative identity element in each ring $(\mathbf{Z}_n, +, \cdot)$. To see that the distributive law is satisfied, we note that for any triple $[x]$, $[y]$, and $[z]$ in $\mathbf{Z}_n$, we have

$$[x] \cdot ([y] + [z]) = [x] \cdot ([y + z]) = [x(y + z)] = [xy + xz]$$
$$= [xy] + [xz] = [x] \cdot [y] + [x] \cdot [z]$$

Tables 8.14 and 8.15 are the operation tables for both $+$ and $\cdot$ on $\mathbf{Z}_4$.

Table 8.14

+	[0]	[1]	[2]	[3]
[0]	[0]	[1]	[2]	[3]
[1]	[1]	[2]	[3]	[0]
[2]	[2]	[3]	[0]	[1]
[3]	[3]	[0]	[1]	[2]

Table 8.15

$\cdot$	[0]	[1]	[2]	[3]
[0]	[0]	[0]	[0]	[0]
[1]	[0]	[1]	[2]	[3]
[2]	[0]	[2]	[0]	[2]
[3]	[0]	[3]	[2]	[1]

In any ring the additive operation $+$ is necessarily commutative, but the multiplicative operation need not be. A ring is called a **commutative ring** if the multiplication operation $*$ is also commutative. The rings $(\mathbf{Z}_n, +, \cdot)$ are all commutative rings. A noncommutative ring is given in the following example.

Example 2. Let M be the set of all 2×2 matrices. Let $+$ denote matrix addition, and let $\times$ denote matrix multiplication. Then $(M, +, \times)$ is a noncommutative ring. (That is to say, the multiplication operation is not commutative.) The additive identity element is the matrix

$$E = \begin{pmatrix} 0 & 0 \\ 0 & 0 \end{pmatrix}$$

and the multiplicative identity is

$$I = \begin{pmatrix} 1 & 0 \\ 0 & 1 \end{pmatrix}$$

Example 3. Let X be a nonempty set and let $\mathscr{P}(X)$ be its power set. The operations of union $\cup$ and intersection $\cap$ satisfy the distributive law on $\mathscr{P}(X)$, and each of the operations has an identity element: the set X is the identity element for intersection, and the empty set is the identity element for union. $\mathscr{P}(X)$ is not a ring, however, because it is not a group under either operation.

We know that, in integer arithmetic, for every integer n we have $n \cdot 0 = 0$ and $(-1) \cdot n = -n$. In the following proposition we prove that similar results hold in every ring.

Proposition 1. Let $(R, +, *)$ be a ring with additive identity 0 and multiplicative identity 1. Let x be any element in the set R. Then $x * 0 = 0 * x = 0$ and $x * (-1) = (-1) * x = -x$.

PROOF: Since $0 = 0 + 0$, we have $x * 0 = x * (0 + 0)$. We use the distributive law to obtain $x * (0 + 0) = x * 0 + x * 0$, and so we have $x * 0 = x * 0 + x * 0$. Adding $-(x * 0)$ to both sides of the last equation, we see that $0 = x * 0$. The proof for $0 * x$ is similar. We have $0 = 1 + (-1)$, $x * 0 = 0$, and $x * 1 = x$. Again using the distributive law, we have $0 = x * 1 + x * (-1)$. Adding $-x$ to both sides, we have $-x = x * (-1)$. The proof for $(-1) * x$ is similar.

As with ordinary arithmetic, we shall denote the sum $x + (-y)$ by $x - y$ in any ring.

If $(R, +, *)$ is a ring and a and b are elements of R, we can always solve the equation $a + x = b$ for x because R is an abelian group under $+$. (We let $x = b - a$.) However, even in the ring of integers, we cannot in general solve the equation $a * x = b$. In the integers we *do* know that if, for instance, $3x = 3y$, then $x = y$. (*Proof:* If $3x = 3y$, then $3(x - y) = 0$. Since $3 \neq 0$, then $(x - y) = 0$ so that $x = y$.) If we proceed similarly in the ring $(\mathbf{Z}_6, +, \cdot)$, we fail. Consider the analogous expression $[3] \cdot x = [3] \cdot y$. If we let $x = [2]$ and $y = [4]$, we find that $[3] \cdot x = [6] = [0]$ and $[3] \cdot y = [6] = [0]$, but $x \neq y$. If we look at the equation $[3] \cdot (x - y) = 0$, we cannot conclude that $(x - y) = [0]$ since in fact $[3] \cdot [2] = [0]$. We call $[3]_6$ and $[2]_6$ **zero divisors** in $\mathbf{Z}_6$ because, although neither is equal to zero, their product is. More generally, we have the following definition.

> **DEFINITION.** Let $(R, +, \cdot)$ be a ring, and let x and y be elements of R with neither x nor y equal to 0. If $x \cdot y = 0$, then x and y are called **zero divisors** of R.

Example 4. In $\mathbf{Z}_4$, $[2]$ is a zero divisor since $[2] \cdot [2] = [0]$. The ring of integers has no zero divisors. In $\mathbf{Z}_{12}$ the zero divisors are $[2]$, $[3]$, $[4]$, $[6]$, $[8]$, and $[10]$. There are no zero divisors in the rings $\mathbf{Z}_5$, $\mathbf{Z}_7$, and $\mathbf{Z}_{19}$.

Proposition 2. Suppose that x is a zero divisor. Then x has no multiplicative inverse.

PROOF: Suppose that x has a multiplicative inverse x^{-1}. Suppose also that $x * y = 0$. Then $(x^{-1} * x) * y = 1 * y = y$ but $x^{-1} * (x * y) = x^{-1} * 0 = 0$. Thus y must equal 0 and x is not a zero divisor.

(The converse of Proposition 2 is not true because, for instance, in the ring of integers, the number 3 has no multiplicative inverse but it is not a zero divisor.)

The following proposition tells us how to find the zero divisors of $\mathbf{Z}_n$.

Proposition 3. Let a be an integer and suppose that $[a] \neq [0]$. Then $[a]$ is a zero divisor in $(\mathbf{Z}_n, +, \cdot)$ if and only if a and n are not relatively prime.

PROOF: Let $d = (a, n)$ and suppose that $d > 1$. The integers n/d and a are relatively prime and $[n/d] \neq [0]$. Since a is divisible by d, we can find an integer q such that $a = qd$, but then $[n/d] \cdot [a] = [nq] = [0]$. Hence $[a]$ is a zero divisor. Conversely, if a and n are relatively prime, we can find integers s and t such that $sa + tn = 1$. Thus $[s] \cdot [a] = [1]$ and $[a]$ has a multiplicative inverse. By Proposition 2, $[a]$ cannot be a zero divisor.

Thus the zero divisors in $\mathbf{Z}_{15}$ are $[3]$, $[5]$, $[6]$, $[9]$, $[10]$, and $[12]$. We verify that, for instance, $[9] \cdot [10] = [90] = [0]$.

Corollary 1. If p is prime, then $(\mathbf{Z}_p, +, \cdot)$ has no zero divisors.

DEFINITION. A commutative ring with no zero divisors is called an **integral domain.**

The ring of integers, the ring of rational numbers, and the ring of real numbers are all integral domains. It follows from Corollary 1 that $(\mathbf{Z}_n, +, \cdot)$ is an integral domain if and only if n is a prime number.

In an integral domain $(R, +, *)$ we can always cancel. Suppose $a \neq 0$ and $a * x = a * y$. Then $a * (x - y) = 0$. Since a is *not* a zero divisor, $(x - y) = 0$ and indeed $x = y$. So, for instance, to solve the equation $[2] \cdot x = [6]$ in $\mathbf{Z}_{17}$, we note that $[6] = [2] \cdot [3]$. Canceling $[2]$ from both sides, we see that the unique solution to this equation is $x = [3]$.

The integral domains $(\mathbf{Z}_p, +, \cdot)$ have a further property not shared by all integral domains: every nonzero element $[a]_p$ in $\mathbf{Z}_p$ has a multiplicative inverse— that is, a solution to the equation $[a]_p \cdot x = [1]$. For example, in $\mathbf{Z}_5$ the multiplicative inverse of $[3]_5$ is $[2]_5$ because $[3]_5 \cdot [2]_5 = [6]_5 = [1]_5$. In $\mathbf{Z}_{11}$ the multiplicative inverse of $[3]_{11}$ is $[4]_{11}$ since $[3]_{11} \cdot [4]_{11} = [1]_{11}$. To see that we can do this in general, we note that if a is not a multiple of the prime p, we can find integers s and t such that the sum $as + pt = 1$. Thus $[a]_p \cdot [s]_p = [1]_p$ and $[s]_p = [a]_p^{-1}$.

Since every nonzero element $[a]_p$ has a multiplicative inverse in $\mathbf{Z}_p$, every equation of the form $[a]_p \cdot x = [b]_p$ has the unique solution $x = [a]_p^{-1} \cdot [b]_p$. For example, in $\mathbf{Z}_7$ we can solve the equation $[5]_7 \cdot x = [4]_7$ by multiplying both sides of the equation by $[5]_7^{-1} = [3]_7$. Thus $x = [3]_7 \cdot [4]_7 = [5]_7$. We check our results: $[5]_7 \cdot [5]_7 = [25]_7 = [4]_7$ in $\mathbf{Z}_7$.

DEFINITION. A **field** is an integral domain in which every nonzero element has a multiplicative inverse.

The rings $(\mathbf{Z}_p, +, \cdot)$ are thus fields for every prime integer p. The ring $(\mathbf{Z}, +, \cdot)$ is not a field since no integers other than 1 and -1 have multiplicative inverses. The set of rational numbers and the set of real numbers are both fields.

Proposition 4. Let $(F, +, *)$ be a field and let $U = F \setminus \{0\}$. Then $(U, *, 1)$ is an abelian group.

PROOF: The set U is the set of nonzero elements of F. It is closed under multiplication, because a field is an integral domain and the product of two nonzero elements must again be a nonzero element of F. The operation $*$ is associative and commutative. Every element in U has a multiplicative inverse since F is a field. Thus $(U, *, 1)$ is an abelian group.

Let U_p denote $\mathbf{Z}_p - \{[0]\}$. By Proposition 4, each $(U_p, \cdot, [1])$ is an abelian group of order $p - 1$. Tables 8.16 and 8.17 are the Cayley tables for U_5 and U_7.

Table 8.16

U_5:	$\cdot$	[1]	[2]	[3]	[4]
[1]		[1]	[2]	[3]	[4]
[2]		[2]	[4]	[1]	[3]
[3]		[3]	[1]	[4]	[2]
[4]		[4]	[3]	[2]	[1]

Table 8.17

U_7:	$\cdot$	[1]	[2]	[3]	[4]	[5]	[6]
[1]		[1]	[2]	[3]	[4]	[5]	[6]
[2]		[2]	[4]	[6]	[1]	[3]	[5]
[3]		[3]	[6]	[2]	[5]	[1]	[4]
[4]		[4]	[1]	[5]	[2]	[6]	[3]
[5]		[5]	[3]	[1]	[6]	[4]	[2]
[6]		[6]	[5]	[4]	[3]	[2]	[1]

Fermat's Theorem. Let p be any prime integer, and let a be any integer. Then $a^p \equiv a \pmod{p}$.

PROOF: The theorem is true if a is a multiple of p since both sides are then congruent to 0 modulo p. Suppose now that a is not divisible by p so that a and p are relatively prime. We can cancel a from both sides of the congruence and restate our theorem as $a^{p-1} \equiv 1 \pmod{p}$. In terms of equivalence classes, we must prove that $[a]_p^{p-1} = [1]_p$ in $(U_p, \cdot, [1]_p)$. But this is true because $p - 1$ is the order of the group U_p and is thus a multiple of the order of $[a]_p$ in $(U_p, \cdot, 1)$.

Example 5. We check Fermat's theorem for some values of p and a.

(a) Let $p = 3$ and $a = 2$. Then $2^3 = 8$ and $8 \equiv 2 \pmod{3}$.

(b) Let $p = 5$ and let $a = 3$. Then $3^5 = 243$ and $243 \equiv 3 \pmod{5}$.

Example 6. We can use Fermat's theorem to find the multiplicative inverse of a non-zero element $[a]$ in $\mathbf{Z}_p$ by simply finding the equivalence class of a^{p-2}. Let us check in $\mathbf{Z}_{11}$. Let $[a] = [5]$. We claim that $[5]^{-1} = [5^9]$. Now $5^9 = 1,953,125$, which has a remainder of 9 after division by 11. Thus $[9] = [5]^{-1}$, and indeed $[5] \cdot [9] = [45] = [1]$ in $\mathbf{Z}_{11}$.

We now show how Fermat's theorem (known for at least two centuries) is the basis for a new and useful scheme for formulating secret codes, or "ciphers." The method was presented in a paper by R. L. Rivest, A. Shamir, and L. Adelman ("A Method for Obtaining Digital Signature and Public-Key Cryptosystem," Communications of the A.C.M., 1978, Vol. 21, #2, page 120). A somewhat simplified version works as follows.

Suppose we wish to encode a set of symbols such as $S = \{A, B, C, D, E, F, G, H\}$ as a series of numbers. First we assign each symbol an integer greater than 1 in some simple fashion, such as $A = 2$, $B = 3$, . . . , $H = 9$. The dictionary for this assignment may be well known and public. We now pick a prime p larger than any of the integers representing the numbers in the set S. (For our example, the smallest available prime is $p = 11$.) Suppose we have chosen p. We next find two positive integers n and m that are relatively prime to $p - 1$ and have the property that their product nm is congruent to 1 modulo $p - 1$. That is, $nm = 1 + (p - 1)q$ for some integer q. (If $p = 11$, for instance, then $p - 1 = 10$ and we may choose $n = 7$ and $m = 3$ since $7 \cdot 3 = 21$ and $21 \equiv 1 \pmod{10}$.)

Let x be an integer such that $0 < x < p$. Then

$$x^{nm} = x^{1+(p-1)q} = x \cdot (x^{p-1})^q$$

From Fermat's theorem we know that $x^{p-1} \equiv 1 \pmod{p}$. Thus we have

$$x^{nm} \equiv x \pmod{p}$$

For example, if $p = 11$, $n = 7$, $m = 3$, and $x = 2$, we have $2^7 \equiv 7 \pmod{11}$ and $7^3 \equiv 2 \pmod{11}$. That is, $(2^7)^3 \equiv 2 \pmod{11}$. This simple result is the basis for our encoding scheme.

The sender of the message (the encoder) and the receiver of the message (the decoder) know the integers p, n, and m. (The spies do not.) Let x be an integer representing a symbol in S in our public dictionary. The encoder replaces x in his message by the integer y, where $0 < y < p$ and $x^n \equiv y \pmod{p}$. In other words, the encoder sends the representative of $[x^n]_p$ that lies between 0 and p. In our example with $p = 11$, $n = 7$, $m = 3$, and $x = 4$, the encoder would replace the symbol C (represented by 4 in the dictionary) by 5 because $4^7 \equiv 5 \pmod{11}$. The complete encoding scheme for our example is given in Table 8.18.

Table 8.18

Symbol	x	Code (x^7 mod 11)
A	2	7
B	3	9
C	4	5
D	5	3
E	6	8
F	7	6
G	8	2
H	9	4

The simple digital representation of the message BAD is 3, 2, 5. The encoder would send the sequence 9, 7, 3.

To decode a message, the decoder receiving the number y finds x such that $0 < x < p$ and $y^m \equiv x \pmod{p}$. Because of Fermat's theorem and because $y = x^n \pmod{p}$, he knows that $y^m \equiv x \pmod{p}$. Thus to decode 9, 7, 3 in our example, the decoder finds that $9^3 \equiv 3 \pmod{11}$, that $7^3 \equiv 2 \pmod{11}$, and that $3^3 \equiv 5 \pmod{11}$. He decodes his message (correctly) as 3, 2, 5 and consults his public dictionary to find that his message spells BAD.

The effectiveness of such a coding scheme lies not in its conceptual complexity but in the difficulties that its arithmetic presents. Even if a potential code breaker knows the modulus prime p, he must also know the factors of $p - 1$ in order to find n and m. If $p - 1$ is very large—with, say, 100 or so digits—it could take centuries of computer time to obtain the factorization. By then the message would be considerably outdated.

Section 8.5 | **Exercises**

---

1. Decide which of the following are rings.

   (a) The set Q^+ of positive rational numbers under addition and multiplication

   (b) Let p be a prime, and let Q_p be the set of rational numbers of the form $\{x : x = m/p^n \text{ for } m \text{ in } \mathbf{Z}\}$ under addition and multiplication.

   (c) The subset $\{[0], [2], [4]\}$ of $(\mathbf{Z}_6, +, \cdot)$

2. Show that the set of real numbers $\mathbf{R}$ under addition and the operation # defined by $x \mathbin{\#} y = 2xy$ is a ring. What is the identity element for the operation #? Find the inverse with respect to # of each of the following: 1, 1/2, and 5/3.

3. Write down the Cayley tables for both $+$ and $\cdot$ for the ring $(\mathbf{Z}_8, +, \cdot)$.

4. Which elements in $(\mathbf{Z}_8, +, \cdot)$ have multiplicative inverses? What elements in $(\mathbf{Z}_8, +, \cdot)$ are zero divisors?

5. Let U be the elements of $(\mathbf{Z}_8, \cdot, +)$ that are neither [0] nor zero divisors. Write down the Cayley table for the multiplication operation $\cdot$ on U, and show that U is a group under $\cdot$.

6. Repeat the instructions of Exercise 5 for the ring $(\mathbf{Z}_{12}, \cdot, +)$.

7. Find two distinct values of x and y in $(\mathbf{Z}_{12}, +, \cdot)$ for which $[8] \cdot x = [8] \cdot y$.

8. Find the multiplicative inverse of each nonzero element in $(\mathbf{Z}_7, +, \cdot)$. Solve each of the following equations (if possible) in $\mathbf{Z}_7$.

(a) $[3] \cdot x = [4]$　　　　　　　(b) $[6] \cdot x = [2]$

(c) $x^2 = [2]$　　　　　　　　　　(d) $x^2 = [3]$

9. Is the ring given in Exercise 2 a field? (Justify your answer.)

10. Verify Fermat's theorem for $p = 5$ and $a = 3$, and also for $p = 7$ and $a = 2$.

11. Use Fermat's theorem to find the multiplicative inverse of $[2]$ in the field $\mathbf{Z}_{13}$.

12. Find another code for the set $\{A, B, \ldots , H\}$, using $p = 11$ again but letting $n = 9$. What value of m is needed for decoding?

Bridge to Computer Science

8.6 More About Coding Theory (optional)

(*Note:* This section assumes the results of the bridge that appears in Chapter 3.)

Suppose that G is a set and that $*$ is a binary operation on G. Let $G^n = G \times G \times \cdots \times G$ be the cartesian product of G with itself n times. The operation $*$ can be extended to G^n as follows: for two n-tuples in G^n, we define $(x_2, x_2, \ldots, x_n) * (y_1, y_2, \ldots, y_n)$ to be $(x_1 * y_1, x_2 * y_2, \ldots, x_n * y_n)$. (We just operate term by term.)

Example 1. If we extend the addition operation on the real numbers $\mathbf{R}$ to the cartesian product $\mathbf{R} \times \mathbf{R}$, we simply obtain the familiar operation of vector addition on the x–y plane.

Example 2. Let us denote the set $\mathbf{Z}_2$ by B and its elements by 0 and 1 (rather than $[0]_2$ and $[1]_2$). The addition and multiplication tables for $B \times B$ are shown in Table 8.19.

Table 8.19

+	(0, 0)	(0, 1)	(1, 0)	(1, 1)		$*$	(0, 0)	(0, 1)	(1, 0)	(1, 1)
(0, 0)	(0, 0)	(0, 1)	(1, 0)	(1, 1)		(0, 0)	(0, 0)	(0, 0)	(0, 0)	(0, 0)
(0, 1)	(0, 1)	(0, 0)	(1, 1)	(1, 0)		(0, 1)	(0, 0)	(0, 1)	(0, 0)	(0, 1)
(1, 0)	(1, 0)	(1, 1)	(0, 0)	(0, 1)		(1, 0)	(0, 0)	(0, 0)	(1, 0)	(1, 0)
(1, 1)	(1, 1)	(1, 0)	(0, 1)	(0, 0)		(1, 1)	(0, 0)	(0, 1)	(1, 0)	(1, 1)

In what follows, we shall let B^n represent the cartesian product of B with itself n times. If we equip B^n with the extended addition and multiplication operations, the n-tuple $(0, 0, \ldots, 0)$ is the identity element for $+$. We shall denote it by $\mathbf{0}$. The n-tuple $(1, 1, \ldots, 1)$ is the identity element for $*$. Then $(B^n, +, \mathbf{0})$ is an abelian group and $(B^n, +, *)$ is a ring. We shall let $e_i = (0, 0, \ldots, 1, 0, \ldots)$ be the element of B^n with a 1 in the ith place and zeros elsewhere. The set of elements $\{e_1, e_2, \ldots, e_n\}$ generates the group $(B^n, +, \mathbf{0})$.

Suppose $f : B^m \to B^n$ is a group homomorphism. We can use an $m \times n$ matrix M_f to represent f as follows. We set the ith row of M_f equal to n-tuple $f(e_i)$; that is, we place the jth coordinate of $f(e_i)$ in the ith row, jth column of M_f. Then, for any m-tuple $x = (x_1, x_2, \ldots, x_m)$ in B^m, we obtain $f(x)$ by the matrix multiplication $(x_1, x_2, \ldots, x_m) \times M_f$.

Example 3. Let $f : B^3 \to B^2$ be defined by the following diagram:

$$(0, 0, 0) \longrightarrow (0, 0) \qquad (0, 1, 1) \longrightarrow (1, 0)$$

$$(0, 0, 1) \longrightarrow (1, 1) \qquad (1, 1, 0) \longrightarrow (1, 1)$$
$$(0, 1, 0) \longrightarrow (0, 1) \qquad (1, 0, 1) \longrightarrow (0, 1)$$
$$(1, 0, 0) \longrightarrow (1, 0) \qquad (1, 1, 1) \longrightarrow (0, 0)$$

Then the matrix M_f is

$$\begin{pmatrix} 1 & 0 \\ 0 & 1 \\ 1 & 1 \end{pmatrix}$$

We check that matrix multiplication yields the correct value of $f(1, 0, 1)$ by evaluating as follows:

$$(1, 0, 1) \begin{pmatrix} 1 & 0 \\ 0 & 1 \\ 1 & 1 \end{pmatrix} = (0, 1)$$

This agrees with the value given in the diagram. The matrix representation of f is much more concise than the diagram originally given.

When a homomorphism f is the encoding function of an (m, n)-encoding scheme, we call M_f an **encoding matrix.** We now investigate how we can use the algebraic structure of B^n in coding theory. So let us suppose that $f : B^m \to B^n$ is an encoding function that is a group homomorphism, and let us denote the set of code words in B^n by C—namely, $C = f(B^m)$. The set C is a subgroup of B^n. (See Exercise 18 of Section 8.4.)

Theorem 1.   Let k be the minimal distance between any two code words in C. We can find a code word w in C such that the weight of w is equal to k.

PROOF: Suppose that x and y are code words in C such that the distance between x and y is k. Then the weight of the word $x + y$ is equal to k. (When we add x and y, we obtain 1's in exactly the places where x and y differ.) Since C is a subgroup of B^n, the word $x + y$ is also a code word and $x + y$ satisfies the theorem.

Thus if we are looking for k, the minimal distance between two code words in the subgroup C, we need only look for the minimal weight of the nonzero code words.

Example 4. Consider the $(3, 6)$-encoding scheme given by the following matrix:

$$\begin{pmatrix} 1 & 0 & 0 & 1 & 1 & 0 \\ 0 & 1 & 0 & 0 & 1 & 1 \\ 0 & 0 & 1 & 1 & 1 & 1 \end{pmatrix}$$

The list of code words generated is

$$000 \longrightarrow 000000 \qquad 001 \longrightarrow 001111$$
$$100 \longrightarrow 100110 \qquad 101 \longrightarrow 101001$$
$$010 \longrightarrow 010011 \qquad 011 \longrightarrow 011100$$
$$110 \longrightarrow 110101 \qquad 111 \longrightarrow 111010$$

Checking the list of code words, we find that, of the nonzero code words, there are 4 of weight 3 and the remainder are of greater weight. Thus the minimal distance between code words is 3. Our previous results tell us that this code can detect all double errors and correct all single errors.

Recall that in order for a code to correct all cases of k or fewer errors, the minimal distance between two code words must be at least $2k + 1$. Even when this criterion is met, however, not all decoding functions correct such errors. In the bridge appearing in Chapter 3, we obtained a k-correcting decoding function by mapping each of the words in B^n to a nearest code word. We had no rules about how to conduct this search. We now show how to use LaGrange's theorem to accomplish this task.

The method we use is called the **coset method.** Let z be a word in B^n and let $C(z)$ be the coset of C determined by z; that is, $C(z) = \{x : x = z + w \text{ for some } w \text{ in } C\}$. Recall that LaGrange's theorem guarantees that each element in B^n is in exactly one such coset. In each coset $C(z)$, we can choose an element y of minimal weight. Since y and z determine the same coset, we can rewrite (if necessary) each element x in $C(z)$ as $x = y + w$ for some w in C. The element y is called the **coset leader.** We construct a decoding table as follows. Suppose that the 2^m elements of C are denoted by $\{0, b_1, b_2, \ldots, b_{2m}\}$. We place this list in the first row of the decoding table. Then we pick an element z of B^n that is not in C. We then find the coset leader of $C(z)$—call it z_1—and list the elements of the coset $C(z)$ as follows: $\{z_1, z_1 + b_1, z_1 + z_2, \ldots\}$. We enter this list as the second row of the decoding table. We continue by picking another element of B^n that has not already appeared on the decoding list; finding its coset and coset leader; constructing a coset list as before; and entering that list as the next row. We terminate when we have exhausted the elements of B^n.

Example 5. We construct a decoding table for the $(2, 4)$-code given by the following encoding matrix:

$$\begin{pmatrix} 1 & 0 & 1 & 0 \\ 0 & 1 & 0 & 1 \end{pmatrix}$$

The code words generated are $\{(0, 0, 0, 0), (1, 0, 1, 0), (0, 1, 0, 1), (1, 1, 1, 1)\}$.

The word $z = (1, 1, 1, 0)$ is not in C. The coset $C(z) = \{(1, 1, 1, 0), (0, 1, 0, 0),$ $(1, 0, 1, 1), (0, 0, 0, 1)\}$. The coset leader of $C(z)$ is the word $(0, 0, 0, 1)$ since it is the word of least weight. We now relist the members of the coset. Adding the members of C to $(0, 0, 0, 1)$ in order, we obtain $\{(0, 0, 0, 1), (1, 0, 1, 1), (0, 1, 0, 0), (1, 1, 1, 0)\}$. We now repeat the process for the element $z = (0, 1, 1, 1)$, which has not yet appeared in either list. We obtain the following coset list: $\{(0, 0, 1, 0), (1, 0, 0, 0), (0, 1, 1, 1), (1, 1, 0, 1)\}$. Finally the coset list for the element $(1, 1, 0, 0)$ is $\{(1, 1, 0, 0), (0, 1, 1, 0), (1, 0, 0, 1), (0, 0, 1, 1)\}$. The decoding table is

$(0, 0, 0, 0)$	$(1, 0, 1, 0)$	$(0, 1, 0, 1)$	$(1, 1, 1, 1)$
$(0, 0, 0, 1)$	$(1, 0, 1, 1)$	$(0, 1, 0, 0)$	$(1, 1, 1, 0)$
$(0, 0, 1, 0)$	$(1, 0, 0, 0)$	$(0, 1, 1, 1)$	$(1, 1, 0, 1)$
$(1, 1, 0, 0)$	$(0, 1, 1, 0)$	$(1, 0, 0, 1)$	$(0, 0, 1, 1)$

To decode a word x in B^n, we first find the column in which x appears in the decoding table. We map x to the code word at the top of the column and decode this code word. Thus, suppose we are using the $(2, 4)$-code given in Example 5 and receive the word 0110. We find the word 1010 at the top of the column in which 0110 appears. Since 1010 is the encodement of the word 10 from B^2, we decode 0110 as 10.

The next theorem shows that if x is a word in B^n, then the code word at the top of the column in which x appears in the decoding table is a code word of minimal distance from x. (Other code words may be as close, but none is closer.) By Theorem 3 of the bridge appearing in Chapter 3, we know that our procedure guarantees that if the minimal distance between any two code words is $2k + 1$, then our decoding process detects all cases of $2k$ or fewer errors and corrects all cases of k or fewer errors. As such, it is an optimal decoding process.

Theorem 2.   Suppose that x is a word in B^n and that z is the code word at the top of the column in which x appears in the decoding table. Then there is no code word of smaller distance to x than z.

PROOF: Let b be any code word with minimal distance to x. Let $e = x - b$. The weight of the word e is exactly the distance between x and b. Now the word x is in the coset $C(e)$ since $(-b)$ is an element of C. Let $y = b_k + e$ be the coset leader of the coset $C(e)$. We can rewrite x in terms of y as follows: $x = (b - b_k) + (b_k + e) = (b - b_k) + y$. So we know that x is in the column of the code word z and that $z = (b - b_k)$. Since $x = z + y$ or $y = x - z$, we know that $w(y) = d(x, z)$. Since y is the coset leader of $C(e)$, the weight of y is less than or equal to the weight of e. Thus we conclude that z is also a code word of minimal distance to x.

Section 8.6   Exercises

---

1. Write out the code defined by the following encoding matrix.

$$\begin{pmatrix} 1 & 0 & 0 & 1 & 1 & 0 & 1 \\ 0 & 1 & 0 & 0 & 1 & 0 & 0 \\ 0 & 0 & 1 & 0 & 0 & 1 & 1 \end{pmatrix}$$

2. Find the encoding matrix for the following $(2, 5)$-encoding function.

$$00 \longrightarrow 00000$$
$$10 \longrightarrow 01101$$
$$01 \longrightarrow 10011$$
$$11 \longrightarrow 11110$$

3. Determine how many errors the coding scheme generated in Exercise 2 can detect and how many it can correct. Define a decoding function that accomplishes optimal error correction.

4. Set up a decoding table (by coset leaders) for the function defined in Exercise 2. Then decode the following words accordingly: 00110, 10101, and 00100.

5. Repeat Exercise 4 for the $(3, 6)$-code defined in Example 2. Decode the words 001100 and 010000.

6. Show that the additive group B^n is generated by $\{e_1, e_2, \dots, e_n\}$.

7. Show that the function defined by an encoding matrix is a group homomorphism.

key concepts

8.1 Binary Operations, Semigroups, and Monoids	binary operation on a set A Cayley table associative operation commutative operation semigroup identity elements: right identity element, left identity element monoid submonoid monoid generated by the set S: $\langle S \rangle$ semigroup of a finite state machine
8.2 The Integers	divisibility, $a \mid b$, factor, prime the division theorem the greatest common divisor of a and b: (a, b) Euclid's algorithm relatively prime unique factorization
8.3 Groups	inverse of the element a for the operation $*$: a^{-1} inverse of the element a for the operation $+$: $-a$ group $(G, *, e)$ abelian group modular groups $(Z_n, +, [0])$ symmetric groups $(S_n, \circ, \mathrm{id}_{X_n})$ order of G order of an element a: $o(a)$ isomorphism homomorphism
8.4 Subgroups	subgroup subgroup generated by a subset S: $\langle S \rangle$ generators cyclic group cosets: right cosets and left cosets LaGrange's theorem
8.5 Rings and Fields	ring, $(R, +, *)$ commutative ring zero divisor integral domain field the finite fields $(Z_P, +, \cdot)$ $(U_P, \cdot, [1])$, the group of nonzero divisors of $(Z_P, +, \cdot)$ Fermat's theorem Fermat's theorem and cryptic codes

| Chapter 8 | Exercises |

1. In parts (a) through (c) that follow, a set with a binary operation is given. Determine which operations are associative and which are commutative. Determine which sets have identity elements. Determine which are monoids.

(a) The real numbers under multiplication

(b) The natural numbers under the operation #, where $a \# b$ denotes the least common multiple of a and b

(c) The set $A = \{a, b, c, d, e, f\}$ under the $*$ operation defined by the following table:

$*$	a	b	c	d	e	f
a	a	a	c	a	b	f
b	a	b	c	b	b	f
c	c	c	c	c	c	c
d	a	b	c	d	d	f
e	a	b	c	d	e	f
f	f	f	c	f	f	f

2. Let $\max(a, b)$ denote the maximum of the numbers a and b. Determine whether the set of natural numbers under max is a monoid. What about the integers under max?

3. Let A denote the set $\{a, b, c\}$, and let S denote the set of functions $\{f : f : A \rightarrow A \text{ and } f(c) = c\}$. Write out the Cayley table for the monoid S under the operation of composition.

4. Let X be a nonempty set, and define the operation on $\mathscr{P}(X)$ as follows: $A \mathbin{\Delta} B = A \setminus B \cup B \setminus A$. Write out the Cayley table for Δ when $X = \{a, b, c\}$. Determine whether $\mathscr{P}(X)$ is, in general, a monoid under Δ.

5. Let $*$ be a binary operation on the set A. Prove that A has at most one identity element.

6. Let B be a nonempty subset of the set X. Show that $\mathscr{P}(B)$ is a monoid under the operation of union. Is $\mathscr{P}(B)$ a submonoid of $\mathscr{P}(X)$ under union?

7. (a) Find the submonoid of $(\mathbf{Z}, +, 0)$ generated by $\{2, -3\}$.

(b) Find the submonoid of the monoid $(\mathbf{N}, \cdot, 1)$ generated by $\{2\}$.

(c) Characterize the submonoid generated by the words ab and b in $(E^*, *, \Lambda)$, the monoid of all words over $\{a, b\}$ under concatenation.

8. Write out the Cayley tables for the semigroups of each of the following finite state machines.

(a)

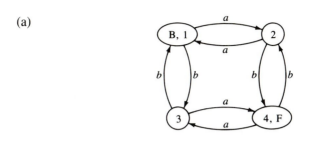

(b)

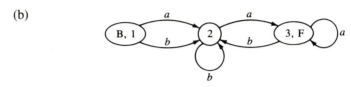

9. Use Euclid's algorithm to find the greatest common divisor of each of the following pairs of integers: 90 and 48; 51 and 143; 102 and 595.

10. Theorem 2 in Section 8.2 says that the greatest common divisor of a pair of integers a and b can be written as $am + bn$. Do this for each pair given in Exercise 9. *Hint*: To avoid "trial and error," use Euclid's algorithm in reverse. For instance, suppose that we had iterated Euclid's algorithm three times before halting and that we have

$$a = qb + r_0$$
$$b = q_1 r_0 + r_1$$
$$r_0 = q_2 r_0 + 0$$

So $r_1 = (a, b)$. Solving for the remainders, we have

(1) $r_1 = b - q_1 r_0$
(2) $r_0 = a - qb$

Substituting for r_0 in Equation (1), we have

(3) $(a, b) = r_1 = b - q_1(a - qb)$ or, regrouping,
(4) $(a, b) = q_1 a + (1 - q)b$

11. Use Theorem 2 in Section 8.2 to prove the following: Let m, n, and q be integers. The equation $mx + ny = q$ has integer solutions for x and y if and only if $(m, n) \mid q$. Use this result to find integer solutions to the following equations, where possible.

(a) $5x + 13y = 1$
(b) $33x + 24y = 6$
(c) $17x + 9y = 5$

12. Prove: $(a, b) = (a, b + qa)$.

13. Prove: $m(a, b) = (ma, mb)$.

14. Solve the following congruences. Express each nonempty solution as the union of equivalence classes.

    (a) $4x \equiv 6 \pmod{12}$

    (b) $3x \equiv 5 \pmod{11}$

    (c) $2x \equiv 6 \pmod 8$

    (d) $6x \equiv 9 \pmod{12}$

    (e) $9x \equiv 6 \pmod{12}$

15. Consider the congruence $ax \equiv y \pmod m$ and let $q = (a, m)$, $q > 0$. Prove that if $q | y$, then the solution set to $ax \equiv y \pmod m$ can be expressed as the union of exactly q equivalence classes modulo m. (Verify this first on the congruences given in Exercise 14.)

16. Prove that the least common multiple of any pair of nonzero integers always exists. (For a definition, see Exercise 16 in Section 8.2.) *Hint*: Use the fact that the natural numbers are well-ordered and the division theorem.

17. Let $(G, *, e)$ be a group, and let a be an element of G. Prove that if $a * a = a$, then a is the identity element. Is this always true of a monoid that is not a group?

18. Find the group T of the symmetries of the equilateral triangle. (It has six elements.) Write out the Cayley table for this group. Is it abelian?

19. Write out the Cayley table for the group of symmetries of a rectangle that is not a square. Is it abelian?

20. (a) Let $(G, *, e)$ be a group, and let y be an element of G. Prove that the function $f_y : G \rightarrow G$ defined by $f(x) = y * x$ is a bijection.

    (b) Write out the bijection described in part (a) explicitly for the group S of the symmetries of the square with $y = V$.

    (c) Write out the bijection for $y = [2]$ in $(\mathbf{Z}_5, +, [0])$.

21. (a) Let $(G, *, e)$ and $(H, \#, i)$ be groups, and let K be the set $G \times H$. Define the operation $\circ$ on H by letting $(a, b) \circ (s, t) = (a * s, b \# t)$. Show that $(K, \circ, (e, i))$ is a group.

    (b) Write out the Cayley table for $\mathbf{Z}_2 \times \mathbf{Z}_2$ under $\circ$.

    (c) Write out the Cayley table for $\mathbf{Z}_4 \times \mathbf{Z}_6$ under $\circ$.

22. (Do Exercise 21 first.)

    (a) Let x be an element in an abelian group $(G, *, e)$ of order n, and let y be an element of an abelian group $(H, \#, i)$ of order m. Prove that the order of (x, y) in the group $(G \times H, \circ, (e, i))$ is the least common multiple of m and n.

(b) Find the order of $([4], [6])$ in $\mathbf{Z}_{12} \times \mathbf{Z}_{12}$.

(c) Show that $\mathbf{Z}_2 \times \mathbf{Z}_2$ is isomorphic to the Klein 4 group.

(d) Show that $\mathbf{Z}_3 \times \mathbf{Z}_2$ is isomorphic to $(\mathbf{Z}_6, +, [0])$.

23. Find an isomorphism from the group of the symmetries of the equilateral triangle (see Exercise 18) to $(S_3, \circ, \mathrm{id}_{X_3})$.

24. Find an isomorphism f from the group of the symmetries of the square to a subgroup of $(S_4, \circ, \mathrm{id}_{X_4})$. What do the permutations of S_3 that are not in the image of f—that is, not in $f(S_4)$—represent geometrically?

25. Prove that every cyclic group of order m is isomorphic to $(\mathbf{Z}_m, +, [0])$.

26. Prove that every group of order 4 is isomorphic to either the Klein 4 group or $(\mathbf{Z}_4, +, [0])$.

27. Prove Proposition 3 in Section 8.3.

28. (a) Find a homomorphism from $(\mathbf{Z}, +, [0])$ to $(S_4, \circ, \mathrm{id}_{X_4})$, and find its kernel.

(b) Find a homomorphism from $(\mathbf{Z}_8, +, [0])$ to $(\mathbf{Z}_2, +, [0])$, and find its kernel.

29. Show that if q is a factor of n, then $(\mathbf{Z}_n, +, [0])$ has an element of order q.

30. Show that the order of any element in $(S_n, \circ, \mathrm{id}_{X_n})$ is the least common multiple of the length of the cycles in its cycle decomposition. Demonstrate your results with a few examples.

31. (a) Find the subgroup of $(S, *, I)$ generated by V and H.

(b) Find the subgroup of $(\mathbf{Z}, +, 0)$ generated by 6 and 9.

(c) Find the subgroup of $(\mathbf{Z}, +, 0)$ generated by 2 and 5.

(d) Find the subgroup of $(\mathbf{Z}_{13}, +, [0])$ generated by $[2]$.

(e) Find the subgroup of $(\mathbf{Z}_{12}, +, [0])$ generated by $[4]$ and $[6]$.

(f) Find all the subgroups of $(\mathbf{Z}_{12}, +, [0])$.

32. Find a pair of generators for the group of symmetries of the equilateral triangle.

33. Let $(G, *, e)$ be a finite group.

(a) Show that for any element a in G, there is a positive integer n such that $a^n = e$. Show also that there is a positive integer m such that $a^m = a^{-1}$.

(b) Show that if S is a subset of G that is closed under $*$, then S is a subgroup of G.

34. Which of the following groups is cyclic?

(a) $(Q, +, 0)$, the set of rational numbers under addition

(b) $\mathbf{Z}_3 \times \mathbf{Z}_5$

(c) $\mathbf{Z}_2 \times \mathbf{Z}_4$

35. Let S be a subset of the integers $\mathbf{Z}$ containing n elements. Prove by induction on n that the subgroup $\langle S \rangle$ is cyclic.

36. Prove that every subgroup of a finite cyclic group is also cyclic.

37. Let $(G, *, e)$ be a group, and let H and K be subgroups of G. Show that the intersection of the sets H and K is also a subgroup of G.

38. (a) Find all the cosets of the subgroup generated by the integer 3 in $(\mathbf{Z}, +, 0)$.

(b) Find the left and right cosets of the subgroup $\{I, R_1, R_2, R_3\}$ in $(S, *, I)$.

(c) Find the right and the left cosets of the subgroup $\{I, R_2\}$ in $(S, *, I)$.

39. Prove that the converse of LaGrange's theorem is true for any finite cyclic group.

40. Write out the multiplication tables for the rings $(\mathbf{Z}_3, +, \cdot)$, $(\mathbf{Z}_5, +, \cdot)$, and $(\mathbf{Z}_9, +, \cdot)$.

41. Find the multiplicative inverse of each element in $\mathbf{Z}_{11} \backslash \{[0]\}$.

42. Find all the zero divisors in $(\mathbf{Z}_{15}, +, \cdot)$.

43. Find a generator of the group $(U_7, \cdot, [1])$, and find an isomorphism from this group to $(\mathbf{Z}_6, +, [0])$.

44. Let m be any positive integer, not necessarily prime. Let U_m be the subset of $\mathbf{Z}_m$ of elements that are not zero divisors.

(a) Prove that $(U_m, \cdot, [1])$ is a group.

(b) Find a value of m for which $(U_m, \cdot, [1])$ is not cyclic.

45. Let M be the set of 2×2 matrices. Prove that a matrix

$$A = \begin{pmatrix} a & b \\ c & d \end{pmatrix}$$

has a multiplicative inverse if and only if $ad - cb \neq 0$. Prove also that the set of invertible matrices is closed under multiplication.

46. (a) Let R be a ring that is an integral domain. Prove that the ring $R \times R$ is not an integral domain. (We define $+$ and $*$ by $(a, b) + (s, t) = (a + s, b + t)$ and $(a, b) * (s, t) = (a * s, b * t)$.)

(b) Find all the zero divisors of the ring $\mathbf{Z}_5 \times \mathbf{Z}_5$.

(c) Find all the zero divisors of the ring $\mathbf{Z}_2 \times \mathbf{Z}_2$.

47. Let $\mathbf{R}$ be the set of real numbers, and let $C = \mathbf{R} \times \mathbf{R}$. Define addition by

$(a, b) + (s, t) = (a + s, b + t)$ and multiplication by $(a, b) \cdot (s, t) = (as - bt, bs + at)$. Show that $(C, +, \cdot)$ is a field.

48. Let p be a prime and suppose that x is not divisible by p. Prove that $[x]^{-1} = [x]^{p-2}$ in $(\mathbf{Z}_p, +, \cdot)$. Verify your result with a few examples.

49. Let $(R, +, *)$ be a ring. Let U be the set of elements of R that have multiplicative inverses. Prove that $(U, *, 1)$ is a group. Find U and its Cayley table for the ring of integers and for the ring $(\mathbf{Z}_{15}, +, \cdot)$.

50. Prove that, in any ring, $(a + b)^2 = a * a + a * b + b * a + b * b$. Give a counterexample to the statement that $(a + b)^2 = a * a + 2(a * b) + b * b$. (*Hint*: Your example must come from a noncommutative ring.)

Appendix

Matrices and Matrix Operations

An **n × m matrix** is a rectangular array of numbers with n rows and m columns. We denote the entry in the ith row, jth column of a matrix A by $a_{i,j}$. We denote the entire matrix A in terms of its entries by $(a_{i,j})_{n \times m}$. The matrix A in Figure A.1 is a 3×4 matrix in which $a_{2,3} = -2$ and $a_{3,2} = 0.5$. All the entries in the first row are equal to 5; that is, $a_{1,j} = 5$ for $j = 1$ to 4. All the entries in the first column are also equal to 5; that is, $a_{i,1} = 5$ for $i = 1$ to 3.

Figure A.1

$$A = \begin{pmatrix} 5 & 5 & 5 & 5 \\ 5 & -0.3 & -2 & 0 \\ 5 & 0.5 & 4 & -1 \end{pmatrix}$$

A $1 \times m$ matrix is called a **row matrix** and an $n \times 1$ matrix is called a **column matrix**. The matrices in Figures A.2(a) and A.2(b) are row (1×4) and column (3×1) matrices, respectively.

Figure A.2

$$(1 \quad 3 \quad 7 \quad 0) \qquad \begin{pmatrix} 1 \\ 3 \\ 0 \end{pmatrix}$$

(a)   (b)

Let A be an $n \times m$ matrix and let c be any constant. We define the **scalar product** cA to be the $n \times m$ matrix $(ca_{i,j})_{n \times m}$. To obtain cA, we multiply each of the entries in A by c. For example, if A is the matrix displayed in Figure A.3(a) and $c = 1/2$, then cA is the matrix displayed in Figure A.3(b).

Figure A.3

$$\begin{pmatrix} 2 & 3 & 4 \\ 9 & -2 & 0 \end{pmatrix} \qquad \begin{pmatrix} 1 & 3/2 & 2 \\ 9/2 & -1 & 0 \end{pmatrix}$$

(a)   (b)

Let $A = (a_{i,j})_{n \times m}$ and $B = (b_{i,j})_{n \times m}$ be $n \times m$ matrices. The **matrix sum** $A + B$ is the $n \times m$ matrix $(a_{i,j} + b_{i,j})_{n \times m}$. We obtain $A + B$ by adding entries in corresponding positions. The difference $A - B$ is defined to be $A + (-1)B$.

Example 1. Let A, B, C, and D be the matrices displayed in Figure A.4.

Figure A.4

$$A = \begin{pmatrix} 1 & -3 \\ 3 & 0 \\ 4 & 3 \end{pmatrix} \qquad B = \begin{pmatrix} 1 & 0 \\ 0 & 1 \end{pmatrix} \qquad C = \begin{pmatrix} 2 & 5 \\ -2 & 6 \\ 0 & 1 \end{pmatrix} \qquad D = \begin{pmatrix} 1 & 4 \\ -1 & 5 \end{pmatrix}$$

(a) $A + C = \begin{pmatrix} 3 & 2 \\ 1 & 6 \\ 4 & 4 \end{pmatrix}$ (b) $A + B$ is not defined.

(c) $-3B = \begin{pmatrix} -3 & 0 \\ 0 & -3 \end{pmatrix}$ (d) $2B - D = \begin{pmatrix} 1 & -4 \\ 1 & -3 \end{pmatrix}$

Matrix addition is both commutative and associative.

Suppose that A is a $1 \times n$ row matrix and that B is an $n \times 1$ column matrix. The **product** $A \times B$ is a 1×1 matrix. Its single entry c is obtained as follows:

$$c = a_{1,1}b_{1,1} + a_{1,2}b_{2,1} + \cdots + a_{1,n}b_{n,1}$$

Example 2. The matrices A and B are displayed in Figure A.5. The single entry of the product $A \times B$ is obtained as follows:

$$1 \cdot 2 + 3 \cdot 5 + 7 \cdot (-3) + 0 \cdot 9 = -4$$

Thus $A \times B = (-4)$. The product $B \times A$ is *not* defined.

Figure A.5

$$A = (1 \quad 3 \quad 7 \quad 0) \qquad B = \begin{pmatrix} 2 \\ 5 \\ -3 \\ 9 \end{pmatrix}$$

We now define matrix multiplication more generally. Let A be an $n \times r$ matrix and let B be an $r \times m$ matrix. The product $A \times B$ is the $n \times m$ matrix $A \times B = (c_{i,j})_{n,m}$ where

$$c_{i,j} = a_{i,1}b_{1,j} + a_{i,2}b_{2,j} + \cdots + a_{i,r}b_{r,j}$$

To find the entry in the ith row and jth column, we take the product of the ith row of A with the jth column of B as defined in the preceding paragraph. We enter the number obtained in the ith row, jth column of $A \times B$. Note that $A \times B$ is defined only when the number of columns of A is equal to the number of rows of B.

Example 3. Let A, B, and C be the matrices displayed in Figure A.6.

Figure A.6

$$A = \begin{pmatrix} 1 & 3 & -2 & 4 \\ -1 & 2 & 5 & 4 \\ 3 & 5 & -1 & 0 \\ 3 & 1 & 1 & 2 \end{pmatrix} \quad B = \begin{pmatrix} 1 & 2 & 3 \\ 0 & 1 & 4 \\ 4 & 2 & -7 \\ 4 & 5 & -1 \end{pmatrix} \quad C = \begin{pmatrix} 1 & 3 & 1 \\ 7 & 6 & 0 \\ 5 & 0 & -1 \end{pmatrix}$$

(a) Since A is a 4×4 matrix and B is a 4×3 matrix, the product $A \times B$ is a 4×3 matrix. The product $A \times B$ is displayed in Figure A.7. To find the entry in the second row, third column of $A \times B$, we take the product of the second row of A with the third column of B as shaded in Figure A.7. That is, we compute

$$-1 \cdot 3 + 2 \cdot 4 + 5 \cdot (-7) + 4 \cdot (-1) = -34$$

Figure A.7

$$\begin{pmatrix} 1 & 3 & -2 & 4 \\ -1 & 2 & 5 & 4 \\ 3 & 5 & -1 & 0 \\ 3 & 1 & 1 & 2 \end{pmatrix} \times \begin{pmatrix} 1 & 2 & 3 \\ 0 & 1 & 4 \\ 4 & 2 & -7 \\ 4 & 5 & -1 \end{pmatrix} = \begin{pmatrix} 9 & 21 & 25 \\ 35 & 30 & -34 \\ -1 & 9 & 36 \\ 15 & 19 & 4 \end{pmatrix}$$

(b) The product $B \times A$ is not defined, because the number of columns of B is 3, but the number of rows of A is 4.

(c) The product $B \times C$ is a 4×3 matrix:

$$\begin{pmatrix} 1 & 2 & 3 \\ 0 & 1 & 4 \\ 4 & 2 & -7 \\ 4 & 5 & -1 \end{pmatrix} \times \begin{pmatrix} 1 & 3 & 1 \\ 7 & 6 & 0 \\ 5 & 0 & -1 \end{pmatrix} = \begin{pmatrix} 30 & 15 & -2 \\ 27 & 6 & -4 \\ -17 & 24 & 11 \\ 34 & 42 & 5 \end{pmatrix}$$

Example 4. Let $A = \begin{pmatrix} 1 & 3 & 6 \\ 3 & 0 & 1 \end{pmatrix}$ and let $B = \begin{pmatrix} -1 & 4 \\ 2 & 1 \\ 0 & 3 \end{pmatrix}$. Then $A \times B$ is the 2×2 matrix

$$\begin{pmatrix} 5 & 25 \\ -3 & 15 \end{pmatrix}$$

$B \times A$ is the 3×3 matrix

$$\begin{pmatrix} 11 & -3 & -2 \\ 5 & 6 & 13 \\ 9 & 0 & 3 \end{pmatrix}$$

We can see from Examples 3 and 4 that matrix multiplication is not commutative. However, it is an associative operation.

Let I_n be an $n \times n$ matrix in which each diagonal entry is 1 and every other entry is 0. That is, $I_n = (a_{i,j})_{n \times n}$ where $a_{i,i} = 1$ and $a_{i,j} = 0$ if $i \neq j$. We call I_n the $n \times n$ **identity matrix**. The matrices I_1, I_2, I_3, and I_4 are displayed in Figure A.8.

Figure A.8

(1)
$$\begin{pmatrix} 1 & 0 \\ 0 & 1 \end{pmatrix} \qquad \begin{pmatrix} 1 & 0 & 0 \\ 0 & 1 & 0 \\ 0 & 0 & 1 \end{pmatrix} \qquad \begin{pmatrix} 1 & 0 & 0 & 0 \\ 0 & 1 & 0 & 0 \\ 0 & 0 & 1 & 0 \\ 0 & 0 & 0 & 1 \end{pmatrix}$$

The matrices I_n are called identity matrices because they have the property that if A is any $n \times m$ matrix, then $A \times I_m = A$ and $I_n \times A = A$. In particular, if A is a square $n \times n$ matrix, then $A \times I_n = I_n \times A = A$.

Suppose that M is a square $(n \times n)$ matrix. We denote the product of M with itself n times by M^n, and we define M^0 to be I_n. The matrix M^n may also be defined inductively as follows:

(1) $M^0 = I_n$
(2) $M^{n+1} = M^n \times M$ for $n \geq 0$

Example 5. Let $M = \begin{pmatrix} 1 & 3 \\ 1 & 2 \end{pmatrix}$. Then $M^0 = I_2$ and $M^1 = M$. Also,

$$M^2 = M \times M = \begin{pmatrix} 4 & 9 \\ 3 & 7 \end{pmatrix} \quad \text{and} \quad M^3 = M^2 \times M = \begin{pmatrix} 13 & 30 \\ 10 & 23 \end{pmatrix}$$

We may extend our matrix operations to matrices with entries taken from any set that is equipped with two operations. We demonstrate this in the next several examples.

Example 6. The entries of the matrices A and B displayed in Figure A.9 are all polynomials in the variable x. We can add and multiply polynomials, and so we can add and multiply the matrices A and B. (Their sizes are compatible.)

Figure A.9
$$A = \begin{pmatrix} x & x - 1 \\ x^2 & 1 \end{pmatrix} \qquad B = \begin{pmatrix} x^3 & x + 1 \\ -x & 2x \end{pmatrix}$$

(a) Let $C = A \times B$. Then

$$c_{1,1} = x \cdot x^3 + (x - 1) \cdot (-x) = x^4 - x^2 + x$$
$$c_{1,2} = x \cdot (x + 1) + (x - 1) \cdot 2x = 3x^2 - x$$
$$c_{2,1} = x^2 \cdot x^3 + 1 \cdot (-x) = x^5 - x$$
$$c_{2,2} = x^2 \cdot (x + 1) + 1 \cdot 2x = x^3 + x^2 + 2x$$

The resulting matrix is

$$\begin{pmatrix} x^4 - x^2 + x & 3x^2 - x \\ x^5 - x & x^3 + x^2 + 2x \end{pmatrix}$$

(b) $A + B = \begin{pmatrix} x + x^3 & 2x \\ x^2 - x & 2x + 1 \end{pmatrix}$

Example 7. Suppose the set $\mathbf{Z}_5 = \{0, 1, 2, 3, 4\}$ and that the two operations $+$ and $\cdot$ are defined on $\mathbf{Z}_5$ by the following tables:

Table A.1

+	0	1	2	3	4
0	0	1	2	3	4
1	1	2	3	4	0
2	2	3	4	0	1
3	3	4	0	1	2
4	4	0	1	2	3

Table A.2

·	0	1	2	3	4
0	0	0	0	0	0
1	0	1	2	3	4
2	0	2	4	1	3
3	0	3	1	4	2
4	0	4	3	2	1

$$\text{Let } A = \begin{pmatrix} 2 & 4 \\ 3 & 3 \\ 1 & 3 \end{pmatrix} \quad \text{and} \quad \text{let } B = \begin{pmatrix} 3 & 1 & 4 \\ 2 & 1 & 0 \end{pmatrix}$$

The product of A and B is a 3×3 matrix. To obtain the entry in the first row, first column, we perform the following computation:

$$2 \cdot 3 + 4 \cdot 2 = 1 + 3 = 4$$

To obtain the entry in the second row, second column, we compute as follows:

$$3 \cdot 1 + 3 \cdot 1 = 3 + 3 = 1$$

The other entries are obtained similarly. The resulting matrix is

$$\begin{pmatrix} 4 & 1 & 3 \\ 0 & 1 & 2 \\ 4 & 4 & 4 \end{pmatrix}$$

Example 8. Suppose we let $B = \{0, 1\}$ and define the operations $+$ and $*$ by the following tables:

Table A.3	+	0	1
	0	0	1
	1	1	1

Table A.4	*	0	1
	0	0	0
	1	0	1

The operations $+$ and $*$ are called *Boolean addition* and *Boolean multiplication*. If we interpret 0 and 1 as truth values, then $+$ is the disjunction operation and $*$ is the conjunction operation. Let

$$M = \begin{pmatrix} 1 & 0 & 1 \\ 0 & 1 & 1 \\ 1 & 0 & 1 \end{pmatrix} \quad \text{and} \quad Q = \begin{pmatrix} 0 & 1 & 0 \\ 1 & 0 & 0 \\ 0 & 1 & 0 \end{pmatrix}$$

Then

$$M^2 = \begin{pmatrix} 1 & 0 & 1 \\ 1 & 1 & 1 \\ 1 & 0 & 1 \end{pmatrix} \qquad M + Q = \begin{pmatrix} 1 & 1 & 1 \\ 1 & 1 & 1 \\ 1 & 1 & 1 \end{pmatrix}$$

Appendix Exercises

Exercise 1 refers to the following matrices:

$$A = \begin{pmatrix} 0 & 1 & 3 \\ 1 & 0 & 1 \\ 5 & 6 & 2 \end{pmatrix} \qquad B = \begin{pmatrix} 1 & 3 & 1 \\ 4 & 1 & 3 \end{pmatrix}$$

$$C = \begin{pmatrix} 1 & 2 \\ 2 & 7 \\ -1 & 3 \\ 0 & 5 \end{pmatrix} \qquad D = \begin{pmatrix} 1 & 2 & 5 \\ 0 & -2 & 3 \\ 3 & 5 & 0 \end{pmatrix}$$

1. Compute where possible.

(a) $(1/3)D$ (b) $A + D$ (c) $2D - 3A$ (d) $A \times D$

(e) $D \times A$ (f) $B \times C$ (g) $C \times B$ (h) A^2

(i) $I_4 \times C$ (j) $C \times I_4$ (k) $(D - I_3)^2$

2. Let $\mathbf{Z}_5 = \{0, 1, 2, 3, 4\}$ and let $+$ and $\cdot$ be defined as in Tables A.1 and A.2 given in Example 7. Compute the following.

(a) $\begin{pmatrix} 1 & 2 & 0 & 4 \\ 4 & 3 & 1 & 2 \end{pmatrix} + \begin{pmatrix} 2 & 4 & 3 & 1 \\ 2 & 3 & 4 & 1 \end{pmatrix}$

(b) $\begin{pmatrix} 1 & 4 & 2 & 3 \\ 2 & 3 & 1 & 4 \end{pmatrix} \times \begin{pmatrix} 2 & 1 & 4 \\ 2 & 3 & 1 \\ 1 & 2 & 4 \\ 0 & 2 & 3 \end{pmatrix}$

3. Let $B = \{0, 1\}$ and let $+$ and $*$ be the Boolean operations defined in Tables A.3 and A.4 in Example 8. Let A and B be the following 3×3 matrices.

$$A = \begin{pmatrix} 1 & 0 & 1 \\ 0 & 1 & 1 \\ 1 & 1 & 1 \end{pmatrix} \qquad B = \begin{pmatrix} 0 & 0 & 0 \\ 1 & 1 & 1 \\ 1 & 0 & 0 \end{pmatrix}$$

Compute the following.

(a) $A + B$ (b) $A \times B$ (c) $B \times A$ (d) $A + A^2$

Chapter 1

Section 1.1 (page 9)

1. (a) yes;   (b) yes;   (c) yes;   (d) no;   (e) yes;   (f) no;   (g) no

2. (a) Mary is not tall.

   (b) The cat is black.

   (c) Some dogs do not bark.

   (d) No birds have feathers.

   (e) Mary is not tall or Burt is not thin.

   (f) John is not smart and Fred is tall.

3. (a) The earth is flat or all birds sing.

   (b) Some birds do not sing and Manhattan is an island.

   (c) It is not the case that all birds sing and that the earth is flat.

   (d) Either Manhattan is an island or it is the case that both the earth is flat and all birds sing.

   (e) The earth is flat and it is not the case that either all birds sing or that Manhattan is an island.

4. (a) 1;   (b) 0;   (c) 1

5. (a) 0;   (b) 1;   (c) 0

6. (a)

P	Q	$\sim Q$	$P \wedge \sim Q$
0	0	1	0
0	1	0	0
1	0	1	1
1	1	0	0

(b)

P	Q	R	$Q \wedge \sim R$	$P \vee (Q \wedge \sim R)$
0	0	0	0	0
0	0	1	0	0
0	1	0	1	1
0	1	1	0	0
1	0	0	0	1
1	0	1	0	1
1	1	0	1	1
1	1	1	0	1

(c)

P	Q	R	$(\sim P \vee Q)$	$(\sim P \vee Q) \wedge \sim R$
0	0	0	1	1
0	0	1	1	0
0	1	0	1	1
0	1	1	1	0
1	0	0	0	0
1	0	1	0	0
1	1	0	1	1
1	1	1	1	0

(d)

P	Q	R	T	$(\sim P \wedge Q)$	$(R \wedge T)$	$(\sim P \wedge Q) \vee (R \wedge T)$
0	0	0	0	0	0	0
0	0	0	1	0	0	0
0	0	1	0	0	0	0
0	0	1	1	0	1	1
0	1	0	0	1	0	1
0	1	0	1	1	0	1
0	1	1	0	1	0	1
0	1	1	1	1	1	1
1	0	0	0	0	0	0
1	0	0	1	0	0	0
1	0	1	0	0	0	0
1	0	1	1	0	1	1
1	1	0	0	0	0	0
1	1	0	1	0	0	0
1	1	1	0	0	0	0
1	1	1	1	0	1	1

7. (a) $P = 0, Q = 0$;   (b) $P = 0, Q = 1, R = 1$;   (c) $P = 0, Q = 1$

8. (a) no;   (b) yes;   (c) no;   (d) yes

9. (a) $\sim(\sim P \wedge \sim Q)$;   (b) $\sim(P \wedge \sim(\sim Q \wedge R))$

10. (a) $\sim(\sim P \vee \sim Q)$;   (b) $\sim(\sim(\sim P \vee \sim Q) \vee R)$

11. Assume that $\sim(P \wedge \sim Q)$ is true. Then $P \wedge \sim Q$ is false so that at least one of P or $\sim Q$ is false. If P is false, then $\sim P$ is true so that $\sim P \vee Q$ is true. If $\sim Q$ is false, then Q is true and $\sim P \vee Q$ is also true. So if $\sim(P \wedge \sim Q)$ is true, so is $\sim P \vee Q$. Conversely, assume that $\sim P \vee Q$ is true. Then at least one of $\sim P$ or Q is true. If $\sim P$ is true then $(P \wedge \sim Q)$ is false so that $\sim(P \wedge \sim Q)$ is true. If Q is true, then $(P \wedge \sim Q)$ is false so that $\sim(P \wedge \sim Q)$ is true. Thus if $\sim P \vee Q$ is true, $\sim(P \wedge \sim Q)$ must also be true. Thus the forms are equivalent.

Section 1.2 (page 15)

1. (a) If John passes calculus, then John is happy.

    (b) If John is happy, then John both passes calculus and has a job.

    (c) John has a job or if John passes calculus, then John is happy.

(d) John either has a job or passes calculus if and only if John is happy.

(e) It is not the case that if John passes calculus then he is happy. (Alternatively, John passes calculus and John is not happy.)

(f) It is not the case that if John passes calculus, then he neither is happy nor has a job.

(g) John passes calculus if and only if he is happy and does not have a job.

(h) It is not the case that John passes calculus if and only if he is happy.

2. (a) false;   (b) true;   (c) false;   (d) false;
   (e) true;   (f) false;   (g) false;   (h) true

3. (a) true;   (b) true;   (c) true;   (d) false;
   (e) false;   (f) false;   (g) true;   (h) false

4. (a)

P	Q	R	$(P \wedge Q) \Rightarrow R$
0	0	0	1
0	0	1	1
0	1	0	1
0	1	1	1
1	0	0	1
1	0	1	1
1	1	0	0
1	1	1	1

(b)

P	Q	$(P \wedge Q) \Leftrightarrow \sim Q$
0	0	0
0	1	1
1	0	0
1	1	0

(c)

P	Q	R	$(P \vee Q) \Leftrightarrow (R \vee Q)$
0	0	0	1
0	0	1	0
0	1	0	1
0	1	1	1
1	0	0	0
1	0	1	1
1	1	0	1
1	1	1	1

5. (a) $P \wedge \sim(Q \wedge R)$;   (b) $(P \wedge Q) \wedge \sim R$;   (c) $(\sim P \vee Q) \wedge \sim R$

6. (a) If Mary loses, then Sally wins.
       If Mary does not lose, then Sally does not win.
       Sally wins and Mary does not lose.

   (b) If the square of 9 is odd, then 9 is odd.
       If the square of 9 is not odd, then 9 is not odd.
       Nine is odd and the square of 9 is not odd.

(c) If some dogs bark, then all cats meow.
   If no dogs bark, then some cats do not meow.
   All cats meow and no dogs bark.

(d) If Mary loses and the school closes, then John wins.
   If Mary does not lose or the school does not close, then John does not win.
   John does not win and either Mary does not lose or the school does not close.

7. (a) contingency;   (b) tautology;   (c) absurdity

8. $(P \wedge \sim Q) \vee (Q \wedge \sim P)$

9. The contrapositive of $\sim P \Rightarrow Q$ is the equivalent form $\sim Q \Rightarrow \sim(\sim P)$ or $\sim Q \Rightarrow P$.

Section 1.3 (page 24)

1. (a) $(\sim P \wedge \sim Q) \vee (P \wedge \sim Q) \vee (P \wedge Q)$

   (b) $(\sim P \wedge \sim Q \wedge \sim R) \vee (\sim P \wedge Q \wedge \sim R) \vee (P \wedge \sim Q \wedge \sim R) \vee (P \wedge Q \wedge R)$

2. (a) $(P \wedge Q) \vee (\sim P \wedge \sim Q)$

   (b) $(\sim P \wedge \sim Q \wedge \sim R) \vee (\sim P \wedge \sim Q \wedge R) \vee (\sim P \wedge Q \wedge \sim R) \vee (\sim P \wedge Q \wedge R)$
      $\vee (P \wedge \sim Q \wedge R) \vee (P \wedge Q \wedge \sim R) \vee (P \wedge Q \wedge R)$

   (c) as is

3. (a) $\{[(P \wedge Q) \vee (P \wedge \sim Q)] \vee (\sim P \wedge Q)\} \Leftrightarrow \{[P \wedge (Q \vee \sim Q)] \vee (\sim P \wedge Q)\}$
      $\Leftrightarrow [P \vee (\sim P \wedge Q)] \Leftrightarrow [(P \vee \sim P) \wedge (P \vee Q)] \Leftrightarrow (P \vee Q)$

   (b) $\{[(P \wedge Q \wedge R) \vee (P \wedge \sim Q \wedge R)] \vee (P \wedge Q \wedge \sim R)\} \Leftrightarrow \{\{P \wedge [(Q \wedge R)$
      $\vee (\sim Q \wedge R)]\} \vee [P \wedge (Q \wedge \sim R)]\} \Leftrightarrow \{P \wedge [(Q \wedge R) \vee (\sim Q \wedge R) \vee (Q \wedge \sim R)]\}$
      $\Leftrightarrow \{P \wedge \{[Q \wedge (R \vee \sim R)] \vee (\sim Q \wedge R)\}\} \Leftrightarrow \{P \wedge [Q \vee (\sim Q \wedge R)]\}$
      $\Leftrightarrow \{P \wedge [(Q \vee \sim Q) \wedge (Q \vee R)]\} \Leftrightarrow [P \wedge (Q \vee R)] \Leftrightarrow [P \wedge (\sim Q \Rightarrow R)]$

4. (a) $(P \mid P) \Leftrightarrow \sim(P \wedge P) \Leftrightarrow (\sim P \vee \sim P) \Leftrightarrow \sim P$;

   (b) $(P \mid Q) \mid (P \mid Q)$;   (c) $(P \mid P) \mid (Q \mid Q)$;

   (d) $P \mid (Q \mid Q)$

5. (a) $\sim P \Leftrightarrow (P \downarrow P)$, $(P \wedge Q) \Leftrightarrow [(P \downarrow P) \downarrow (Q \downarrow Q)]$,
      $(P \vee P) \Leftrightarrow [(P \downarrow Q) \downarrow (P \downarrow Q)]$

   (b) $[(P \downarrow P) \downarrow (Q \downarrow Q)] \downarrow [(P \downarrow P) \downarrow (Q \downarrow Q)]$

   (c) $\{P \downarrow [(Q \downarrow Q) \downarrow (R \downarrow R)]\} \downarrow \{P \downarrow [(Q \downarrow Q) \downarrow (R \downarrow R)]\}$

6. (a) $PQ \vee PQ' \vee P'Q'$;   (b) $PQR \vee PQ'R \vee P'QR \vee P'Q'R$;

   (c) $PQRT \vee PQ'RT \vee P'Q'R'T' \vee P'QR'T$

7. (a) $P' \vee Q'$;   (b) $PQ' \vee P'Q$ (the original expression was minimal)

8. (a) $PQ'R \vee QR' \vee P'Q$;   (b) $Q' \vee PR'$;   (c) Q

9. (a) $P'QRT \vee Q'RT' \vee PQ'T'$

   (b) $P'Q'RT \vee P'QT \vee PQ'R' \vee R'T$

   (c) $P'QR \vee PR'T' \vee Q'R' \vee P'RT'$ or $P'QR \vee PR'T' \vee Q'R' \vee P'Q'T'$

Section 1.4 (page 29)

1. (a) $2 - 3 = 0$;   (b) $3 - 3 = 0$;   (c) $y - x = 0$;

   (d) For each x, there exists a y such that $x - y = 0$.

   (e) There is a value of x such that, for all values of y, $y - x = 0$.

2. (a) $8 \cdot (0.5) = 4$, true
   (b) There exists a value of y such that $2 \cdot y = 4$. True
   (c) For each value of x, there is some value of y such that $x \cdot y = 4$. False
   (d) There is some value of x such that for all values of y, $x \cdot y = 4$. False
   (e) For every value of x and y, if $x \cdot y = 4$ then $x > y$. False
   (f) There exists a value of x and a value of y such that $x \cdot y = 4$ and $x > y$. True

3. (a) $\forall x[x^2 \geq 0]$
   (b) $\forall x \exists y[x \cdot y = 1]$
   (c) $\exists x \exists y[(x > 0) \wedge (y > 0) \wedge (x \cdot y > 0)]$
   (d) $\exists x \forall y[(y > 0) \Rightarrow (x + y < 0)]$

4. (a) $\exists x[x^2 \leq 0]$
   (b) $\forall x[x \cdot 2 \neq 1]$
   (c) $\exists x \forall y[x + y \neq 1]$
   (d) $\exists x \exists y[(x > y) \wedge (x^2 \leq y^2)]$

5. (a) true;   (b) false;   (c) true;   (d) true

Section 1.5 (page 37)

1. *modus ponens*: $[(P \Rightarrow Q) \wedge P] \Rightarrow Q$
   modus tollens: $[(P \Rightarrow Q) \wedge \sim Q] \Rightarrow \sim P$
   hypothetical syllogism: $[(P \Rightarrow Q) \wedge (Q \Rightarrow R)] \Rightarrow (P \Rightarrow R)$

   To prove validity, use truth tables to show that each of the above forms is always true.

2. "Affirming the consequent": If $P = 0$ and $Q = 1$, then $[(P \Rightarrow Q) \wedge Q] \Rightarrow P$ is false.
   "Denying the antecedent": If $P = 0$ and $Q = 1$, then $[(P \Rightarrow Q) \wedge \sim P] \Rightarrow \sim Q$ is false.

3. (a) G: John graduates. $G \Rightarrow J$
       J: John gets a job. $\sim J$
       _____
                                       $\therefore \sim G$ valid, modus tollens

   (b) G: Mary wears the green hat. $G \Rightarrow B$
       B: Mary leads the band. $\sim G$
       _____
                                       $\therefore \sim B$ invalid

   (c) F: Fred will sing. $F \vee C$
       C: Cat will bark. $\sim C$
       _____
                                       $\therefore F$ valid, disjunctive syllogism

4. (a) valid;   (b) invalid
5. (a) valid;   (b) valid

Section 1.6 (page 41)

1. Let x and y be odd integers. Then we can find integers n and m such that $x = 2n + 1$ and $y = 2m + 1$. Then $xy = 4nm + 2m + 2n + 1 = 2(2nm + n + m) + 1$ which is also odd.

2. Suppose that x and y are not odd and are therefore even. Then we can find integers m and n such that $x = 2m$ and $y = 2n$. Then $xy = 4mn = 2(2nm)$ which is even. Thus if x and y are not odd, their product is not odd.

3. Use Exercises 1 and 2.

4. Suppose that x is a rational number and that y is an irrational number. Let $z = x + y$ and suppose that z is rational. Solving for y, we find that $y = z - x$. Since z and x are rational and the difference of two rational numbers is again rational, we must conclude that y is rational, contradicting our assumption that y is irrational. Thus z must be irrational.

5. (a) The number 9 is an integer but its square root 3 is not irrational.

    (b) The number 7 is odd but not divisible by 3.

    (c) The number 7 is an odd prime, but $7 + 2 = 9$ and 9 is not a prime number.

Chapter 2   ## Section 2.1   (page 57)

1. (a) $\{5, 10, 15, 20, \ldots\}$;   (b) $\{-1, -3, -5, \ldots\}$;   (c) $\{2\}$;
    (d) $\{\ldots, -3, 1, 5, 9, \ldots\}$;   (e) $\{\ldots, -10, 0, 10, 20, \ldots\}$;
    (f) $\{$MATH, MAHT, MTAH, MTHA, HAMT, $\ldots\}$ (24 elements);
    (g) $\{$BOOT, TOOB, OTOB, $\ldots\}$ (12 elements)

2. (a) $\{x : x = -2n \text{ for } n \in \mathbf{N}\}$;   (b) $\{x : x = 3n \text{ for } n \in \mathbf{Z}\}$;
    (c) $\{x : x = 2n \text{ or } x = 3n \text{ for } n \in \mathbf{N}\}$;   (d) $\{x : x = 1/n \text{ for } n \in \mathbf{N}\}$;
    (e) $\{x : 0 \le x \le 100 \text{ and } x \in \mathbf{Z}\}$

3. (a) true;   (b) false;   (c) true;   (d) true

4. (a) $S \ne Q$;   (b) $X = T$

Section 2.2   (page 59)

1. (a) no;   (b) yes, proper;   (c) no;   (d) yes, not proper;
    (e) no;   (f) yes, proper;   (g) no

2. (a) yes, proper;   (b) no;   (c) yes, proper;   (d) yes, proper;
    (e) yes, not proper;   (f) no;   (g) yes, proper

3. (a) proper;   (b) proper;   (c) not proper

4. (a) Let x be any element of A. Then $x = 10q$ for some q in $\mathbf{Z}$. Since $10 = 2 \cdot 5$, $x = 5(2 \cdot q)$. Since $2 \cdot q$ is an integer, x is an element of B. Thus $A \subset B$.

    (b) Let x be any element of A. Then $x = 2n$ for some n in $\mathbf{Z}$. Thus $x^2 = 4n^2$ which is also even. Thus $A \subset B$. To show that $B \subset A$, we need to show that if $x \in B$, then $x \in A$. We use the contrapositive of this statement and show that if x is not in A, then x is not in B. So suppose that x is an integer that is not in A; that is, suppose x is odd. Then $x = 2n + 1$ for some integer n. So $x^2 = 2(2n^2 + 2n) + 1$, which is also odd. Thus x^2 is not an element of B. Thus $B \subset A$. Since $A \subset B$ and $B \subset A$, $A = B$.

5. Suppose x is an element of B and that $x = y^2$ for some integer y. Suppose that y

$= p_1 p_2 \cdots p_n$ where each p_i is a prime. Then x is divisible by p_i^2 for each i and no other prime numbers divide x. So x is an element of A. The containment is proper since 8 is an element of A but not of B.

6. Let x be any element of A. Since $A \subset B$, x is an element of B. Since $B \subset C$, x is also an element of C. Thus $A \subset C$.

Section 2.3 (page 64)

1. (a) $\{h, i, j, k, l\}$;   (b) $\{a, b, c, \ldots, j, k, l, o, p, q, \ldots, z\}$;
   (c) $\{h, i, j, k, l\}$;   (d) $\{h, i, j, k, l, o, p, q, \ldots, z\}$;   (e) $\{r, s, t, \ldots, z\}$;
   (f) $\{a, b, c, d, e, f, g, m, n, o, p, \ldots, z\}$;   (g) $\{a, b, c, d, e, f, g\}$;
   (h) $\{m, n, o, p, q\}$;   (i) $\{a, b, c, d, e, f, g\}$;   (j) $\{a, b, c, \ldots, l\} = A$

2. (a) $\{\ldots, -6, 0, 6, 12, 18, \ldots\}$
   (b) $\{\ldots, -5, -3, 0, 3, 5, 6, 9, 10, \ldots\}$
   (c) $\{\ldots, -15, -12, -6, 0, 6, 12, 15, 18, \ldots\}$
   (d) $\{\ldots, -10, -6, -5, 0, 5, 6, 10, 12, \ldots\}$
   (e) $\{\ldots, -11, -7, -5, -1, 1, 5, 7, 11, \ldots\}$
   (f) $\{\ldots, -3, -2, -1, 1, 2, 3, 4, 5, 7, 8, \ldots\}$
   (g) $\{\ldots, -9, -3, 3, 9, 15, \ldots\}$
   (h) $\{\ldots, -4, -2, 2, 4, 8, 10, 14, \ldots\}$
   (i) $(A \cap B') \cup (A \cap C)$
   (j) $(A \cap C') \cup (A \cap B)$

3. (a) false;   (b) true;   (c) true;   (d) false;   (e) false

4. Any element of A or A' is in U. So $A \cup A' \subset U$. Let x be any element of U. If x is not in A, then x must be in A'. Thus x is in $A \cup A'$ and $A \cup A' \subset U$. An element x is in A' if and only if x is not in A. Thus $A \cap A' = \emptyset$.

5. Let x be an element of $A \backslash B$. Then x is not an element of B and hence not an element of $B \cap A'$. Thus x is not an element of $B \backslash A$. So $A \backslash B$ and $B \backslash A$ are disjoint.

6. $(A \backslash B) \cup (B \backslash A) \cup (A \cap B) = [(A \cap B') \cup (B \cap A')] \cup (A \cap B)$
$$= \{[(A \cap B') \cup B] \cap [(A \cap B') \cup A']\} \cup (A \cap B)$$
$$= \{[(A \cup B) \cap (B' \cup B)] \cap [(A \cup A') \cap (B' \cup A')]\} \cup (A \cap B)$$
$$= [(A \cup B) \cap (B \cap A)'] \cup (A \cap B)$$
$$= [(A \cup B) \cup (A \cap B)] \cap [(A \cap B)' \cup (A \cap B)]$$
$$= A \cup B$$

7. $A \cup B = \{x : 0 < x \le 2\} \cup \{x : 5 \le x < 8\} \cup \{x : 2 < x < 5\}$

Section 2.4 (page 67)

1. (a) $A_4 = \{x : -4 \le x \le 4\}$;   (b) A_5;   (c) A_7;   (d) $\mathbf{R}$;   (e) A_1

2. $\bigcup\limits_{r \in \mathbf{R}^+} A_r = \mathbf{R}^+$ and $\bigcap\limits_{r \in \mathbf{R}^+} A_r = \emptyset$

3. (a) $\{x : x > 1\}$; (b) $\{x : x < -1\} \cup \{x : x > 0\}$; (c) same as (b);
   (d) same as (a)

4. (a) $\{x : 0 < x \leq \frac{1}{3}\}$; (b) same as (a); (c) $\{x : x \leq \frac{1}{2}\}$; (d) same as (c)

Section 2.5 (page 69)

1. (a) $\mathcal{P}(A) = \{\varnothing, \{a\}, \{b\}, \{2\}, \{3\}, \{a, b\}, \{a, 2\}, \{a, 3\}, \{b, 2\}, \{b, 3\}, \{2, 3\}, \{a, b, 2\}, \{a, b, 3\}, \{b, 2, 3\}, \{a, 2, 3\}, \{a, b, 2, 3\}\}$

   (b) $\mathcal{P}(B) = \{\varnothing, \{cat\}, \{dog\}, \{mouse\}, \{cat, dog\}, \{dog, mouse\}, \{cat, mouse\}, \{cat, dog, mouse\}\}$

   (c) $\mathcal{P}(C) = \{\varnothing, \{\{a\}\}, \{\{b\}\}, \{\{a\}, \{b\}\}\}$

   (d) $\mathcal{P}(D) = \{\varnothing, \{\varnothing\}, \{\{\varnothing\}\}, \{\varnothing, \{\varnothing\}\}\}$

2. $\varnothing, \{1, 2, 3, 4, 5\}, \{x : x = 3n - 1 \text{ for } n \text{ in } \mathbf{N}\}, E, \mathbf{N}, \mathbf{W},$
   $\{x : x = 5n \text{ for } n \text{ in } \mathbf{Z}\}$

3. $A \cap B = \{a, b\}$; $C \cap D = B$; $A \cap D = B$;
   $C \cap \mathcal{P}(A) = C$; $D \cap \mathcal{P}(A) = B$
   (a) true; (b) true; (c) false; (d) true; (e) true; (f) true

4. Let X be any element of $\mathcal{P}(A)$. Then X is a subset of A. Since A is contained in B, X is also a subset of B. Thus X is an element of $\mathcal{P}(B)$ and $\mathcal{P}(A) \subset \mathcal{P}(B)$.

Section 2.6 (page 70)

1. $A \times C = \{(1, x), (1, y), (2, x), (2, y)\}$;
   $B \times B = \{(a, a), (a, b), (a, c), (b, a), (b, b), (b, c), (c, a), (c, b), (c, c)\}$;
   $A \times B \times C = \{(1, a, x), (1, a, y), (1, b, x), (1, b, y), (1, c, x), (1, c, y), (2, a, x), (2, a, y), (2, b, x), (2, b, y), (2, c, x), (2, c, y)\}$;
   $(A \times A) \times (C \times C) = \{((1, 1), (x, x)), ((1, 1), (x, y)), ((1, 1), (y, x)), ((1, 1), (y, y)), ((1, 2), (x, x)), ((1, 2), (x, y)), ((1, 2), (y, x)), ((1, 2), (y, y)), \ldots, ((2, 2), (y, y))\}$

2. (a) $\{(1, -1), (1, -2), \ldots, (4, -5), \ldots\} = A \times B$ where A is the set of positive integers and B is the set of negative integers.

   (b) $\{(0, a), (0, b), \ldots, (9, y), (9, z)\} = A \times B$ where $A = \{0, 1, 2, \ldots, 9\}$ and B is the set of letters of the alphabet.

   (c) $\{(1, 2), (2, 3), (3, 4), \ldots\}$. This set cannot be expressed as a cartesian product because 2 is both a first coordinate and a second coordinate, but $(2, 2)$ is not in the set.

   (d) $\{\ldots, (0.3, 7.1), (0.333, 45.7), \ldots\} = A \times B$ where $A = \{x : 0 \leq x \leq 1\}$ and $B = \{x : x \geq 0\}$.

   (e) $\{\ldots, (0.2, 0.04), (-3, 9), (10,100), \ldots\}$. This set cannot be expressed as a cartesian product because $(2, 4)$ is in the set and $(4, 16)$ is in the set, but $(4, 4)$ is not in the set.

3. (a) $\{\ldots, (0, 0, -5), (1, 2, 19), (9, 8, -39), \ldots\} = D \times D \times \mathbf{Z}$ where
$D = \{0, 1, 2, \ldots, 9\}$.

(b) $\{\ldots, (0, 0, 1), (1, 0, 0), (0, 1, 0), \ldots, (3^{-1/2}, 3^{-1/2}, 3^{-1/2}), \ldots\}$. This set
cannot be expressed as a cartesian product because $(1, 1, 1)$ is not in the set.

(c) $\{(0.2, 0.3, 0.5, 0.13), (0.3335, 0.3335, 0.44, 0.12), \ldots\} = A \times A \times A \times A$
where $A = \{x : x \text{ is real and } 0 < x < 1\}$.

Section 2.7 (page 74)

1. $\{2, 6, 18, 54, 162, 486, \ldots\}$
2. $\{5, 10, 0, 15, -10, -5, \ldots\}$
3. $\{?, !, ??, !!, !?, !!?, !??, !!!, ???, \ldots\}$
4. $\{3, 5, 8, 2, -2, 6, 10, 1, 0, \ldots\} = \mathbf{Z}$
5. (1) 1 is in S. (2) If x is in S then $2x$ is in S. (3) No other elements are in S.
6. (1) ab is in S. (2) If x is in S then xab is in S. (3) No other elements are in S.
7. (1) 2 is in S. (2) If x is in S then x^2 is in S. (3) No other elements are in S.
8. (1) 2 is in S. (2) If x and y are elements in S then $x + y$ and $x - y$ are in S.
   (3) No other elements are in S.
9. (1) 0 is in S. (2) If x is in S then $10x01$ is in S.
   (3) No other words are in S.

Section 2.8 (page 78)

1. $\{a, aa, aaa, abab, baaabb, bababb, baaabbb, bb, aaabb, ababbbbb, \ldots\}$
2. $\{aaaa, aaab, aabb, abbb, aaba, abab, abba, abaa, baaa, baab, \ldots\}$
3. $\{aaa, a, aa, aabaaab, aaaaabaaaaab, aabaabaabaaa, aaaabaabaaaa, aaab,$
   $aaaabaab, aabaabaabaaaaaa, \ldots\}$
4. $\{aa, ab, aab, babb\} = R$
5. $\{abbb, aabbbbbb, aaabbbbbbbbbbbb, aaaabbbbbbbbbbbbbbbbbbbbbbbb, \ldots\}$
6. (1) ba is in S.
   (2) If x is in S then $bbxaa$ is in S.
   (3) Only words so formed are in S.
7. (1) 000 is in A.
   (2) If x is in A, then $1x$ and $0x$ are in A.
   (3) Only words so formed are in A.
8. (1) w is in L.
   (2) If x is in L, then xw is in L.
   (3) Only words so formed are in L.
9. (1) The empty word is in B.

(2) If x is a word in B, then xb is in B and xab is in B.

(3) Only words so formed are in B.

10. (a) $\{\Lambda, 11, 1111, 111111, \ldots\}$

(b) $\{\Lambda, 00, 1, 100, 001, 0000, 11, 100100, \ldots\}$

(c) $\{\Lambda, 10, 100, 1010, 10100, 101000, 100100, 10010, \ldots\}$

11. $LS = L$

$SL = \{x : x = 1110y, x = 0010y, \text{ or } x = 10y \text{ for some } y \text{ in } E^*\}$

12. (a) $L \cap S = \{x : x = ayb \text{ for some } y \text{ in } E^*\}$

$L \cup S = \{x : x = ay \text{ or } x = yb \text{ for some } y \text{ in } E^*\}$

(b) $LS = L \cap S$ and $SL = \{x : x = ybaz \text{ for some } y \text{ and } z \text{ in } E^*\}$

Section 2.9 (page 82)

1. $P(5): 1 + 2 + 3 + 4 + 5 = 5(5 + 1)/2$

$P(n + 2): 1 + 2 + \cdots + n + (n + 1) + (n + 2) = (n + 2)\,[(n + 2) + 1]/2$

$P(2n + 1): 1 + 2 + \cdots + (2n + 1) = (2n + 1)\,[(2n + 1) + 1]/2$

2. (a) $Q(1): 2 = 1(1 + 1)$

$Q(4): 2 + 4 + 6 + 8 = 4(4 + 1)$

$Q(n + 1): 2 + 4 + \cdots + 2(n + 1) = (n + 1)\,[(n + 1) + 1]$

$Q(2n): 2 + 4 + \cdots + 2(2n) = (2n)(2n + 1)$

(b) $Q(1)$ is true since $2 = 2$. Assume $Q(n)$ is true. We shall prove that $Q(n + 1)$ is true. Now $Q(n + 1)$ states that

$$[2 + 4 + \cdots + 2n] + 2(n + 1) = (n + 1)(n + 2)$$

By the induction hypothesis we can substitute $n(n + 1)$ for the expression in the square brackets to obtain: $n(n + 1) + 2(n + 1) = (n + 1)(n + 2)$ which is a true statement. Thus $Q(n)$ is true for all $n > 0$.

3. (a) $P(1): 1 = 1^2$

$P(3): 1 + 3 + 5 = 3^2$

$P(n + 1): 1 + 3 + \cdots + (2n - 1) + [2(n + 1) - 1] = (n + 1)^2$

(b) $P(1)$ is true since $1 = 1^2$. Assume that $P(n)$ is true. We shall prove that $P(n + 1)$ is true, namely that

$$[1 + 3 + \cdots + (2n - 1)] + (2n + 1) = (n + 1)^2$$

By the induction hypothesis we can substitute n^2 for the expression in the square brackets to obtain $n^2 + (2n + 1) = (n + 1)^2$, which is true. Thus $P(n)$ is true for all $n > 0$.

4. See the note for Exercise 4 and the solution to Exercise 3(b).

5. For $n = 0$ we have $1 = 2 - 1$, which is true. Assume that $1 + 2 + \cdots + 2^n = 2^{n+1} - 1$. We shall prove that $[1 + 2 + \cdots + 2^n] + 2^{n+1} = 2^{n+2} - 1$. We may use the induction hypothesis to substitute for the expression in square brackets and obtain $[2^{n+1} - 1] + 2^{n+1} = 2^{n+2} - 1$. Simplifying, we find we need to show that $2(2^{n+1}) = 2^{n+2}$, which is indeed true. So the statement is true for all $n \geq 0$.

6. For $n = 1$ we need to show that $1 = 1(2)(3)/6$, which is true. Assume the statement is true for n. We must show that $[1 + 2^2 + 3^2 + \cdots + n^2] + (n + 1)^2 = (n + 1)(n + 2)(2n + 3)/6$. By the induction hypothesis, we may substitute $n(n + 1)(2n + 1)/6$ for the expression in the square brackets. So we then need to show that $n(n + 1)(2n + 1)/6 + (n + 1)^2 = (n + 1)(n + 2)(2n + 3)/6$. This may be done by simplifying the left side.

7. For $n = 1$, there is nothing to prove. For $n = 2$, this is simply the distributive law as proved in Chapter 2. Assume that the statement is true for n. We must show that:

$$B \cap ([A_1 \cup A_2 \cup \cdots \cup A_n] \cup A_{n+1}) = (B \cap A_1) \cup \cdots \cup (B \cap A_{n+1})$$

We may distribute and find that the left side is:

$$\{B \cap [A_1 \cup A_2 \cup \cdots \cup A_n]\} \cup (B \cap A_{n+1})$$

We now use the induction hypothesis to substitute $[(B \cap A_1) \cup (B \cap A_2) \cup (B \cap A_n)]$ for the expression in the curly brackets to finish the proof.

8. The empty word has an even number of a's and b's and so the statement is true for the words given in the basis clause. Now suppose x is a word in A and that x has an even number of a's and b's. Then axa and bbx also have an even number of a's and b's. So the statement is true for all words obtained by iterating step (2), that is, for all words in A.

Chapter 3 | **Section 3.1 (page 106)**

---

1. (a) and (e)

2. There are eight functions from A to B:

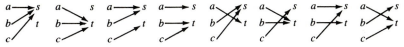

There are nine functions from B to A:

3. $f(2) = 2.5, f(1.5) = 2.166. \ldots, f(-1)$ is not defined.
   $f(z + 2)$ is defined for $z \geq -1$.

4. $g(3, 2) = 8, g(2, 3) = 7, g(x, x) = 3x, g(s + t, s - t) = 3s + t$.
   $g(x, y) = 0$ for (x, y) on the line $y = -2x$.
   $g(x, y) > 1$ for (x, y) above, but not on the line $y = -2x + 1$.

5. $f(x) = 2$ for $2 \leq x < 3; f(x) = -3$ for $-3 \leq x < -2; f(x) = n$ for $n \leq x < n + 1$

6. $g(4.0001) = 5;$ $\quad g(-3.2) = -3;$ $\quad g(0) = 0;$
   $g(x) \neq \lfloor x \rfloor + 1$ for any integer value of x, e.g., $g(3) = 3 = \lfloor 3 \rfloor$.

7. (a) $f(1) = 1, f(2) = 3, f(3) = 7, f(4) = 15, f(5) = 31, f(6) = 63, f(7) = 127$
   (b) $g(1) = 1, g(2) = 1, g(3) = 2, g(4) = 6, g(5) = 24, g(6) = 120, g(7) = 720$
   (c) $f(1) = 1, f(2) = 3, f(3) = 4, f(4) = 7, f(5) = 11, f(6) = 18, f(7) = 29$

8. Let $n = 1$. Since $2^n - 1 = 1$ and $f(1) = 1$, the statement is true for $n = 1$. Suppose that $f(n) = 2^n - 1$. We prove that $f(n + 1) = 2^{n+1} - 1$. Now $f(n + 1) = 2f(n) + 1$. Using the induction hypothesis, we substitute for $f(n)$ to obtain $f(n + 1) = 2(2^n - 1) + 1$. Simplifying, we obtain $f(n + 1) = 2^{n+1} - 1$.

9. (a) $f(1) = 3$ and $f(n + 1) = 3f(n)$ for $n + 1 > 1$.
   (b) $f(1) = 5$ and $f(n + 1) = 3 + f(n)$ for $n + 1 > 1$.

10. The initial pair of rabbits breeds in the first month. So at the end of the first month, there are 2 pairs $[= f(2)]$ of rabbits. Only the initial pair will breed during the second month. After two months, there will be $1 + 2$ pairs $[= f(3)]$. During month n, only those pairs existing at the end of month $n - 2$ (there are $f(n - 1)$ of these) will breed. So after n months there will be $f(n + 1) = f(n - 1) + f(n)$ pairs of rabbits. In particular, after one year, there are $f(13)$ pairs.

Section 3.2 (page 109)

1. (a) bijective;   (b) injective;   (c) neither;   (d) surjective;   (e) neither

2. (a) $f(n) = 2n$;   (b) $f(n) = \lfloor n/2 \rfloor + 1$;
   (c) $f(n) = n + 1$ for n odd and $f(n) = n - 1$ for n even.

3. (a) Suppose $f(x, y) = f(s, t)$. Then $x + y = s + t$ and $2x - y = 2s - t$. Adding, we see that $3x = 3s$ or $x = s$. Thus $y = t$ and $(x, y) = (s, t)$ and we proved injectivity. Let (u, v) be an element of the codomain. We must find (x, y) such that $x + y = u$ and $2x - y = v$. Solving simultaneously, we find that $x = (u + v)/3$ and $y = (2u - v)/3$. Since $((u + v)/3, (2u - v)/3)$ is in the domain of f, we have surjectivity.
   (b) Similar to (a).

4. This function is injective, but it is not surjective. If $(s, t) = f(n, m)$, then s and t are either both even or both odd.

5. (a) neither injective nor surjective;   (b) both injective and surjective

6. Define $f : \mathbf{N} \times \mathbf{N} \to A$ by $f(m, n) = a^{m-1} b^{n-1}$.

Section 3.3 (page 113)

1.

2. $f \circ g(x) = (3x^2 + 1)/10$, $g \circ f(x) = (9x^2 - 6x + 26)/50$

3. No, e.g., $f \circ g(-3.2) = 3$, but $g \circ f(-3.2) = 4$

4. $f \circ g : \mathbf{R} \to \mathbf{R} \times \mathbf{R}$ and $f \circ g(t) = (3t + 2t^2, t^2 - 3t)$

$h \circ f : \mathbf{R} \times \mathbf{R} \to \mathbf{R}$ and $h \circ f(x, y) = 4y - x$
$h \circ (f \circ g) : \mathbf{R} \to \mathbf{R}$ and $h \circ (f \circ g)(t) = 4t^2 - 3t$

5. $f^0 = \mathrm{id}_A$ and $f^1 = f$.

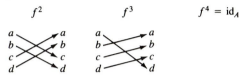

$f^2 \qquad\qquad f^3 \qquad\qquad f^4 = \mathrm{id}_A$

6. $f^0 = \mathrm{id}_{\mathbf{R}}$, $f^1 = f$, $f^2(x) = 9x + 8$, $f^3(x) = 27x + 26$

7. See answer to Exercise 9.

8. $f(n) = n$ and $g(n) = |n|$

9. $f(n) = n + 1$; $\quad g(n) = n - 1$ for $n > 1$ and $g(1) = 1$

Section 3.4 (page 116)

1. (a) $f^{-1}(x) = (1 - 3x)/2$

   (b) $g^{-1}(x) = (3x - 1)^{1/3}$

   (c) $f^{-1}(x) = (2x - 1)^{1/2}$

   (d) $g^{-1}(x, y) = (x - y, 2y - x)$

   (e) $f^{-1}(x) = (x)^{1/2}$ of $x \geq 4$ and $f^{-1}(x) = x - 2$ if $x < 4$

2. (a) $f : E \to O$ where $f(n) = n - 1$, and $f^{-1}(n) = n + 1$

   (b) $f : \mathbf{N} \to S$ where $f(n) = (n - 1)^2$, and $f^{-1}(n) = n^{1/2} + 1$

   (c) $f : A \to B$ where $f(x) = 2x + 1$, and $f^{-1}(x) = (x - 1)/2$

   (d) $f : E \to \mathbf{N}^-$ where $f(n) = -n - 1$ if $n \geq 0$, and $f(n) = n$ if $n < 0$
      $f^{-1}(n) = -n - 1$ if n is odd, and $f^{-1}(n) = n$ if n is even

3. (a) $f(a^m b^n) = (m, (n + 1)/2)$, and $f^{-1}(m, n) = a^m b^{2n-1}$

   (b) $g(a^m b^n) = (m + 1, (n + 1)/2)$, and $g^{-1}(m, n) = (a^{m-1}, b^{2n-1})$

4. $\varnothing \longrightarrow \varnothing$
    $\{a\} \longrightarrow \{c\}$
    $\{b\} \longrightarrow \{b\}$
    $\{a, b\} \longrightarrow \{b, c\}$

5. Let $f : \mathbf{W} \to \mathbf{Z}$ where $f(n) = n$ and let $g : \mathbf{Z} \to \mathbf{W}$ where $g(m) = |m|$. Then $g \circ f = \mathrm{id}_{\mathbf{W}}$. Since g is not injective, it has no inverse. Since f is not surjective, it has no inverse.

6. Suppose that $(f^{-1})^{-1}(x) = y$. Then $f^{-1}(y) = x$. Since $f^{-1}(y) = x$, $f(x) = y$. So $(f^{-1})^{-1}(x) = f(x)$.

Section 3.5 (page 120)

1. $f(A) = \{x : -7 \leq x \leq 9\}; \quad f^{-1}(B) = \left\{x : \frac{1}{2} \leq x \leq \frac{3}{2}\right\}; \quad f(\mathbf{R}^+) = \{x : x > -1\};$

   $f^{-1}(\mathbf{R}^+) = \left\{x : x > \frac{1}{2}\right\}$

2. $f(A) = \{x : 1 \le x \le 17\}$ and $f^{-1}(A) = \{x : -3^{1/2} \le x \le 3^{1/2}\}$

3. $f^{-1}(\{n\}) = \{x : n \le x < n + 1\}$. Since f is not a bijection it has no inverse. Hence $f^{-1}(n)$ is not defined. (Note the absence of braces.)

4. $f(S) = \{4, 0, 2\};$ $f^{-1}(\{0\}) = \{n : n \le 0\};$ $f(\mathbf{Z}) = \{0, 2, 4, 6, \ldots\}$

5. $L(A) = \{0, 1, 2, 3, \ldots, 7\}$. There are $2^4 = 16$ words in $L^{-1}(\{4\})$ and 2^n words in $L^{-1}(\{n\})$. The number of words in $L^{-1}(X_n) = 2 + 2^2 + \cdots + 2^n = 2(2^n - 1)$.

6. Let $f : \mathbf{R} \to \mathbf{R}$ be defined by $f(x) = x^2$ and let $A = \{x : -1 < x < 4\}$. Then $f^{-1}(A) = \{x : 0 \le x < 2\}$ and $f(f^{-1}(A)) = \{x : 0 \le x < 4\}$. Equality holds for surjective functions.

| **Chapter 4** | **Section 4.1 (page 140)** |

1. $|A \cup C| = 8;$ $|A \times B| = 20;$ $|A \cup B| = 7;$ $|\mathscr{P}(A)| = 2^5 = 32;$
$|\mathscr{P}(A) \cup \mathscr{P}(C)| = 39;$ $|\mathscr{P}(A) \cup \mathscr{P}(B)| = 44$

2.
```
a        1
b    ✕   2
c    ✕   3
d        4
s ────→  5
t    ✕   6
u        7
```

3.
$$\varnothing \longrightarrow \varnothing$$
$$\{a\} \longrightarrow \{s\}$$
$$\{a, b\} \longrightarrow \{s, t\}$$
$$\{a, d\} \longrightarrow \{s, u\}$$
$$\{a, b, d\} \longrightarrow \{s, t, u\}$$
$$\{b, d\} \longrightarrow \{t, u\}$$
$$\{b\} \longrightarrow \{t\}$$
$$\{d\} \longrightarrow \{u\}$$

4. 36 outfits

5. 1900 bytes

6. 30 like both; 40 like oranges but not kiwis.

7. 7^5; 2520; $7 \cdot 6^4$

8. 2520; 210; $2520 + 840 + 210 + 42 + 7 = 3619$

9. 32; 24

10. 36; 9; 6

11. 265

12. 1800

13. 308

14. 772

15. 1539

Section 4.2 (page 145)

1. (a) $4^3 = 64$; (b) $4 \cdot 3 \cdot 2 = 24$; (c) $3^4 = 81$;
   (d) 0; (e) 2^9; (f) $8 \cdot 7 \cdot 6 = 336$

2. $A \cup B = \{a_1, \ldots, a_m, b_{r+1}, \ldots, b_n\}$. Define $f : A \cup B \to X_{m+n-r}$ as follows:
   $$f(a_i) = i$$
   $$f(b_i) = i - r + m$$

3. (a) $aaa \longrightarrow 1,$ $aab \longrightarrow 2,$ $aba \longrightarrow 3,$ $abb \longrightarrow 4$
   $baa \longrightarrow 5,$ $bab \longrightarrow 6,$ $bba \longrightarrow 7,$ $bbb \longrightarrow 8$

   (b) $f : A \to X_{501}$ where $f(n) = n/2 + 1$

   (c) $f : A \to X_{68}$ where $f(n) = n/2 + 1$ for n even and $f(n) = 51 + n/3$ for n divisible by 3 but not 2.

   (d) $f : A \to X_{24}$

$1234 \longrightarrow 1,$	$1243 \longrightarrow 2,$	$1324 \longrightarrow 3,$	$1342 \longrightarrow 4$
$1423 \longrightarrow 5,$	$1432 \longrightarrow 6,$	$2134 \longrightarrow 7,$	$2143 \longrightarrow 8$
$2314 \longrightarrow 9,$	$2341 \longrightarrow 10,$	$2413 \longrightarrow 11,$	$2431 \longrightarrow 12$
$3124 \longrightarrow 13,$	$3142 \longrightarrow 14,$	$3214 \longrightarrow 15,$	$3241 \longrightarrow 16$
$3412 \longrightarrow 17,$	$3421 \longrightarrow 18,$	$4123 \longrightarrow 19,$	$4132 \longrightarrow 20$
$4213 \longrightarrow 21,$	$4231 \longrightarrow 22,$	$4312 \longrightarrow 23,$	$4321 \longrightarrow 24$

4. $256!/220!$; $128!/92!$ if each string must have an even number of 1's.

5. 100^{50}; $100!/50!$; $2 \cdot 50!$ (Assume each person has a different number.)

6. $n > 13$

7. 6001

8. 21; 11

9. 360; 1260

10. $13!/[(4!)(6!)(3!)]$

11. $10!$; $9!$

12. $10!/7! = 720$; $10!/[(7!)(3!)] = 120$

Section 4.3 (page 152)

1. $P(6, 4) = 360$, $C(6, 4) = 15$, $P(7, 1) = 7$, $C(7, 1) = 7$
   $P(8, 6) = 20{,}160$, $C(8, 6) = 28$, $C(8, 2) = 28$

2. (a) $C(n, 0) = n!/[(0!)(n!)] = 1$
   (b) $C(n, 1) = n!/[(1!)(n - 1)!] = n!/(n - 1)! = n$
   (c) $C(n, n - 1) = n!/[(n - 1)!(1!)] = n$

3. $P(30, 2) = 30 \cdot 29 = 870$

4. $P(20, 5) = 20 \cdot 19 \cdot 18 \cdot 17 \cdot 16 = 1{,}860{,}480$

5. $P(8, 3) = 8 \cdot 7 \cdot 6 = 336$

6. $C(50, 5) = 2,118,760$

7. $C(30, 9) = 14,307,150$

8. $C(10, 3) = 120$

9. $C(20, 6) \cdot C(14, 3) \cdot C(11, 5)$

10. $C(25, 10) \cdot C(15, 8) \cdot C(7, 7)$

11. 6^6; $C(6, 3) \cdot 5^3$; $C(6, 6) + C(6, 4) \cdot 5^2 + C(6, 2) \cdot 5^4 + C(6, 0) \cdot 5^6$

12. $C(5, 3) \cdot 4^2 + C(5, 4) \cdot 4 + C(5, 5)$; $C(8, 3) \cdot C(5, 2) \cdot 3^3$; $3 \cdot 5^2 = 75$

13. $C(52, 7)$; $C(13, 4) \cdot C(13, 3)$

14. $2 \cdot C(8, 4)$

15. $C(14, 4)$; $C(14, 4)/5!$

16. 8

17. $C(17, 2)$

Section 4.4 (page 156)

1. (a) $x^6 + 6x^5y + 15x^4y^2 + 20x^3y^3 + 15x^2y^4 + 6xy^5 + y^6$

   (b) $x^5 - 5x^4y + 10x^3y^2 - 10x^2y^3 + 5xy^4 - y^5$

   (c) $x^4 - 12x^3 + 54x^2 - 108x + 81$

   (d) $x^4 + 12x^3y + 54x^2y^2 + 108xy^3 + 81y^4$

2. There is a one-to-one correspondence between the subsets of cardinality 1 and the subsets of cardinality $n - 1$ obtained by corresponding a singleton subset to its complement. Thus $C(n, 1) = C(n, n - 1)$.

3. Let A be a set of cardinality n. Suppose we choose a subset B from A of cardinality m, and then, from B, choose a subset C of cardinality k. There are $C(n, m) \cdot C(m, k)$ ways to do this. Now we can also choose B and C as follows. First choose the k elements of C from A and then choose $m - k$ elements from the remaining $n - k$ elements of A to adjoin to C to form B. There are $C(n, k) \cdot C(n - k, m - k)$ ways to do this.

4. $n = 2$:

$$C(2, 0) \cdot C(2, 2) + C(2, 1) \cdot C(2, 1) + C(2, 2) \cdot C(2, 0) = 1 + 4 + 1 = 6 = C(4, 2)$$

$n = 3$:

$$C(3, 0) \cdot C(3, 3) + C(3, 1) \cdot C(3, 2) + C(3, 2) \cdot C(3, 1) + C(3, 3) \cdot C(3, 0)$$
$$= 1 + 9 + 9 + 1 = 20 = C(6, 3)$$

$n = 4$:

$$C(4, 0) \cdot C(4, 4) + \cdots + C(4, 4) \cdot C(4, 0) = 1 + 16 + 36 + 16 + 1 = 70$$
$$= C(8, 4)$$

Let A be a set of cardinality $2n$ with disjoint subsets B and C of cardinality n. Each subset of A of cardinality n has k elements from B and $n - k$ elements from C for some k, $k = 0, 1, \ldots, n$. There are $C(n, k) \cdot C(n, n - k)$ ways to choose a subset of cardinality n with k elements in B and $n - k$ elements in C. Thus the total number of subsets of A of cardinality n is given by the sum of $C(n, k) \cdot C(n, n - k)$ over $k = 0, 1, \ldots, n$.

5. (a) $|S| = 2^n$

   (b) Let $A = \{a_1, \ldots, a_n\}$. Define $f : S \rightarrow \mathscr{P}(A)$ as follows:

      $a_i \in f(s)$ if and only if the ith position in the string s is 1.

Section 4.5 (page 165)

1. (a) $f : A \rightarrow B$ where $f(x) = 2x + 1$

   (b) First, let $f : \mathbf{N} \rightarrow \mathbf{Z}$ be defined by:

$$f(n) = \begin{cases} n/2 \text{ if } n \text{ is even} \\ -(n - 1)/2 \text{ if } n \text{ is odd} \end{cases}$$

   Then let $g : \mathbf{N} \times \mathbf{N} \rightarrow \mathbf{Z} \times \mathbf{Z}$ be defined by:

$$g(i, j) = (f(i),\ f(j))$$

   (c) $f : A \rightarrow B$ where $f(m) = 3m/2$

   (d) $f : S \rightarrow T$ where $f(x) = -x$

2. (a) $f(i) = 2i - 1$;   (b) $f(i) = 5i - 4$;

   (c) $f(i) = 3i/2$ for i even and $f(i) = 3(1 - i)/2$ for i odd;

   (d) $f(i) = 2^i$ (Note: In each case, the domain is $\mathbf{N}$.)

3. (a) The words of S are of the form $a^n b^n$ where n is a whole number. Define a bijection $f : S \rightarrow \mathbf{N}$ by $f(a^n b^n) = n + 1$.

   (b) The words of S are of the form $b(ab)^n$. Define $f : S \rightarrow \mathbf{N}$ by $f(b(ab)^n) = n + 1$.

   (c) The words of S are of the form $a^{2n} b^{3n}$. Define $f : S \rightarrow \mathbf{N}$ by $f(a^{2n} b^{3n}) = n + 1$.

4. (a) S is a subset of the set of rational numbers Q. Since Q is countable and any subset of a countable set is countable, S must be countable.

   (b) $\mathbf{Z}^4 = \mathbf{Z} \times \mathbf{Z} \times \mathbf{Z} \times \mathbf{Z}$ is countable since it is the cartesian product of a finite number of countable sets. Define $f : \mathbf{Z}^4 \rightarrow M$ by $f(a, b, c, d) = \begin{pmatrix} a & b \\ c & d \end{pmatrix}$.

   Since f is a bijection, M is countable.

   (c) Let P be the set of polynomials. Let P_n be the set of polynomials of degree n. P is the countable union of the P_n. Hence, if P_n is countable, P is also. There is a bijection from the countable set Q^{n+1} to P_n given by $(a_1, \ldots, a_{n+1}) \longrightarrow a_1 x^n + \cdots + a_{n+1}$. Thus P_n is countable and so is P.

5. (a) uncountable;   (b) countable;   (c) uncountable;

   (d) countable;   (e) uncountable

6. Let $f : A \rightarrow B$ be a bijection and let B be countable. If B is finite, we can find a bijection $g : B \rightarrow S_n$ for some n. Then $g \circ f$ is a bijection from A to S_n, and A is finite. If B is infinite, we can find a bijection g from B to $\mathbf{N}$. Then $g \circ f$ is a bijection from A to $\mathbf{N}$, and A is countably infinite.

Chapter 5

Section 5.1 (page 185)

1. (a)

(b)

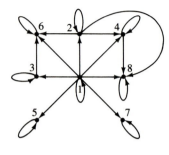

(c)

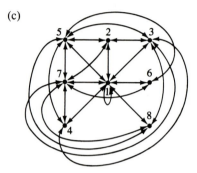

2. (a) Symmetric;   (b) Reflexive, transitive, antisymmetric;   (c) Symmetric

3. This relation is reflexive, transitive, and antisymmetric.

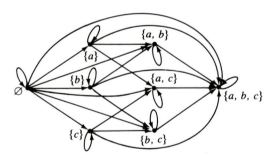

4. R is neither reflexive nor transitive.

5.

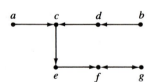

6. (a) No;   (b) Yes

7. (a)

	a	b	c	d
a	0	1	0	1
b	1	0	0	1
c	0	0	1	0
d	1	1	0	0

(b)

	1	2	3	4	5	6	7	8
1	1	1	1	1	1	1	1	1
2	0	1	0	1	0	1	0	1
3	0	0	1	0	0	1	0	0
4	0	0	0	1	0	0	0	1
5	0	0	0	0	1	0	0	0
6	0	0	0	0	0	1	0	0
7	0	0	0	0	0	0	1	0
8	0	0	0	0	0	0	0	1

(c)

	1	2	3	4	5	6	7	8
1	1	1	1	1	1	1	1	1
2	1	0	1	0	1	0	1	0
3	1	1	0	1	1	0	1	1
4	1	0	1	0	1	0	1	0
5	1	1	1	1	0	1	1	1
6	1	0	0	0	1	0	1	0
7	1	1	1	1	1	1	0	1
8	1	0	1	0	1	0	1	0

8.

9. There should be only 0's on the diagonal of the matrix of an irreflexive relation. There should be at least one 0 and at least one 1 on the diagonal of the matrix of a relation that is neither reflexive nor irreflexive.

Section 5.2 (page 189)

1. $R \circ R = \{(a, a), (a, b), (a, c), (a, d), (b, b), (b, d), (c, c), (d, d)\}$

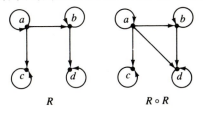

2. For $k > 6$, $R_k = R_6$.

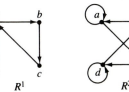

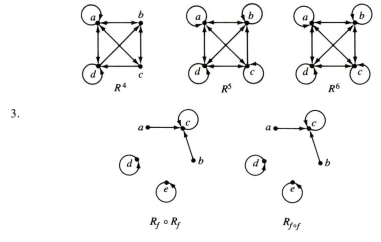

$$R^4 \qquad R^5 \qquad R^6$$

3.

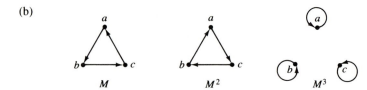

$$R_f \circ R_f \qquad\qquad R_{f \circ f}$$

4. Suppose that (a_1, a_2) is in $R_{f \circ f}$. Then $f \circ f(a_1) = a_2$, and we can find a' such that $f(a_1) = a'$ and $f(a') = a_2$. So (a_1, a') is in R_f and (a', a_2) is in R_f. Thus (a_1, a_2) is in $R_f \circ R_f$.

Conversely, suppose that (a_1, a_2) is in $R_f \circ R_f$. We can find a' such that (a_1, a') and (a', a_2) are in R_f so that $f(a_1) = a'$ and $f(a') = a_2$. Thus $f \circ f(a_1) = a_2$, and (a_1, a_2) is in $R_{f \circ f}$.

5. The digraphs for R^2 and R^3 are the same as that for R. In fact, $R^n = R$ for all $n > 0$.

6. (a)

$$\text{(b)} \quad M \times M = \begin{pmatrix} 0 & 1 & 0 & 1 \\ 0 & 0 & 0 & 1 \\ 0 & 0 & 0 & 1 \\ 1 & 1 & 0 & 1 \end{pmatrix}$$

(c)

$$\text{(d)} \quad \begin{pmatrix} 0 & 1 & 0 & 1 \\ 0 & 0 & 0 & 1 \\ 0 & 0 & 0 & 1 \\ 1 & 1 & 0 & 1 \end{pmatrix}. \text{ The matrices are the same.}$$

7. (a) If $n = 3q$, $q \in \mathbf{W}$, then $M^n = I$.
   If $n = 3q + 1$, $q \in \mathbf{W}$, then $M^n = M$.
   If $n = 3q + 2$, $q \in \mathbf{W}$, then $M^n = M^2$.

   (b)

   $$M \qquad\qquad M^2 \qquad\qquad M^3$$

8. (a) $R^T = \{(a, b), (a, a), (a, c), (a, d), (c, c), (c, d), (c, a), (b, a), (c, b), (b, b),$
   $(b, c), (b, d), (d, a), (d, b), (d, c), (d, d), (a, e), (b, e), (c, e), (d, e)\}$

(b) $R^T = R$

(c) $R^T = \{(a, b), (a, d), (a, f), (a, g), (a, e), (a, c), (a, a), (g, g), (b, b), (b, d),$
$(b, f), (b, g), (b, e), (b, c), (b, a), (d, b), (d, d), (d, f), (d, g), (d, e), (d, c),$
$(d, a), (f, g), (f, c), (f, e), (g, c), (g, e), (e, c), (e, e), (e, g), (c, e), (c, c), (c, g)\}$

9. Let (a, b) and (b, c) be in R^T. Then we can find m and n such that $a\, R^m\, b$ and $b\, R^n\, c$. Thus $a\, R^{m+n}\, c$ and (a, c) is in R^T, and R^T is transitive.

Section 5.3 (page 195)

1. (a) Yes;   (b) No;   (c) No

2. (a) Yes;   (b) No;   (c) No

3. (a) Reflexivity: $(x, y)\, R\, (x, y)$ because both $x - x$ and $y - y$ are integers.

   (b) Symmetry: If $(x, y)\, R\, (s, t)$, then $x - s$ and $y - t$ are integers. Thus $s - x$ and $t - y$ are both integers and we have $(s, t)\, R\, (x, y)$.

   (c) Transitivity: Suppose $(x, y)\, R\, (s, t)$ and $(s, t)\, R\, (u, v)$. Then $x - s$, $y - t$, $s - u$, and $t - v$ are all integers. Now $x - u = (x - s) + (s - u)$ and $y - v = (y - t) + (t - v)$. Thus $x - u$ and $y - v$ are integers and $(x, y)\, R\, (u, v)$.

4. (a) Reflexivity: $n \equiv n \pmod{m}$ since $n - n = 0 = 0m$.

   (b) Symmetry: If $n \equiv k \pmod{m}$, then $n - k = rm$ for some integer r. Thus $k - n = (-r)m$ and $k \equiv n \pmod{m}$.

   (c) Transitive. If $p \equiv q \pmod{m}$ and $q \equiv r \pmod{m}$, then $p - q = sm$ and $q - r = tm$ for some integers s and t. Thus $p - r = p - q + q - r = (s + t)m$ and hence $p \equiv r \pmod{m}$.

5. (a) $x \equiv 5 \pmod{7}$ or $\{\ldots, -2, 5, 12, 19, \ldots\}$

   (b) $x \equiv 8 \pmod{9}$ or $\{\ldots, -1, 8, 17, 26, \ldots\}$

   (c) This congruence has no solution.

   (d) $x \equiv 4 \pmod{5}$ or $\{\ldots, -1, 4, 9, 14, \ldots\}$

   (e) $x \equiv 1, 3,$ or $5 \pmod{6}$ or $\{\ldots, -5, 1, 7, \ldots\} \cup \{\ldots, -3, 3, 9, \ldots\}$
   $\cup \{\ldots, -1, 5, 11, \ldots\}$

6. Let (a, b) and (b, c) be in $\mathcal{R}_f$. Then $b = f^m(a)$ and $c = f^n(b)$ for some nonnegative integers n and m. Thus $c = f^{m+n}(a)$ and (a, c) is in $\mathcal{R}_f$.

7. (a)

   (b)

 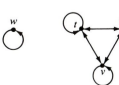

Section 5.4 (page 199)

1. $E(3) = \{\ldots, -11, -4, 3, 10, 17, \ldots\}$
   $E(-1) = \{\ldots, -10, -1, 8, 17, \ldots\}$

2. (a) 2;   (b) m

3. Let E_m be the equivalence class of words of length m. Then E_m has n^m elements. Let a be a symbol in the alphabet E. A complete set of representatives is the set $\{\lambda, a, a^2, a^3, \ldots\}$.

4. Assume that $E(x) = E(y)$. Thus x is in $E(y)$ and y is in $E(x)$. So $x\ R\ y$ and $y\ R\ x$. Now assume that $x\ R\ y$. By symmetry we also have $y\ R\ x$. Let z be any element of $E(x)$ so that we have $x\ R\ z$. By transitivity, we have $y\ R\ z$. Thus z is in $E(y)$ and $E(x) \subset E(y)$. The proof that $E(y) \subset E(x)$ is similar.

5. $E((1, 2)) = \{(1, y) : y \text{ is a real number}\}$
   $E\left(\left(0, \frac{1}{2}\right)\right) = \{(0, y) : y \text{ is a real number}\}$
   The set $\{(x, 0) : x \text{ is a real number}\}$ is a complete set of representatives of R.

6. $E((-1, 2)) = \{(x, y) : x = 2m - 1 \text{ and } y = 3n + 2 \text{ for some } m \text{ and } n \text{ in } \mathbf{Z}\}$. The set $\{(x, y) : 0 \le x < 2, 0 \le y < 3\}$ is a complete set of representatives.

7. The orbits of f are $\{a, b, c, d, e\}$ and $\{f, g, h\}$. A complete set of representatives of $\mathcal{R}_f$ is $\{a, f\}$. The cycle decomposition of f is $(a, b, c, d, e)(f, g, h)$.

8. $(a_1)(a_2) \cdots (a_n)$

9.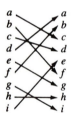

10. $(a, d, c)(b)(e, i, g)(f)(h)$

Section 5.5 (page 209)

1. The digraphs in (b) and (c) define partial orders. The digraph in (a) does not.

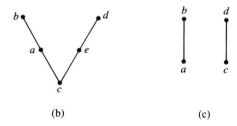

(b)   (c)

2. (a)

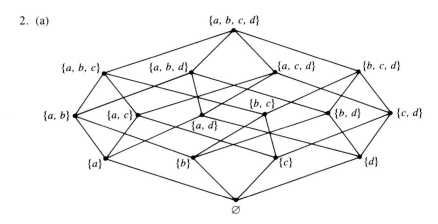

(b) 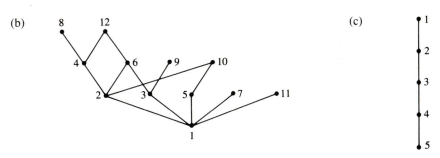 (c)

3. Only the relation described in (c) is a poset. It is also totally and well ordered.

4. (a) @@@@, @##$, @#$$, #@$$, ##, $, $$@@##
   (b) $, ##, @@@@, @##$, @#$$, #@$$, $$@@##
   The empty word is the least element of E^*.

5. (a) lub = c, glb does not exist
   (b) lub = f, glb = a
   (c) lub = g, glb = e
   (d) lub = g, glb = e

6. (a) lub = 2, glb = -1;   (b) lub = {a, b, c}, glb = {b};
   (c) lub = 11111, glb = 00000

7. The Hasse diagram in part (a) of the figure is not a lattice. There is no lower bound for {f, g}. The Hasse diagram in part (b) of the figure is a lattice.

8. $f \wedge g = g$, $d \wedge e = d$, $d \wedge b = a$, $a \wedge (b \wedge c) = a$, $a \wedge (b \vee c) = a$
   $f \vee g = f$, $d \vee e = e$, $d \vee b = e$, $a \vee (b \vee c) = e$, $a \vee (b \wedge c) = a$

9. Let x and y be elements in a totally ordered set. Either $x \leq y$, $y \leq x$, or $x = y$. If $x \leq y$ then $x \wedge y = x$ and $x \vee y = y$. If $y \leq x$ then $x \wedge y = y$ and $x \vee y = x$. If $x = y$ then $x \wedge y = x \vee y = x$. Thus in each case the condition for being a lattice is satisfied.

10. (a) The element a is a lower bound for the set {$a, a \vee b$} because $a \leq a$ and

$a \lesssim a \vee b$. If x is any other lower bound, then $x \lesssim a$. Thus $a = \text{glb}\{a, a \vee b\}$ $= a \wedge (a \vee b)$.

(b) The element a is an upper bound for the set $\{a, a \wedge b\}$ because $a \lesssim a$ and $a \wedge b \lesssim b$. If y is any other upper bound, then $a \lesssim y$. Thus $a = \text{lub}\{a, a \wedge b\}$ $= a \vee (a \wedge b)$.

Chapter 6   # Section 6.1 (page 232)

---

1. (a) Not simple

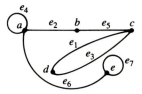

(b) Simple

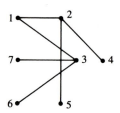

(c) Not simple

2. (a) $\deg(a) = 4$, $\deg(b) = 2$, $\deg(c) = 3$, $\deg(d) = 2$, $\deg(e) = 3$. The sum of degrees is 14 and the number of edges is 7. Parts (b) and (c) are similar.

3. (a)

(b)

(c) can't be done;

(d)

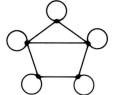

(e) can't be done;   (f)

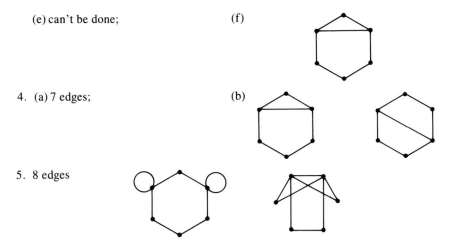

4. (a) 7 edges;   (b)

5. 8 edges

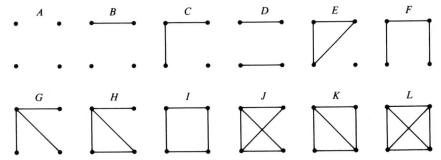

6. No, each node of degree 5 is connected to each of the other nodes and the two nodes of degree 3 are connected to each other. Any bijection which corresponds nodes of the same degree will establish an isomorphism.

7. There are 4 nodes of degree 5 and 2 nodes of degree 2.

8. 21

9. (a) isomorphic;   (b) not isomorphic

10. Each simple graph with 4 nodes is isomorphic to one of the following:

11. None exists. Use 102 nodes to represent the students and 35 nodes to represent the computers. If we place an edge between each computer and the student assigned to it, there will be 102 edges. Each computer will have degree 1 or 3. Let x be the number of computers of degree 1 and y be the number of computers of degree 3. Then $x + y = 35$ and $x + 3y = 102$. If we solve these equations simultaneously, we find that $y = 67/2$, which is impossible.

Section 6.2 (page 242)

1. *abce, abcde, abcdge, abcdfhge, abde, abdce, abdge, abdfhge, afde, afdce, afdge, afdbce, afhge, afhgde, afhgdce, afhgdbce.* The length of the shortest path is 3, e.g., *abce* has length 3. The length of the longest path is 7, e.g., *afhgdbce.*

2. *abde*

3. If we ignore the orientation counterclockwise or clockwise and also ignore the starting point we have:

Length 3: *bdcb*, *dced*, *degd*

Length 4: *abdfa*, *fdghf*, *dbced*, *dcegd*   (There are other proper circuits.)

4. The minimal number of edges in a proper circuit is 1 (a self-loop). A graph with two circuits can have as few as two edges:

5. Each simple connected graph with 4 nodes and no proper circuits has 3 edges and is isomorphic to one of the following:

Each simple connected graph with 5 nodes has 4 edges and is isomorphic to one of the following:

A simple connected graph with *n* nodes and no proper circuits has $n - 1$ edges. For a proof, see the proof of Theorem 1, part (b) of Section 6.4.

6. If *G* does not contain a proper circuit, there is at most one path between any two nodes.

7. (a) 3;   (b) 2;   (c) 3;   (d) 2 components, not simple

8. (a) 2;   (b) 4

9. $\lfloor 16/5 \rfloor = 3$

Met   Not met

10. (a) *acbdaebfa*;   (b) None exists because deg(*a*) = 3;
    (c) None exists.

11. (a) While there is an Euler path, namely *acbdaebf*, there is none from *a* to *b*.
    (b) *afbdecdgcab*
    (c) None exists because there are 4 nodes of odd degree.

12. Any connected graph in which each node has degree 2 like:

13. *AFBCDEA*

14. There is no Hamiltonian circuit, but *abcdefghij* is a Hamiltonian path.

15. There is no Hamiltonian circuit in (b). $K_{n,m}$ has a Hamiltonian circuit if and only if $n = m$.

Section 6.3 (page 248)

1. (a) 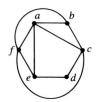 (b)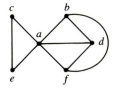

2. (a) $|V| = 6$, $|E| = 10$, $R = 6$, and $6 = 10 - 6 + 2$.
   (b) $|V| = 6$, $|E| = 9$, $R = 5$, and $5 = 9 - 6 + 2$.

3. A copy of K_5 is found as indicated by the circled nodes.

4. To find a copy of $K_{3,3}$ in the graph, ignore the nodes of degree 2.

5. 6 regions:

6. 6 regions:

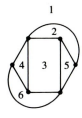 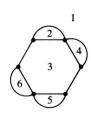

7. $|E| = 10$:

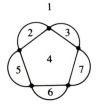

8.

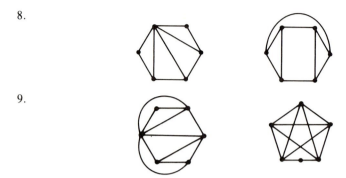

9.

10. Self-loops and parallel edges do not change planarity. There is no maximal number of edges if we do not require the graph to be simple. There is no maximal number of regions because the addition of a self-loop increases the number of regions by 1.

11. (Sketch the possible configurations as you read this argument.) Let G be a simple connected graph with 7 nodes, $a_1, a_2, \ldots, a_7$, and suppose that each node has degree at least 5. The number of nodes in G is odd, so at least one node, say a_1, has degree 6. Suppose that a_2 has degree 5 and that there is no edge from a_2 to a_7. Because the degree of a_3 is at least 5, there must be edges from a_3 to at least three of the nodes a_4, a_5, a_6, and a_7. Assume that we have edges (a_3, a_4) and (a_3, a_5). (We must also have one or both of the edges (a_3, a_6) and (a_3, a_7), but these will not matter for the rest of this discussion.) Now consider a_4. It must be connected by an edge to at least two of the nodes a_5, a_6, and a_7. If there is an edge from a_4 to a_5, then we can find a copy of K_5 in the nodes a_1, a_2, a_3, a_4, and a_5 and the edges between each pair of them. If there is no edge from a_4 to a_5, then there must be an edge from a_4 to a_6 and another from a_5 to a_6. In this case we can find a copy of $K_{3,3}$ in the nodes a_1, a_2, a_3, a_4, a_5, and a_6 and the edges between them. In either case we have shown that G cannot be planar.

Section 6.4 (page 259)

1.

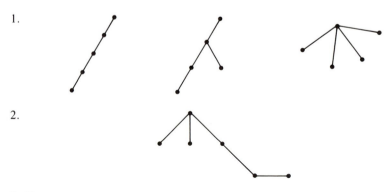

2.

3. No

4.

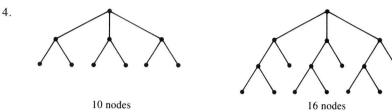

10 nodes   16 nodes

Let x and y be the number of nodes of degree 1 and 3 respectively. A tree with 11 nodes has 10 edges. So $2 \cdot 10 = x + 3y$ and $11 = x + y$. Solving simultaneously, we have $y = 9/2$, which is impossible. A tree with 15 nodes has 14 edges. So $2 \cdot 14 = x + 3y$ and $x + y = 15$, and we have $y = 13/2$, which is impossible. We can draw such a tree when the number of nodes is even and greater than or equal to 2.

5. If T is such a tree with exactly 2 nodes, then both are of degree 1. Assume it is true of a tree with m nodes, $2 \leq m \leq n$, and let T be a tree with $n + 1$ nodes. Remove an edge e of T. The resulting graph has two components C_1 and C_2, both of which are trees and at least one of which, say C_2, has 2 or more nodes. Let x and y be nodes in C_2 of degree 1. If C_1 has only 1 node z, then it is an endnode of e. So z and at least one of x or y has degree 1 in T. If C_1 has at least 2 nodes s and t, e is incident to at most one of these, say t. Then s and at least one of x or y has degree 1 in T. Thus T has at least two nodes of degree 1. The maximal number of nodes of degree 1 is $n - 1$.

6.

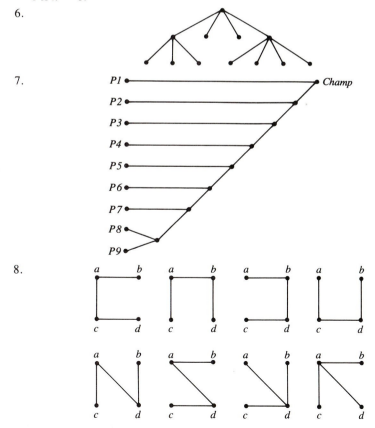

7.

8.

9. Yes, there is a path.

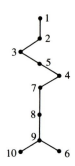

10. Node 1: see answer to Exercise 9.

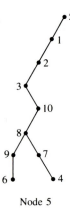

Node 5

11. Shortest path: (1, 2, 3, 10)

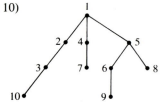

12. The weight for both is 12.

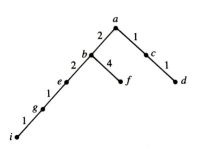 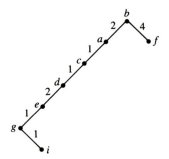

Section 6.5 (page 271)

1. Each rooted tree will be isomorphic to one of the following:

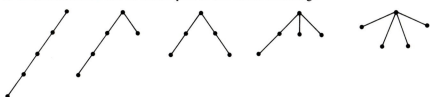

2. Each binary tree will be isomorphic to one of the following:

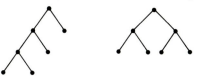

3. Each binary tree will be isomorphic to one of the following:

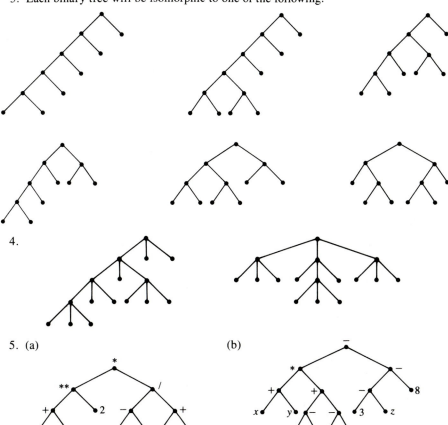

6. (a) 11;   (b) 333

7. (a) 11;   (b) 31;   (c) 61

8. No, there is no integer solution to $100 = 4i + 1$.

9. Let q = the number of leaves. Then $q = n - i = (3i + 1) - i = 2i + 1$. This number is always odd.

10. $q = (2i + 1) - i = i + 1$. We can always solve this for i.

11. In terms of n, $i = (n - 1)/m$ and $q = ((m - 1)n + 1)/m$. For $m > 1$, $q > i$.

12. (a) 5;   (b) 4;   (c) 6;   (d) 4

13. Maximal number of nodes: 511; minimal number of nodes: 256

14.

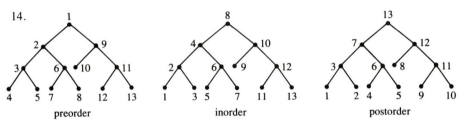

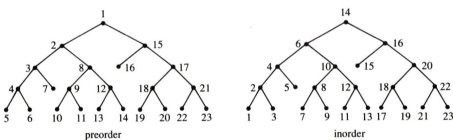

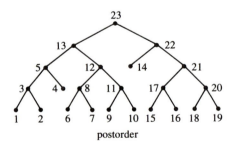

15. (a) $*, **, +, 2, 3, 2, /, -, 3, 9, +, 7, 2$

   (b) $-, *, +, x, y, +, -, z, w, -, x, 5, -, -, 3, z, 8$

16. (a) $[(3 + 5) * (7 - 1)] + [(8 - 2) - 5]$

   (b) $[(x + y) / (7 + z)] - [(5 + z) / (3 + 7)]$

Chapter 7   **Section 7.1 (page 298)**

1. (a)

Node	a	b	c	d	e
In-degree	0	3	1	1	1
Out-degree	2	0	1	2	1
Total degree	2	3	2	3	2

(b) There are 6 edges. The sum of the in-degrees is $0 + 3 + 1 + 1 + 1 = 6$. The sum of the out-degrees is $2 + 0 + 1 + 2 + 1 = 6$.

2.

Node	aa	bb	cc	dd	ee	ff	gg
In-degree	2	1	2	1	1	1	1
Out-degree	1	2	3	1	0	2	0
Total degree	3	3	5	2	1	3	1

The sum of the total degrees is $18 = 2 \cdot 9$. There are 9 edges.

3. (a)

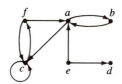

(b)

Node	a	b	c	d	e	f
In-degree	3	1	3	1	0	1
Out-degree	2	1	2	0	2	2
Total degree	5	2	5	1	2	3

(c)

	a	b	c	d	e	f
a	0	1	1	0	0	0
b	1	0	0	0	0	0
c	0	0	1	0	0	1
d	0	0	0	0	0	0
e	1	0	0	1	0	0
f	1	0	1	0	0	0

4. The maximal number of edges is n^2. The digraph that follows has 3 nodes and 9 edges.

5.

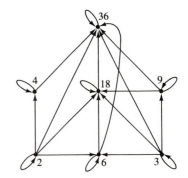

The in-degree of node x is the number of factors of x that appear in the list of nodes. The out-degree of x is the number of multiples of x that appear in the list of nodes.

6. In-degree of 12 = 2.

   In-degree of 36 = 2.

   In-degree of 105 = 3.

   Out-degree of 5 = $\lfloor 200/5 \rfloor$ = 40.

   Out-degree of 6 = 0. (6 is not a prime.)

   Out-degree of 19 = $\lfloor 200/19 \rfloor$ = 10.

   The in-degree of x is the number of distinct primes in the prime factorization of x. The out-degree of x is 0 if x is not a prime and is equal to the number of multiples of x less than or equal to 200 if x is a prime.

7. (a)

Node	Out-degree
a	3
d	1
f	3

   (b)

Node	In-degree
a	4
e	2
d	2

   (c) The number of edges is 15, which is the number of 1's appearing in the adjacency matrix of D.

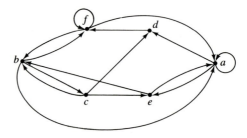

8.

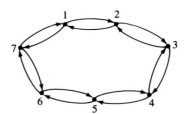

9. There are 10 edges.

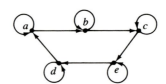

10. D has 5 nodes with out-degree equal to 1 and 1 node with out-degree equal to 2.

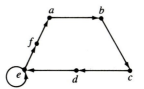

11. (a) Let the states be:

    1: No *b* detected

    2: *b* detected

    3: *ba* detected

    4: *bab* detected

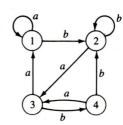

    (b) Let the states be as in (a).

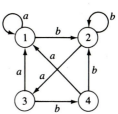

In (a) and (b) the number of times the pattern appears is the number of times node 4 is visited. All paths should begin at node 1.

Section 7.2 (page 308)

1. (a, b, e, f), (a, b, c, d, f), (a, b, c, e, f)

2. (a, b, c, a), (a, b, e, f, c, a)

3. (a, b, c, a), (a, b, e, f, c, a), (a, b, c, d, a), (a, b, c, e, f, d, a), (a, b, e, f, d, a), (a, b, e, f, c, d, a), (e, f, c, e), (d, f, d), (c, d, f, c)

4. (b, e)

5. The longest simple path is (a, b, c, d, f, g). No, there are paths of arbitrarily long length if we do not require simplicity.

6. (d, a, b, e, f, c)

7. There is a directed path from *a* to *g* but none from *g* to *a*.

8. (e, g) and (f, g)

9. D_1 and D_3 are strongly connected but D_2 is not.

10. D_1 is strongly connected but D_2 is not.

11. If the adjacency matrix has a row or column of zeros, then some node either has an in-degree or an out-degree of zero. Either no edge ends at that node or no edge begins at that node. In the first case that node can not be reached from any other node. In the second case no other node may be reached from that node. In either case the graph is not strongly connected.

The fact that the adjacency matrix has no rows or columns of zeros is not enough to ensure strong connectivity. For example the digraph with the following adjacency matrix is not strongly connected.

$$
\begin{array}{c c}
 & \begin{array}{c c} a & b \end{array} \\
\begin{array}{c} a \\ b \end{array} & \left|\begin{array}{c c} 1 & 0 \\ 0 & 1 \end{array}\right.
\end{array}
$$

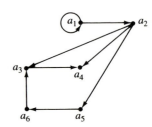

12. D_1 has a directed Euler circuit but no simple directed Euler circuit. D_2 has a simple directed Euler circuit.

13. The maximal number and the minimal number of edges are the same. They both equal n.

14. The edges 001, 010, 101, 011, 111, 110, 100, 000 determine the sequence 00101110.

15. (a)

$$
M^3 = \begin{pmatrix}
1 & 1 & 1 & 1 & 1 & 1 \\
0 & 0 & 1 & 0 & 0 & 0 \\
0 & 0 & 0 & 0 & 0 & 0 \\
0 & 0 & 0 & 0 & 0 & 0 \\
0 & 0 & 0 & 1 & 0 & 0 \\
0 & 0 & 0 & 0 & 0 & 0
\end{pmatrix}
$$

There is a directed path of length 3 between each of the following pairs of nodes: a_1 and itself, a_1 and a_2, a_1 and a_3, a_1 and a_4, a_1 and a_5, a_1 and a_6, a_2 and a_3, a_5 and a_4.

(b) The reachability matrix $M^R = M + M^2 + M^3 + M^4$ equals:

$$
\begin{pmatrix}
1 & 1 & 1 & 1 & 1 & 1 \\
0 & 0 & 1 & 1 & 1 & 1 \\
0 & 0 & 0 & 1 & 0 & 0 \\
0 & 0 & 0 & 0 & 0 & 0 \\
0 & 0 & 1 & 1 & 0 & 1 \\
0 & 0 & 1 & 1 & 0 & 0
\end{pmatrix}
$$

(c) No directed path exists from a_4 to any node.

16. (a)

$$
W_0 = \begin{pmatrix}
1 & 1 & 0 & 0 & 0 & 0 \\
0 & 0 & 1 & 1 & 1 & 0 \\
0 & 0 & 0 & 1 & 0 & 0 \\
0 & 0 & 0 & 0 & 0 & 0 \\
0 & 0 & 0 & 0 & 0 & 1 \\
0 & 0 & 1 & 0 & 0 & 0
\end{pmatrix}
$$

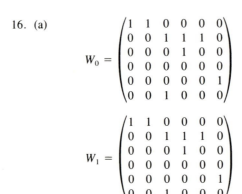

$$
W_1 = \begin{pmatrix}
1 & 1 & 0 & 0 & 0 & 0 \\
0 & 0 & 1 & 1 & 1 & 0 \\
0 & 0 & 0 & 1 & 0 & 0 \\
0 & 0 & 0 & 0 & 0 & 0 \\
0 & 0 & 0 & 0 & 0 & 1 \\
0 & 0 & 1 & 0 & 0 & 0
\end{pmatrix}
$$

$$W_2 = \begin{pmatrix} 1 & 1 & 1 & 1 & 1 & 0 \\ 0 & 0 & 1 & 1 & 1 & 0 \\ 0 & 0 & 0 & 1 & 0 & 0 \\ 0 & 0 & 0 & 0 & 0 & 0 \\ 0 & 0 & 0 & 0 & 0 & 1 \\ 0 & 0 & 1 & 0 & 0 & 0 \end{pmatrix}$$

$$W_3 = \begin{pmatrix} 1 & 1 & 1 & 1 & 1 & 0 \\ 0 & 0 & 1 & 1 & 1 & 0 \\ 0 & 0 & 0 & 1 & 0 & 0 \\ 0 & 0 & 0 & 0 & 0 & 0 \\ 0 & 0 & 0 & 0 & 0 & 1 \\ 0 & 0 & 1 & 1 & 0 & 0 \end{pmatrix}$$

(b) There is, for example, a 1 in the (1, 5) position of W_3 and there is a path with interior nodes in the set $\{a_1, a_2, a_3\}$, namely, (a_1, a_2, a_5).

(c)

$$W_4 = \begin{pmatrix} 1 & 1 & 1 & 1 & 1 & 0 \\ 0 & 0 & 1 & 1 & 1 & 0 \\ 0 & 0 & 0 & 1 & 0 & 0 \\ 0 & 0 & 0 & 0 & 0 & 0 \\ 0 & 0 & 0 & 0 & 0 & 1 \\ 0 & 0 & 1 & 1 & 0 & 0 \end{pmatrix}$$

$$W_5 = \begin{pmatrix} 1 & 1 & 1 & 1 & 1 & 1 \\ 0 & 0 & 1 & 1 & 1 & 1 \\ 0 & 0 & 0 & 1 & 0 & 0 \\ 0 & 0 & 0 & 0 & 0 & 0 \\ 0 & 0 & 0 & 0 & 0 & 1 \\ 0 & 0 & 1 & 1 & 0 & 0 \end{pmatrix}$$

$$W_6 = \begin{pmatrix} 1 & 1 & 1 & 1 & 1 & 1 \\ 0 & 0 & 1 & 1 & 1 & 1 \\ 0 & 0 & 0 & 1 & 0 & 0 \\ 0 & 0 & 0 & 0 & 0 & 0 \\ 0 & 0 & 1 & 1 & 0 & 1 \\ 0 & 0 & 1 & 1 & 0 & 0 \end{pmatrix}$$

W_6 is the reachability matrix, and it is the same matrix as M^R obtained in Exercise 15.

Section 7.3 (page 319)

1. 9

2. 22

3.

Path	Weight
(a, g, f, b, d, e)	29
(a, g, f, b, c, d, e)	25
(a, h, f, b, c, d, e)	17
(a, h, f, b, d, e)	21

Iteration 7:

Node	Label	Path
a	$PL = 0$	$\{a\}$
b	$PL = 4$	$\{a, h, f, b\}$
c	$PL = 7$	$\{a, h, f, b, c\}$
d	$PL = 12$	$\{a, h, f, b, c, d\}$
e	$PL = 17$	$\{a, h, f, b, c, d, e\}$
f	$PL = 3$	$\{a, h, f\}$
g	$PL = 5$	$\{a, g\}$
h	$PL = 2$	$\{a, h\}$

$V = e$

$PL(e) = 17$. The shortest path from a to e is $\{a, h, f, b, c, d, e\}$.

13. Step 1 (Initial labeling):

Node	Label	Path
a	$TL = \infty$	$\emptyset$
b	$TL = \infty$	$\emptyset$
c	$TL = \infty$	$\emptyset$
d	$TL = \infty$	$\emptyset$
e	$PL = 0$	$\{e\}$
f	$TL = \infty$	$\emptyset$
g	$TL = \infty$	$\emptyset$
h	$TL = \infty$	$\emptyset$

$V = e$

Iteration 1:

Node	Label	Path
a	$TL = \infty$	$\emptyset$
b	$TL = \infty$	$\emptyset$
c	$TL = \infty$	$\emptyset$
d	$TL = \infty$	$\emptyset$
e	$PL = 0$	$\{e\}$
f	$PL = 4$	$\{e, f\}$
g	$TL = \infty$	$\emptyset$
h	$TL = \infty$	$\emptyset$

$V = f$

Iteration 2:

Node	Label	Path
a	$TL = \infty$	$\emptyset$
b	$PL = 5$	$\{e, f, b\}$
c	$TL = \infty$	$\emptyset$
d	$TL = \infty$	$\emptyset$
e	$PL = 0$	$\{e\}$
f	$PL = 4$	$\{e, f\}$
g	$TL = \infty$	$\emptyset$
h	$TL = \infty$	$\emptyset$

$V = b$

Iteration 3:

Node	Label	Path
a	$TL = \infty$	$\emptyset$
b	$PL = 5$	$\{e, f, b\}$
c	$PL = 8$	$\{e, f, b, c\}$
d	$TL = 16$	$\{e, f, b, d\}$
e	$PL = 0$	$\{e\}$
f	$PL = 4$	$\{e, f\}$
g	$TL = \infty$	$\emptyset$
h	$TL = \infty$	$\emptyset$

$V = c$

Iteration 4:

Node	Label	Path
a	$TL = \infty$	$\emptyset$
b	$PL = 5$	$\{e, f, b\}$
c	$PL = 8$	$\{e, f, b, c\}$
d	$PL = 13$	$\{e, f, b, c, d\}$
e	$PL = 0$	$\{e\}$
f	$PL = 4$	$\{e, f\}$
g	$TL = \infty$	$\emptyset$
h	$TL = \infty$	$\emptyset$

$V = d$

Iteration 5:

Node	Label	Path
a	$TL = \infty$	$\emptyset$
b	$PL = 5$	$\{e, f, b\}$
c	$PL = 8$	$\{e, f, b, c\}$
d	$PL = 13$	$\{e, f, b, c, d\}$
e	$PL = 0$	$\{e\}$
f	$PL = 4$	$\{e, f\}$
g	$TL = \infty$	$\emptyset$
h	$TL = \infty$	$\emptyset$

The algorithm stops at this point. Since a is not permanently labeled, there is no path from e to a.

14. Step 1:

Node	Label	Path
a	$PL = 0$	(a)
b	$TL = \infty$	$\emptyset$
c	$TL = \infty$	$\emptyset$
d	$TL = \infty$	$\emptyset$
e	$TL = \infty$	$\emptyset$
f	$TL = \infty$	$\emptyset$

$V = a$

Iteration 1:

Node	Label	Path
a	$PL = 0$	(a)
b	$PL = 1$	(a, b)
c	$TL = \infty$	$\emptyset$
d	$TL = 1$	(a, d)
e	$TL = 1$	(a, e)
f	$TL = \infty$	$\emptyset$

$V = b$

Iteration 2:

Node	Label	Path
a	$PL = 0$	(a)
b	$PL = 1$	(a, b)
c	$TL = 3$	(a, b, c)
d	$PL = 1$	(a, d)
e	$TL = 1$	(a, e)
f	$TL = \infty$	$\emptyset$

$V = d$

Iteration 3:

Node	Label	Path
a	$PL = 0$	(a)
b	$PL = 1$	(a, b)
c	$TL = 3$	(a, b, c)
d	$PL = 1$	(a, d)
e	$PL = 1$	(a, e)
f	$TL = 2$	(a, d, f)

$V = e$

Iteration 4:

Node	Label	Path
a	$PL = 0$	(a)
b	$PL = 1$	(a, b)
c	$TL = 3$	(a, b, c)
d	$PL = 1$	(a, d)
e	$PL = 1$	(a, e)
f	$PL = 2$	(a, d, f)

The shortest path from a to f is (a, d, f). It has length 2.

Section 7.4 (page 326)

1. D_a and D_c are acyclic and D_b is not.
2. The reachability matrix of D_1 is:

$$\begin{pmatrix} 1 & 1 & 1 & 1 & 0 \\ 1 & 1 & 1 & 1 & 0 \\ 0 & 0 & 0 & 0 & 0 \\ 1 & 1 & 1 & 1 & 0 \\ 1 & 1 & 1 & 1 & 0 \end{pmatrix}$$

The reachability matrix of D_2 is:

$$\begin{pmatrix} 0 & 1 & 1 & 1 & 1 & 1 \\ 0 & 0 & 1 & 0 & 0 & 0 \\ 0 & 0 & 0 & 0 & 0 & 0 \\ 0 & 0 & 1 & 0 & 0 & 0 \\ 0 & 0 & 1 & 1 & 0 & 1 \\ 0 & 0 & 0 & 0 & 0 & 0 \end{pmatrix}$$

D_2 is acyclic.

3. In D_a, nodes a and g are sources and node e is a sink.
   In D_b, e is a source and g is a sink.
   In D_c, a is a source and c and e are sinks.

4. In D_1, e is a source and c is a sink.
   In D_2, 1 is a source and 3 and 6 are sinks.

5.

Node	Number
a	0
b	1
f	2
h	3
d	4
c	5
e	6

6.

Node	Number
f	0
a	1
h	2
b	3
d	4
e	5
c	6

7.

	a	b	f	h	d	c	e
a	0	1	0	0	0	0	1
b	0	0	0	0	1	0	0
f	0	0	0	1	1	0	0
h	0	0	0	0	0	1	0
d	0	0	0	0	0	1	1
c	0	0	0	0	0	0	0
e	0	0	0	0	0	0	0

	f	a	h	b	d	e	c
f	0	0	1	0	1	0	0
a	0	0	0	1	0	1	0
h	0	0	0	0	0	0	1
b	0	0	0	0	1	0	0
d	0	0	0	0	0	1	1
e	0	0	0	0	0	0	0
c	0	0	0	0	0	0	0

Note that all the 1's are above the diagonal in these matrices.

8.

Node	Number
d	0
a	1
e	2
b	3
i	4
j	5
c	6
h	7
g	8
f	9

9.

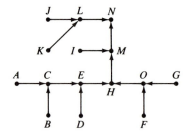

Name	Number
A	0
B	1
C	2
D	3
E	4
F	5
G	6
O	7
H	8
J	9
K	10
L	11
I	12
M	13
N	14

10.

Program	Earliest Time	Latest Time
A	0	0.2
B	0	0
C	0.5	0.7
D	0.3	0.3
E	1.0	1.0
F	0.3	1.3
G	0.5	1.1
H	0.3	1.1
I	0.9	1.7
J	1.6	1.6

Least time to run all programs: 2.1 hours.

Section 7.5 (page 339)

1. (a)

Word	Path	Accepted?
abbaaababa	(B, 1, 2, 3, 4, 5, 2, 3, 4, 3, 4)	no
bbaaa	(B, 1, 2, 2, 2, 2)	no
aabbabb	(B, 1, 2, 3, 3, 4, 3, 3)	no
ababbaa	(B, 1, 2, 2, 3, 3, 4, 5)	yes
bbbbaaa	(B, 1, 2, 3, 3, 4, 5, 2)	no

  (b) Only *ababbaa* is accepted.

  (c) 5

  (d) Yes

  (e) *aabaa, abababbaa, babaa, bbbbbbaa, baaaaabbaa*
  This machine accepts a word if and only if it ends in *baa* and has length of at least 5.

2. (a) They are all accepted.

  (b) This machine does not accept any word in which the pattern *bbaa* appears. It accepts all other words.

3. This machine accepts all words in which the number of *a*'s is even and the number of *b*'s is odd, or in which the number of *a*'s and the number of *b*'s are even.

4. This machine accepts words that end in *aaa* or *bab*.

5. This machine accepts a word if it has at least four *b*'s occurring in such a way that the first *b* is followed somewhere later in the word by the pattern *bbb*.

6. This machine accepts any word in which the pattern *aba* does not appear.

7.

8.

9.

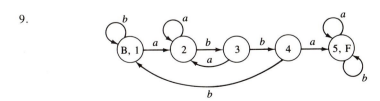

10.

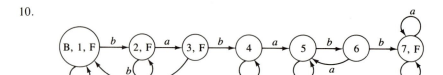

11.

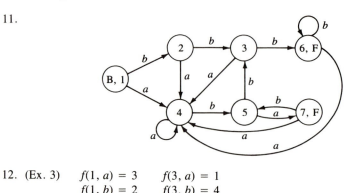

12. (Ex. 3) $f(1, a) = 3$ $f(3, a) = 1$
             $f(1, b) = 2$ $f(3, b) = 4$
             $f(2, a) = 4$ $f(4, a) = 2$
             $f(2, b) = 1$ $f(4, b) = 3$

    (Ex. 4) $f(1, a) = 2$ $f(4, b) = 5$
             $f(1, b) = 5$ $f(5, a) = 6$
             $f(2, a) = 3$ $f(5, b) = 5$
             $f(2, b) = 5$ $f(6, a) = 3$
             $f(3, a) = 4$ $f(6, b) = 7$
             $f(3, b) = 5$ $f(7, a) = 6$
             $f(4, a) = 4$ $f(7, b) = 5$

    (Ex. 5) $f(1, a) = 1$ $f(3, b) = 4$
             $f(1, b) = 2$ $f(4, a) = 2$
             $f(2, a) = 2$ $f(4, b) = 5$
             $f(2, b) = 3$ $f(5, a) = 5$
             $f(3, a) = 2$ $f(5, b) = 5$

    (Ex. 6) $f(1, a) = 2$ $f(3, a) = 4$
             $f(1, b) = 1$ $f(3, b) = 1$
             $f(2, a) = 2$ $f(4, a) = 4$
             $f(2, b) = 3$ $f(4, b) = 4$

13. (a)

Word	Production	Accepted?
abbaabb	1, 2, 3, 3, 5, 4, 3, 3	no
ababaa	1, 2, 3, 5, 6, 5, 4	yes
ababab	1, 2, 3, 5, 6, 5, 6	yes
bbbbaaaa	1, 3, 3, 3, 3, 5, 4, 4, 4	yes
aaaaaabab	1, 2, 4, 4, 4, 4, 4, 3, 5, 6	yes

(b)

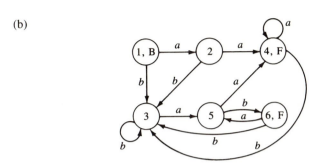

(c) This machine accepts any word that ends in *aa* and also words that end in *ab* and in which *b* occurs somewhere before *ab*.

14.

Input	Output
10001011	2023
100100	210
111111	333

15.

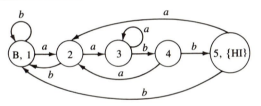

Chapter 8   Section 8.1 (page 374)

1. The operation ∗ is associative and commutative. The element *c* is both a left identity element and a right identity element.

2. The operation # is not associative because (*b* # *c*) # *a* = *a*, but *b* # (*c* # *a*) = *b*. It is not commutative since *c* # *b* = *c* but *b* # *c* = *a*. The element *a* is a left identity element. There is no right identity element.

3. (a) Both commutative and associative

   (b) Neither commutative nor associative

4. The operation is neither commutative nor associative. The empty set is a right identity element.

5. (a) None;   (b) *m* = 0 is a left identity element. There is no right identity element.

6. Associativity follows from the associativity of the conjunction operation ∧. The element *e* = 111 . . . 11, the string of *n* 1's, is the identity element. The Cayley table for (B^2, ∧) is as follows:

∧	00	01	10	11
00	00	00	00	00
01	00	01	00	01
10	00	00	10	10
11	00	01	10	11

7.

∘	id	f_1	f_2	f_3	f_4	f_5
id	id	f_1	f_2	f_3	f_4	f_5
f_1	f_1	f_5	f_3	f_4	f_2	id
f_2	f_2	f_4	id	f_5	f_1	f_3
f_3	f_3	f_2	f_1	id	f_5	f_4
f_4	f_4	f_3	f_5	f_1	id	f_2
f_5	f_5	id	f_4	f_2	f_3	f_1

The operation is not commutative because $f_5 \circ f_4 = f_3$, but $f_4 \circ f_5 = f_2$.

8. The primes are not closed under either multiplication or addition.

9. Let $x = 3m + 1$ and $y = 3n + 1$. Then $xy = 3(3mn + m + n) + 1 = 3p + 1$. So it is closed. Since $1 = 3 \cdot 0 + 1$, the identity is included and the set is a submonoid.

10. S is closed under $\cap$ since the intersection of any two sets containing b must also contain b.

11. (a) $\{0, 6, 9, 12, 15, 18, 21, \ldots\}$
    (b) **Z**

12. (a)

*	g_Λ	g_a
g_Λ	g_Λ	g_a
g_a	g_a	g_Λ

(b)

*	g_Λ	g_a	g_b	g_{aa}	g_{ba}
g_Λ	g_Λ	g_a	g_b	g_{aa}	g_{ba}
g_a	g_a	g_{aa}	g_b	g_{aa}	g_{aa}
g_b	g_b	g_{ba}	g_b	g_{aa}	g_{ba}
g_{aa}	g_{aa}	g_{aa}	g_b	g_{aa}	g_{ba}
g_{ba}	g_{ba}	g_{aa}	g_b	g_{aa}	g_{ba}

Section 8.2 (page 382)

1. For $a = 40$, $q = 2$ and $r = 2$. For $a = -10$, $q = -1$ and $r = 9$. For $a = -10001$, $q = -527$ and $r = 12$. For $a = 400592$, $q = 21083$ and $r = 15$.

2. $(9, 15) = 3$ and $3 = 2 \cdot 9 + (-1) \cdot 15$
   $(7, 5) = 1$ and $1 = 3 \cdot 5 + (-2) \cdot 7$
   $(24, 33) = 3$ and $3 = 3 \cdot 33 + (-4) \cdot 24$

3. (a) $45 = 1 \cdot 33 + 12$
$33 = 2 \cdot 12 + 9$
$12 = 1 \cdot 9 + 3$
$9 = 3 \cdot 3 + 0$ $(45, 33) = 3$

  (b) $1573 = 43 \cdot 36 + 25$
$36 = 1 \cdot 25 + 11$
$25 = 2 \cdot 11 + 3$
$11 = 3 \cdot 3 + 2$
$3 = 1 \cdot 2 + 1$
$2 = 1 \cdot 2 + 0$ $(1573, 36) = 1$

  (c) $91 = 1 \cdot 70 + 21$
$70 = 3 \cdot 21 + 7$
$21 = 3 \cdot 7 + 0$ $(91, 70) = 7$

4. All pairs are relatively prime.

5. $9 \mid 45$ but $9 \nmid 3$ and $9 \nmid 15$

  $6 \mid 18$ but $6 \nmid 2$ and $6 \nmid 9$

6 (a) $\{\ldots, -3, 10, 23, \ldots\} = [10]_{13}$

  (b) $\{\ldots, -3, 2, 7, \ldots\} = [2]_5$

  (c) $\{\ldots, -3, 3, 9, 15, \ldots\} = [3]_{12} \cup [9]_{12}$

  (d) No solution

7. (a) Solution set is one equivalence class mod 17.

  (b) The solution set is the union of two equivalence classes mod 24.

  (c) No solution

8. For $a = 1, x = [1]_5$
$a = 2, x = [3]_5$
$a = 3, x = [2]_5$
$a = 4, x = [4]_5$

9. $p = 3: a = 1, x = [1]_3$
$a = 2, x = [2]_3$

  $p = 7: a = 1, x = [1]_7$
$a = 2, x = [4]_7$
$a = 3, x = [5]_7$
$a = 4, x = [2]_7$
$a = 5, x = [3]_7$
$a = 6, x = [6]_7$

10. We shall use Theorem 2 of this section. Since $(a, b) = 1$ we can find m and n such that $1 = ma + nb$. Since $(a, c) = 1$ we can find s and t such that $1 = sa + tc$. Multiplying and factoring, we see that $1 = (sm + snb + mtc)a + (nt)bc$. Thus $1 = (a, bc)$.

11. By Proposition 6, the solution set is nonempty if and only if $(a, m) \mid 1$. Thus (a, m) must equal 1; that is to say, a and m must be relatively prime.

12. Suppose that the solution set is not empty and that x and y are solutions. Then we can find integers s and t such that $ax - 1 = sm$ and $ay - 1 = tm$. Subtracting, we have $a(x - y) = (s - t)m$. Since $(a, m) = 1$ (see Exercise 11), $m \mid (x - y)$. Thus $x \equiv y \pmod m$. The solution set is thus $[x]_m$, exactly one equivalence class modulo m.

13. $243 = 3 \cdot 3 \cdot 3 \cdot 3 \cdot 3$, $1800 = 2 \cdot 2 \cdot 2 \cdot 3 \cdot 3 \cdot 5 \cdot 5$, $1155 = 3 \cdot 5 \cdot 7 \cdot 11$, $5831 = 7 \cdot 7 \cdot 7 \cdot 17$, $9797 = 97 \cdot 101$

14. 2, 3, 5, 7, 11, 13, 17, 19, 23, 29, 31, 37, 41, 43, 47, 53, 57, 59, 61, 67

15. If m is not prime, then m can be properly factored as $m = a \cdot b$. If both $a > m^{1/2}$ and $b > m^{1/2}$, then $a \cdot b > m$, which is impossible. Thus one of a or b and hence one of the prime factors of m must be less than $m^{1/2}$.

16. (a) lcm of 5 and 7 is 35; of 15 and 10 is 30; of 15 and 14 is 210.

    (b) Let p be a prime and m and n be integers such that m is the highest power of p that divides x and n is the highest power of p that divides y. Let q be the larger of m and n. Then q is the highest power of p that divides the lcm of x and y.

Section 8.3 (page 392)

1. @

2. The identity is c. We have $a^{-1} = a$, $b^{-1} = d$, $c^{-1} = c$, and $d^{-1} = b$. We have $b * a = d$ so $x = a$.

3. $b^2 = a$, $b^3 = d$, $b^4 = c$

4.

$1 \longrightarrow 1$	$1 \longrightarrow 2$
$2 \longrightarrow 2$	$2 \longrightarrow 1$
id_{X_2}	f

$\circ$	id	f
id	id	f
f	f	id

It is abelian.

5. Let x be an element of G and consider the row labeled by the element a. We know that there is exactly one solution to the equation $a * y = x$. Thus in the row labeled by a, x appears only in the column labeled by y. (The proof for columns is similar.)

6.

Element	Order
f_0	1
f_1	3
f_2	3
f_3	2
f_4	2
f_5	2

    (a) $g = f_5$

    (b) $g = f_3$

7.

f g $o(f) = 4$ and $o(g) = 2$

8. In $(S_3, \circ, \mathrm{id}_{X_3})$, we have $f_2 \circ f_4 = f_5$ and $f_5^{-1} = f_5$. However, $f_2^{-1} \circ f_4^{-1} = f_1 \circ f_4 = f_3$.
    Proof: $(x * y) * (y^{-1} * x^{-1}) = x * (y * y^{-1}) * x^{-1} = x * e * x^{-1} = x * x^{-1} = e$.

9. Define $f : \mathbf{Z}(3) \to E$ by $f(x) = 2x/3$.

10.

$$
\begin{array}{ll}
e \longrightarrow & [0] \\
a & [1] \\
b & [2] \\
c & [3] \\
d & [4]
\end{array}
$$

11. No, there is no element of order 6 in $(S_3, \circ, \mathrm{id}_{X_3})$.

12. Define $f : \mathbf{Z} \to \mathbf{Z}_5$ by $f(x) = [x]_5$.

13. Let s and t be elements of H and suppose that $s = f(x)$ and $t = f(y)$. Since f is an isomorphism, $s \circ t = f(x * y)$. Since G is abelian, $x * y = y * x$, and $s \circ t = f(y * x)$. Since $f(y * x) = f(y) * f(x)$, $s \circ t = t \circ s$.

14.

+	[0]	[1]	[2]	[3]	[4]	[5]	[6]	[7]
[0]	[0]	[1]	[2]	[3]	[4]	[5]	[6]	[7]
[1]	[1]	[2]	[3]	[4]	[5]	[6]	[7]	[0]
[2]	[2]	[3]	[4]	[5]	[6]	[7]	[0]	[1]
[3]	[3]	[4]	[5]	[6]	[7]	[0]	[1]	[2]
[4]	[4]	[5]	[6]	[7]	[0]	[1]	[2]	[3]
[5]	[5]	[6]	[7]	[0]	[1]	[2]	[3]	[4]
[6]	[6]	[7]	[0]	[1]	[2]	[3]	[4]	[5]
[7]	[7]	[0]	[1]	[2]	[3]	[4]	[5]	[6]

15.

Element	Order	Inverse
[0]	1	[0]
[1]	12	[11]
[2]	6	[10]
[3]	4	[9]
[4]	3	[8]
[5]	12	[7]
[6]	2	[6]
[7]	12	[5]
[8]	3	[4]
[9]	4	[3]
[10]	6	[2]
[11]	12	[1]

16. (a) $x = [9]$; (b) $x = [10]$

17. Note that since G is abelian, $(x * y)^p = x^p * y^p$. Let $p =$ the lcm of m and n. We can find positive integers s and t such that $p = sm$ and $p = tn$. Thus $(x * y)^p = (x^m)^s * (y^n)^t = e^s * e^t = e$.

18. Let $S_n = \{1, 2, 3, \dots, n\}$. Consider the following functions:

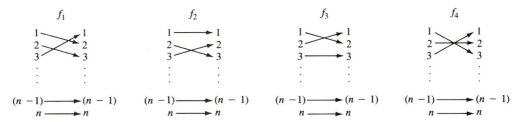

For each f_i, $f_i(x) = x$ for $x > 3$. Now $f_1 \circ f_2 = f_3$ but $f_2 \circ f_1 = f_4$.

Section 8.4 (page 398)

1. (a) No;   (b) Yes;   (c) Yes;   (d) Yes;   (e) No
2. $\{f_0, f_3\}, \{f_0, f_4\}, \{f_0, f_5\}, \{f_0, f_1, f_2\}$
3. $\{[0], [6]\}, \{[0], [4], [8]\}, \{[0], [3], [6], [9]\}, \{[0], [2], [4], [6], [8], [10]\}$
4. $\{[0], [8], [16]\}$
5. $\{a, b\}, \{a, c\}, \{a, d\}$
6. $\mathbf{Z}(3) = \{x : x = 3m \text{ for } m \text{ in } \mathbf{Z}\}$
7. $\{[0], [3], [6], [9]\}$
8. $\{[6], [9]\}$ generates the whole group.
9. $\{b, c\}$ generates the whole group. It is not cyclic.
10. The whole group is generated.
11. $[2], [3], [4], [6], [8], [9], [10]$
12. Each nonidentity element generates a proper cyclic subgroup. Generating pairs are:
    $\{f_1, f_3\}, \{f_1, f_4\}, \{f_1, f_5\}, \{f_2, f_3\},$
    $\{f_2, f_4\}, \{f_2, f_5\}, \{f_3, f_4\}, \{f_3, f_5\},$
    $\{f_4, f_5\}$.
13. $\{[0], [3], [6], [9]\}$
    $\{[1], [4], [7], [10]\}$
    $\{[2], [5], [8], [11]\}$
14. (a) $f_0 A = \{f_0, f_3\}$ (b) $A f_0 = \{f_0, f_3\}$
        $f_1 A = \{f_1, f_5\}$ $A f_1 = \{f_1, f_4\}$
        $f_2 A = \{f_2, f_4\}$ $A f_2 = \{f_2, f_5\}$
15. If n is divisible by d, then we can find k such that $n = dk$. The subgroup $\{[0], [k], [2k], \ldots, [(d-1)k]\}$ has order d.
16. Let S be the subgroup generated by $[x]$. Let d be the smallest positive integer such that $d = sx + tn$ so that $d = (x, n)$. Then $S = \langle [d] \rangle$; we see that $S = \mathbf{Z}_n$ if and only if $d = 1$.
17. (a) $K_f = \mathbf{Z}(n)$; that is, the set of all multiples of n.
    (b) Suppose that x and y are in K_f. Then $f(x * y) = f(x) \circ f(y) = i \circ i = i$. Thus $x * y$ is also in K_f. Also, $f(x * x^{-1}) = f(e) = i$ and $f(x * x^{-1}) = f(x) \circ f(x^{-1})$. So $f(x^{-1}) = i$ and we have that x^{-1} is also in K_f.
18. (a) $\{[0], [2], [4], [6]\}$
    (b) Let x and y be elements of $f(G)$. Find s and t in G such that $f(s) = x$ and $f(t) = y$. Then $f(s * t) = x \circ y$ so that $x \circ y$ is in $f(G)$. Also, $f(s^{-1}) = x^{-1}$ since $i = f(s * s^{-1}) = f(s) \circ f(s^{-1}) = x \circ f(s^{-1})$. Thus x^{-1} is in $f(G)$.

Section 8.5 (page 406)

1. (a) No;   (b) Yes;   (c) No
2. It is an abelian group under $+$. The operation $\#$ is associative: $x \# (y \# z) = 2x(2yz) = 2(2xy)z$. The number $x = 1/2$ is the multiplicative identity. The

distributive law holds: $x \# (y + z) = 2x(y + z) = 2xy + 2xz = x \# y + x \# z$. The inverse of 1 is 1/4. The inverse of 1/2 is 1/2. The inverse of 5/3 is 3/20.

3. For $+$, see Exercise 14, Section 8.3.

·	[0]	[1]	[2]	[3]	[4]	[5]	[6]	[7]
[0]	[0]	[0]	[0]	[0]	[0]	[0]	[0]	[0]
[1]	[0]	[1]	[2]	[3]	[4]	[5]	[6]	[7]
[2]	[0]	[2]	[4]	[6]	[0]	[2]	[4]	[6]
[3]	[0]	[3]	[6]	[1]	[4]	[7]	[2]	[5]
[4]	[0]	[4]	[0]	[4]	[0]	[4]	[0]	[4]
[5]	[0]	[5]	[2]	[7]	[4]	[1]	[6]	[3]
[6]	[0]	[6]	[4]	[2]	[0]	[6]	[4]	[2]
[7]	[0]	[7]	[6]	[5]	[4]	[3]	[2]	[1]

4. The elements [1], [3], [5], and [7] have multiplicative inverses. The elements [2], [4], and [6] are zero divisors.

5. $U = \{[1], [3], [5], [7]\}$

·	[1]	[3]	[5]	[7]
[1]	[1]	[3]	[5]	[7]
[3]	[3]	[1]	[7]	[5]
[5]	[5]	[7]	[1]	[3]
[7]	[7]	[5]	[3]	[1]

We know that · is associative. The table shows us that U is closed under ·, that [1] is the identity element, and that each element has an inverse.

6.

·	[1]	[5]	[7]	[11]
[1]	[1]	[5]	[7]	[11]
[5]	[5]	[1]	[11]	[7]
[7]	[7]	[11]	[1]	[5]
[11]	[11]	[7]	[5]	[1]

7. $x = [3]$ and $y = [6]$

8.

Element	Inverse
[1]	[1]
[2]	[4]
[3]	[5]
[4]	[2]
[5]	[3]
[6]	[6]

(a) $x = [6]$;   (b) $x = [5]$;   (c) $x = [3]$ and $x = [4]$;   (d) No solution

9. Yes. Multiplication is commutative since $2xy = 2yx$. There are no zero divisors since $2xy = 0$ if and only if x or y is 0. For any nonzero element x, $x^{-1} = 1/(4x)$ since $2x(1/(4x)) = 1/2$ and 1/2 is the multiplicative identity.

10. $3^5 = 243$ and $243 \equiv 3 \pmod 5$
$2^7 = 128$ and $128 \equiv 2 \pmod 7$

11. $[2]^{-1} = [2]^{11} = [2^{11}] = 2048$ and $2048 \equiv 7 \pmod{13}$. So $[7] = [2]^{-1}$.

12. For $p = 11$ and $n = 9$, $m = 9$ since $9 \cdot 9 = 81$ and $81 \equiv 1 \pmod{10}$.

Symbol	x	Code [x^9 (mod 11)]
A	2	6
B	3	4
C	4	3
D	5	9
E	6	2
F	7	8
G	8	7
H	9	5

Appendix   Matrices and Matrix Operations (page 426)

1. (a) $\begin{pmatrix} 1/3 & 2/3 & 5/3 \\ 0 & -2/3 & 1 \\ 1 & 5/3 & 0 \end{pmatrix}$

(b) $\begin{pmatrix} 1 & 3 & 8 \\ 1 & -2 & 4 \\ 8 & 11 & 2 \end{pmatrix}$

(c) $\begin{pmatrix} 2 & 1 & 1 \\ -3 & -4 & 3 \\ -9 & -8 & -6 \end{pmatrix}$

(d) $\begin{pmatrix} 9 & 13 & 3 \\ 4 & 7 & 5 \\ 11 & 8 & 43 \end{pmatrix}$

(e) $\begin{pmatrix} 27 & 31 & 15 \\ 13 & 18 & 4 \\ 5 & 3 & 14 \end{pmatrix}$

(f) Not defined

(g) $\begin{pmatrix} 9 & 5 & 7 \\ 30 & 13 & 23 \\ 11 & 0 & 8 \\ 20 & 5 & 15 \end{pmatrix}$

(h) $\begin{pmatrix} 16 & 18 & 7 \\ 5 & 7 & 5 \\ 16 & 17 & 25 \end{pmatrix}$

(i) C

(j) Not defined

(k) $\begin{pmatrix} 15 & 19 & 1 \\ 9 & 24 & -12 \\ -3 & -14 & 31 \end{pmatrix}$

2. (a) $\begin{pmatrix} 3 & 1 & 3 & 0 \\ 1 & 1 & 0 & 3 \end{pmatrix}$　(b) $\begin{pmatrix} 2 & 3 & 0 \\ 1 & 1 & 2 \end{pmatrix}$

3. (a) $\begin{pmatrix} 1 & 0 & 1 \\ 1 & 1 & 1 \\ 1 & 1 & 1 \end{pmatrix}$　(b) $\begin{pmatrix} 1 & 0 & 0 \\ 1 & 1 & 1 \\ 1 & 1 & 1 \end{pmatrix}$

(c) $\begin{pmatrix} 0 & 0 & 0 \\ 1 & 1 & 1 \\ 1 & 0 & 1 \end{pmatrix}$　(d) $\begin{pmatrix} 1 & 1 & 1 \\ 1 & 1 & 1 \\ 1 & 1 & 1 \end{pmatrix}$

INDEX

Abelian group, 384
Absurdity, 14
Accepting states, 330
Acceptword program, 335–337
Acyclic digraphs, 320–329
Addition principle, 134–135
Addition rule, 33, 36
Adjacency matrix, 395
Affirming the consequent, 33–34
Algebra, 366–419
  binary operations, 367–375
  bridge to computer science
    (coding theory), 408–412
  groups, 383–393
  the integers, 375–383
  monoids, 371–375
  rings and fields, 400–407
  semigroups, 369–375
  subgroups, 394–400
Algorithms
  analysis of, 167–171
  Bubblesort, 267
  Dykstra's, 312–319
  Euclid's, 378
  Karnaugh map and, 21
  parentheses checker (algorithm I),
    86–88, 90–91
  Prim's, 256–258

Algorithms—*Cont.*
  TREESORT, 273–280
  Warshall's, 307–308
    modified, 358
  Wff checker (algorithm II), 88–90,
    91–93
Almost binary tree, 273
Alphabet, 75
AND gate, 40
Antisymmetric relation, 183–184
Appendix (matrices and matrix
    operations), 419–425
Arithmetic trees, 263
Associative binary operations, 368
Associativity, 15, 64

Balanced rooted tree, 265
Basis clause (of inductive definition), 72
Biconditional connective, 13–16
Biconditional proof, 39–40
Bijective function (one-to-one
    correspondence), 107–108
Binary operations, 367–375
Binary tree, 261
Binding a variable, 26
Binomial theorem, 155–156
Boolean addition, 190, 426
Boolean multiplication, 188, 424

Breadth-first search tree, 254
Breadth-first spanning tree, 254–255
Bubblesort, 267–271

Cardinality of a finite set, 133–134
Cartesian products, 69–71
Cayley table, 368
Ceiling function, 106
Circuit, 234
Codes (ciphers), Fermat's theorem and, 405–406
Coding theory, 121–126, 408–412
Codomain, 102
COINCIDENCE gate, 44
Column matrix, 419
Combinations
    permutations and, 146–153
    of r objects chosen from
        n (n choose r), 147–148
Combinatorial arguments, 153–157
Combinatorial proof, 154–156
Commutative binary operation, 368
Commutative ring, 401
Commutativity, 14, 64
Complement (set operation), 60
Complete bipartite graph $K_{m,n}$, 285
Complete graph K_n, 227–228
Complete set of representatives, 198–199
Complete tournament, 297–298
Components of a graph, 235–237
Composition of functions, 110–114
Composition of relations, 186–191
Computer science
    algorithm analysis and, 167–171
    coding theory and, 121–126, 408–412
    formal languages (set theory) and, 84–93
    Kleene's theorem and, 342–343
    logic circuits and, 42–47
    TREESORT, 273–280
    two's-complement arithmetic, 212–214
Concatenation, 75
Conditional connective, 11–16
Congruence, 192
Conjunction, 4, 33, 36
Connected digraph, 302

Connected graph, 235–237
Connectivity
    digraphs and, 300–310
    paths and, 233–244
Constructive dilemma, 33, 36
Containment relation, 57
Context-free grammar, 85
Contingency, 14
Contradiction (absurdity), 14, 40
Contraposition, law of, 32, 36
Contrapositive, 12, 15
Contrapositive proof, 39
Converse, 12
Coset leader, 410
Cosets, 396–397
    coding theory and, 410
Countability, infinite sets and, 157–166
Countable set, 159–165
Countably infinite sets, 157, 159
Counterexample, 40
Counting, functions and, 142–146
Counting principles, 133–141
Cryptic codes, Fermat's theorem and, 405–406
Cycle decomposition of a permutation, 197–198
Cyclic group, 396

Degree of a node, 226
DeMorgan's laws, 7, 8–9, 15
    infinite collections of sets and, 66
    set operations and, 62–63
Denying the antecedent, 34
Depth-first search tree, 251–252
Depth-first spanning tree, 254–255
Descendants, 261
Detachment, law of, 31–32, 36
Diagonalization argument, 164–165
Digraphs, 292–364
    acyclic digraphs, 320–329
    bridge to computer science
        (Kleene's theorem), 342–354
    finite state machines, 329–341
    paths and connectivity and, 300–310
    of a relation, 180
    weighted digraphs, 310–320
Directed circuit, 300–301
Directed edge, 180, 294
Directed graph, 180

Directed path, 300
Direct image of a set, 117
Direct proof, 39
Disjoint sets, 61
Disjunction, 4
Disjunctive normal form, 16–25
Disjunctive syllogism, 32, 36
Distance (coding theory), 123
Distributive laws, 9, 15, 64
   rings and, 400
   set operations and, 61–62
Divisibility, 375–376
Division theorem, 376
Domain, 100
Double negation, 14
Dykstra's algorithm, 312–319

Edge connectivity, 236–237
Edge, 225
Edges incident to a node, 226
Elements in a set, 55
Empty set (Ø), 60
Empty word (Λ), 75
Encoding matrix (M_f), 409
Endnodes, 225
End points of the path, 233
Enumerated tree, 269–271
Equivalence classes, 196–200
Equivalence relations, 191–196
Equivalent propositional forms, 7
Euclid's algorithm, 378
Euler circuit, 238
Euler path, 238
Euler's formula, 246–247
Euler's theorem for digraphs,
   302–303
Euler's theorem for graphs, 238
EXCLUSIVE–OR gate, 43
Existential quantification, 27

$f : A \rightarrow B$, 102, 107
f^{-1} (inverse of f), 114
Factor, 375
Factorization, unique, 381–382
Fermat's theorem, 404–406
Fibonacci sequence, 104
Fields, 400–407
Finite set, cardinality of, 133–134

Finite state machines, 329–341
   Kleene's theorem and, 342–354
   semigroups of, 373
Floor function, 103
Formal languages, 75–78
   computer science and, 84–93
Free variables, 26
Full-adder, 45–46
Functions, 101–106
   bijective, 107–108
   bridge to computer science
      (coding theory), 121–126
   composition of, 110–114
   and counting, 142–146
   injective, 107–110
   inverse, 114–117
   and set operations, 117–120
   surjective, 107–110

Gates, 42–45
Generalized transition graph, 345
Generator, 395
Gödel, Kurt, 3
Grammar, 85
Graph, 223, 225
   directed, 180
   generalized transition, 345
Graph theory, 222–290
   basic concepts, 225–233
   bridge to computer science
      (TREESORT), 273–280
   paths and connectivity, 233–244
   planar graphs, 244–249
   rooted trees, 260–272
   trees, 249–260
Gray Code, 122
Greatest common divisor (gcd), 377
Greatest element of a set, 203
Greatest-integer function, 103
Greatest lower bound (glb), 206
Groups, 383–393

Half-adder, 44–45
Hamiltonian circuit, 240–242
Hamiltonian path, 240–242
Hasse diagram, 201–202
Height of a rooted tree, 265
Hilbert, David, 3

Homomorphism, 389
Hypothetical syllogism, 32, 36

Idempotent laws, 14, 64
Identity elements, 369–370
Identity function, 103
Implication, 15
Inclusion-exclusion principle, 137–138
In-degree of a node, 295
Index set, 65
Induction, 79–83
Inductive clause (of inductive definition), 72
Inductive definition of a set, 71–72
Inductively defined function, 103
Inductively defined sets, 71–74
Inductive procedure, 55
Inference, rules of, 32–36
Infinite collection of sets, 65–67
Infinite sets, countability and, 157–166
Injective function (one-to-one), 105–108
Inorder, 269
Integers, 375–383
Integral domain, 403
Interior nodes, 233, 261
Intersection (set operation), 60
Intersection of a collection of sets, 65–66
Invalid arguments, 33–34
Inverse functions, 114–117
Inverse image of a set, 117
Inverse of a group element, 383
Inverter (NOT gate), 42
Irreflexive relation, 183
Isomorphic graphs, 229
Isomorphism, 388

Join of x and y, 207

Karnaugh map method, 18–21
    algorithm using, 21
Kernel of a homomorphism, 399
Kleene-star operation, 77
Kleene's theorem, 342–354
Konigsberg Bridge Problem, 224, 238

LaGrange's theorem, 398
Language acceptor, 330

Languages
    defined or accepted by a finite state machine, 331, 334
    formal, 75–78
        computer science and, 84–93
    regular, 342
Lattice, 208
Law of contraposition, 32, 36
Law of detachment, 31–32, 36
Leaf, 261
Least common multiple (lcm), 383
Least element of a set, 203
Least-integer function, 106
Least upper bound (lub), 206
Left coset, 396–397
Left identity element, 370
Lexicographic order, 203
Literal, 18
Logic, 3–52
    bridge to computer science (logic circuits), 42–47
    the conditional and the biconditional, 11–16
    disjunctive normal form, 16–25
    important logical identities, 14–15
    predicates, 25–30
    proofs in mathematics, 38–41
    propositions, 4–11
    valid arguments, 30–37
Logical argument, 30
Logical connectives, 4
Logical equivalence, 7
Logical identity, 7–9
Logical variables, 4
Logic circuits, 42–47
Longest path from a to z (path of greatest weight), 312
Lower bound, 205

m-ary tree, 263
(m, n)-code, 123
M_R (matrix of a relation R), 184–185
Mathematical proofs, 38–41
Matrices
    adjacency, 395
    column, 419
    encoding (M_f), 409

Matrices—*Cont.*
  identity, 422
  $n \times m$, 419
  reachability, 306–307
  of a relation R (M_R), 184–185
  row, 419
  weighted adjacency, 311
Matrix addition, 420
Matrix multiplication, 420
Matrix operations, 419–425
Matrix sum, 419
Meet of x and y, 207
Members of a set, 55
Message (coding theory), 122
Minimal spanning tree, 256
Minimal expression, 19
Miniterm, 17
Modular groups ($\mathbf{Z}_n$, +, $[0]_n$), 390
Modulus of the relation, 192–193
Modus ponens (law of detachment), 31–32, 36
Modus tollens (law of contraposition), 32, 36
Monoids, 371–375
  generated by a subset, 372
Multiplication principle, 139–140
Mutually disjoint sets, 135

$n \times m$ matrix, 419
$n \times n$ identity matrix, 422
NAND gate, 43
Negation of the proposition, 6
Nodes, 182
  in-degree of, 295
  interior, 235, 261
  out-degree of, 295
  total degree of, 295
Nonplanar graphs, 245–246
NOT gate, 42

One's complement, 212
Operation-preserving function, 388
Orbits of the permutation f, 197
Order of a group, 384
Order of a group element, 387
Order relations, 200–211
OR gate, 42
Out-degree of a node, 295

Parallel edges, 226
Parentheses checker (algorithm I), 86–88, 90–91
Parenthesis-free notation (Polish notation), 271
Parity-bit code, 123
Partially ordered set (poset), 200
Partial orders, 200–201
Path defined by a word, 330
Path P of length n, 233
Paths and connectivity, 233–244
  digraphs and, 300–310
Permutation (bijection), 144
Permutations
  and combinations, 146–153
  cycle decomposition of, 197–198
Pigeon-hole principle, 143
Planar graphs, 244–249
Polish notation, 271
Poset, 200
Postorder, 269
Power sets, 68–69
Predicates, 25–30
Prefix, 75
Premises, 30
Preorder, 269
Prime number, 375
Prim's algorithm, 256–258
Principle of Induction, 79
  Second Principle of Induction, 81
Production defined by a word, 334
Product language LS, 77
Proofs
  biconditional, 39–40
  contradiction, 14, 40
  contrapositive, 39
  counterexample, 40
  direct, 39
  by induction, 79–83
  in mathematics, 38–41
Proper circuit, 234
Proper directed circuit, 301
Proper subset, 58
Propositional form, 4
Propositions, 4–11
Pumping theorem, 337–338

Quantification
  existential, 27
  universal, 26

R^n (composition of a relation with itself
    n times), 187
Reachability matrix, 306–307
Reccurrence relation, 101
Recursively defined function, 103
Reflexive relation, 183, 191
Regions defined by a planar graph, 246–
  247
Regular expressions, 342–343
Regular languages, 342
Relation, 180
Relations, 178–220
  bridge to computer science (two's-
    complement arithmetic), 212–
    214
   composition of, 186–191
   equivalence classes, 196–200
   equivalence relations, 191–196
   order relations, 200–211
   properties of, 182–184
   set relations, 57–59
Relatively prime integers, 379
Right coset, 396–397
Right identity element, 369
Rings, 400–407
Rooted trees, 260–272
Row matrix, 419
Rules of inference, 32–36

Same cardinality for infinite sets, 158–
  159
Scalar products, 419
Second Principle of Induction, 81
Secret codes, Fermat's theorem and,
  405–406
Self-loop, 181, 226
Semigroups, 369–375
  of finite state machines, 373
Set-builder notation, 56
Set difference, 63
Set equality, 55
Set identities, 64
Set operations, 59–65
  functions and, 115–118
Set relations, 57–59

Sets, 55–56
  countable, 159–165
  countably infinite, 157, 159
Set theory, 54–99
  bridge to computer science, 84–93
  Cartesian products, 69–71
  formal languages, 75–78
  inductively defined sets, 71–74
  infinite collection of sets, 65–67
  power sets, 68–69
  proofs by induction, 79–83
  set operations, 59–65
  set relations, 57–59
  sets, 55–56
Shortest path from a to z
    (path of least weight),
    311, 312–319
Simple circuit, 234
Simple directed path, 300
Simple graph, 226
Simple path, 233
Simplification, 33, 36
Singleton set, 56
Sinks, 321
Sources, 321
Spanning trees, 254–256
Stack, 86–87, 275–276
Standard order, 218
State diagram, 333
States, 330
String (word), 75
Strongly connected digraph, 302
Subgroup generated by S, 395
Subgroups, 394–400
Submonoid, 371–375
Subset, 57–58
Subtree, 261
Suffix of z, 75
Sum-of-products expression, 18
Surjective function (onto),
    107–110
Syllogism
  disjunctive, 32, 36
  hypothetical, 32, 36
Symmetric difference, 96
Symmetric groups (S_n, $\circ$, id_{X_n}), 385
Symmetric order, 269
Symmetric relation, 183, 191

Tautology, 14
Teleprinter's Problem, 303
Terminal clause (of inductive definition), 72
Ternary tree, 234
Three Utilities Problem, 224, 245
Topological sorting, 321–323
Total degree of a node, 295
Totally ordered set, 202–203
Transition function of the finite state machine, 332
Transition graph T, 344–345
Transition table, 349–350
Transitive closure of R, 190
Transitive relation, 183, 191
Traveling Sales Representative Problem, 241
Trees, 249–260
 rooted, 260–272
TREESORT, 273–280
Triple-repetition code, 123
Truth table, 6–7
Truth value, 4
Two's-complement arithmetic, 212–214
Two's complement, 212

Uncountable infinite sets, 164–165
Union (set operation), 60

Union of a collection of sets, 65
Unique factorization, 381–382
Universal quantification, 26
Universal set, 60
Universe of discourse, 26
Upper bound, 205

Valid arguments, 30–37
Variables
 binding, 26
 free, 26
 logical, 4
Vertex, 182

Warshall's algorithm, 307–308
 modified, 358
Weight (coding theory), 123
Weighted adjacency matrix, 311
Weighted digraphs, 311–320
Weighted graph, 256
Weight of a path, 311
Weight of the edge, 311
Well-ordered set, 204
Wff checker (algorithm II), 88–90, 91–93
Word (string), 75

Zero divisors, 402